Album of Porous Media

Ezequiel F. Médici • Alejandro D. Otero
Editors

Album of Porous Media

Structure and Dynamics

English and Spanish Edition

InterPore

Springer

Editors
Ezequiel F. Médici
Michigan Technological University
Houghton, MI, USA

Alejandro D. Otero
Universidad de Buenos Aires and CONICET
Buenos Aires, Argentina

ISBN 978-3-031-23802-4 ISBN 978-3-031-23800-0 (eBook)
https://doi.org/10.1007/978-3-031-23800-0

This Springer imprint is published by the registered company Springer Nature Switzerland AG
The registered company address is: Gewerbestrasse 11, 6330 Cham, Switzerland

The editors sincerely thank people, and institutions in the porous media community that made the realization of this album possible. We are especially grateful for the institutional and logistical support provided by The International Society for Porous Media, Interpore, which sponsored and advertised the development of this initiative.

None of this would have been possible without the dedication of S. Majid Hassanizadeh, to whom we are deeply grateful. From the beginning, he encouraged us to execute this project and gave us continuous support through the different stages of producing this album.

We also want to thank the Springer staff for their commitment and support, especially Alexis Vizcaino, the editor of this book, whose dedication and enthusiasm made this non-standard publication possible.

Needless to say, this work would not exist without the wholehearted participation of the authors, who, through their contributions, show us the diversity of subjects and the aesthetic beauty of porous media. We also want to thank the associate editors who devoted time to reviewing and selecting the submitted works.

We also want to highlight the efforts of Leyre Alegre-Figuero, who undertook the complex task of translating this multi-subject album. We are grateful to Roxana Bergonzi for the style and design suggestions, and to Alex Mayer, who provided essential support in defining the focus of the album's content.

Finally, both co-editors are deeply grateful to the public education system of the Argentine Republic, of which we are both university graduates.

Ezequiel F. Medici wishes first to thank his parents Laura and Alberto, for their unconditional support. He also thanks all the family members and friends who provided encouragement and professional support from the beginning, especially Jeffrey S. Allen, Rachel Hetherington, and Marcelino Viera.

Alejandro D. Otero wishes to thank his family, especially Mariana, Tomás, and Guillermina, for their support and help while working on this book during the difficult times of the COVID-19 pandemic. He also offers a big thank you to his colleagues from the Computational Simulation Center of CONICET and the Faculty of Engineering at the University of Buenos Aires, and to all his dear friends.

To everyone... Thank you very much!

Los editores deseamos expresar nuestro más sincero agradecimiento a todas las personas e instituciones de la comunidad de medios porosos que hicieron posible la materialización de este álbum. En especial queremos agradecer el apoyo institucional y logístico provisto por La Sociedad Internacional de Medios Porosos, Interpore, la cual patrocinó el desarrollo y difusión de este esfuerzo.

Nada de esto hubiera sido posible sin la dedicación de S. Majid Hassanizadeh, a quien estamos profundamente agradecidos. Desde el comienzo él nos animó a llevar adelante este proyecto, y nos brindó un apoyo continuo a través las diferentes etapas de producción de este álbum.

También queremos agradecer al personal de Springer por su buena disposición y ayuda, principalmente a Alexis Vizcaino, editor de este libro, quien con su dedicación y entusiasmo ha hecho posible esta publicación fuera de lo convencional.

De más está decir que este trabajo no existiría sin la incondicional participación de los autores, quienes a través de sus contribuciones nos muestran la diversidad temática y belleza estética de los medios poros. Además queremos agradecer a los editores asociados que dedicaron tiempo a la revisión y selección de los trabajos enviados.

Queremos destacar también el esfuerzo de Leyre Alegre-Figuero quien abordó la compleja tarea de traducir este álbum multitemático. Agradecemos las sugerencias de estilo y diseño de Roxana Bergonzi, y a Alex Mayer quien brindó un apoyo fundamental en la definición del enfoque del contenido del álbum.

Por último, ambos coeditores agradecemos profundamente al sistema educativo público de la República Argentina de los cuales ambos somos graduados universitarios.

Ezequiel F. Medici desea agradecer en primer lugar a sus padres, Laura y Alberto, por su apoyo incondicional. También desea agradecer a familiares y amistades que brindaron estímulo y apoyo profesional desde el inicio, especialmente Jeffrey S. Allen, Rachel Hetherington y Marcelino Viera.

Alejandro D. Otero desea agradecer a su familia, especialmente a Mariana, Tomás y Guillermina, por su apoyo y ayuda mientras trabajaba en este libro durante los difíciles momentos de la pandemia de COVID-19. También agradece enormemente a sus colegas del Centro de Simulación Computacional del CONICET y de la Facultad de Ingeniería de la Universidad de Buenos Aires, y a todas sus queridas amistades.

A todos... ¡Muchas gracias!

Ezequiel F. Médici – Alejandro D. Otero
October 2022

Preface

I was born in Tuiserkan, a small town at the foothills of Alvand Mountains in Western Iran. In the summertime, the whole family would go picnicking in the countryside, always next to a body of water or a running stream. My earliest memories are from playing with water; making side channels to a stream, building dams, and trying to get water go up a slope, in vain! I have always found water to be such an intriguing material. It has no odor, but one can smell its freshness; it has no sound, but one can hear it flowing; it is colorless, but one can see many colors in it; it has no taste, yet it is so delectable; it is so soft, yet it can make impressions in stone. Once I got a wristwatch as a birthday gift, and my father showed me how to measure the speed of water flowing in the stream. It was so much fun! I was also fascinated by natural springs. I found them so mysterious as the water seemed to appear from nowhere. My father explained to me that there is water everywhere in the rock behind the spring. I remember I had asked him whether the water flows as fast in the rock as outside, whether it flows on a straight line, and how we can measure its speed using my wristwatch. I also wanted to know whether there is a way we can go inside the rock and see the water moving around. He didn't have answers to those questions, and he said it is impossible to see how water moves inside the rock. That was 65 years ago. Now, here we have a book full of revealing images that show fluids flowing in the intricate pore spaces of porous media.

For a long time, a porous medium was a black box when it came to flow and transport processes. We could measure what goes in and what comes out and thus get an idea of average flow velocity or total bulk concentration inside a porous sample. In early 1950's, with the advent of nuclear scattering technique, one could measure average moisture content (or saturation) along a column. However, these were all average properties and did not provide information about fluid occupancies, flow lines and velocity distribution, or particle trajectories inside the pores and their dynamics. As beautifully demonstrated in this book, we can now get detailed information about the distribution of fluids inside pores, the movement of fluid-fluid interfaces and contact lines, velocity distribution and its magnitude, spreading of solutes and movement of particles. We not only get qualitative images, but we can obtain quantitative data. This has opened up unthinkable and priceless opportunities for increasing our understanding of various flow and transport phenomena. It has provided an immense impetus for improving current theories and advancing new theories of flow and transport in porous media. Moreover, pore-scale images and experiments have been instrumental to improving and advancing pore-scale as well as macroscale numerical models and making it possible to test them. Numerical models have in turn produced greatly valuable insights about the dynamics of flow and transport processes in intricate pore spaces. They have been extremely valuable to us in learning about many situations that cannot be investigated in pore-scale experiments or in quantifying properties that cannot be measured on the pore scale. Many examples of powerful images produced by numerical models can be found in this book. The insight provided by numerical models has in turn helped with refining and designing pore-scale experiments.

I know that porous media researchers will find it fascinating to go through this book and examine scores of beautiful images and insightful descriptions that go with them. We can all learn from imaging and/or modelling approaches that are employed in wholly different porous media applications. Finally, I am also really hoping that this book will fascinate the younger generation and they can see how wonderful the hidden world of porous media is.

International Society for Porous Media (InterPore) is pleased to have embraced the idea of compiling this album of porous media proposed by Ezequiel F. Médici. Ezequiel and his co-editor, Alejandro D. Otero have done a fabulous job and the contributions provided by researchers are all highly valuable. InterPore has supported the whole process in various ways, including communications, advertising, and promoting.

S. Majid Hassanizadeh
Stuttgart Center for Simulation Science (SIMTECH), Stuttgart University, Germany
Department of Earth Sciences, Utrecht University, The Netherlands

Nací en Tuiserkan, un pequeño pueblo al pie de las montañas Alvand en el oeste de Irán. En verano, toda la familia iba de picnic al campo, siempre junto a un cuerpo de agua o un arroyo. Mis primeros recuerdos son de jugar con agua; hacer canales laterales a un arroyo, construir presas y tratar de hacer subir el agua por una pendiente ¡en vano! El agua me ha resultado siempre un material tan intrigante. No tiene olor, pero se puede oler su frescura; no tiene sonido, pero uno puede oírla fluir; es incolora, pero se pueden ver muchos colores en ella; no tiene sabor, pero es tan deliciosa; es tan suave, pero puede hacer impresiones en piedra. Una vez recibí un reloj de pulsera como regalo de cumpleaños y mi padre me mostró cómo medir la velocidad del agua que fluye en un arroyo. ¡Fue muy divertido! También me fascinaron los manantiales naturales. Los encontré tan misteriosos dado que el agua parecía surgir de la nada. Mi padre me explicó que había agua por todas partes en la roca detrás del manantial. Recuerdo haberle preguntado si fluye tan rápido en la roca como afuera, si lo hace en línea recta y cómo podemos medir su velocidad usando mi reloj de pulsera. También quería saber si había alguna forma de entrar en la roca y verla moverse. No tenía respuestas para esas preguntas y dijo que era imposible ver cómo se mueve el agua dentro de la roca. Eso fue hace 65 años. Ahora, aquí tenemos un libro lleno de imágenes reveladoras que muestran fluidos que fluyen en los intrincados espacios de los medios porosos.

Durante mucho tiempo, un medio poroso fue una caja negra en lo que respecta a los procesos de flujo y transporte. Podíamos medir lo que entra y lo que sale y así tener una idea de la velocidad de flujo promedio o la concentración global total dentro de una muestra porosa. A principios de la década de 1950, con el advenimiento de la técnica de dispersión nuclear, se podía medir el contenido de humedad promedio (o saturación) a lo largo de una columna. Sin embargo, todas estas eran propiedades promedio y no proporcionaban información sobre la ocupación de fluidos, las líneas de flujo y la distribución de velocidades, o las trayectorias de partículas dentro de los poros y su dinámica. Como se demuestra maravillosamente en este libro, ahora podemos obtener información detallada de la distribución de fluidos dentro de los poros, el movimiento de las interfaces fluido-fluido y las líneas de contacto, la distribución de velocidades y su magnitud, la dispersión de solutos y el movimiento de partículas. No solo obtenemos imágenes cualitativas, sino que podemos obtener datos cuantitativos. Esto ha abierto oportunidades impensables e invaluables para mejorar nuestra comprensión de varios fenómenos de flujo y transporte. Ha proporcionado un inmenso impulso para mejorar las teorías actuales y proveer nuevas teorías de flujo y transporte en medios porosos. Además, las imágenes y los experimentos a escala de poro han sido fundamentales para mejorar y hacer avanzar los modelos numéricos a escala de poro y en la macroescala y hacer posible probarlos. A su vez, los modelos numéricos han producido información muy valiosa sobre la dinámica de los procesos de flujo y transporte en espacios porosos intrincados. Han sido extremadamente valiosos para nosotros en el aprendizaje de muchas situaciones que no se pueden investigar en experimentos a escala de poro o en la cuantificación de propiedades que no se pueden medir en la escala de poro. En este libro se pueden encontrar muchos ejemplos de poderosas imágenes producidas por modelos numéricos. La información proporcionada por los modelos numéricos ha ayudado a su vez a refinar y diseñar experimentos a escala de poro.

Sé que a los investigadores de los medios porosos les resultará fascinante leer este libro y examinar decenas de bellas imágenes y descripciones perspicaces que las acompañan. Todos podemos aprender de la técnicas de visualización y/o modelado que se emplean en aplicaciones de medios porosos completamente diferentes. Finalmente, también espero que este libro fascine a la generación más joven y que puedan ver lo maravilloso que es el mundo oculto de los medios porosos.

La Sociedad Internacional de Medios Porosos (InterPore) se complace en haber abrazado la idea de recopilar este álbum de medios porosos propuesta por Ezequiel F. Médici. Ezequiel y su coeditor, Alejandro D. Otero han hecho un trabajo fabuloso y las contribuciones de los investigadores son muy valiosas. InterPore ha apoyado todo el proceso de varias maneras, incluidas las comunicaciones, la publicidad y la promoción.

Organization

Editors

Ezequiel F. Médici	Michigan Technological University, USA
Alejandro D. Otero	Universidad de Buenos Aires & CONICET, Argentina

Associate Editors

Martin Blunt	Imperial College London, UK
Jeff Gostick	University of Waterloo, Canada
Daniel S. Hussey	National Institute of Standards and Technology, USA
Ruben Juanes	Massachussets Institute of Technology, USA
Knut Jørgen Måløy	University of Oslo, Norway
Krishna Pillai	University of Wisconsin-Milwaukee, USA
Karl Peterson	University of Toronto, Canada
Marc Prat	Institut de Mécanique des Fluides de Toulouse, France

Translator

Leyre Alegre-Figuero	Michigan Technological University, USA

The International Society for Porous Media (InterPore) has supported this project from the beginning in various ways. It has facilitated contacts with the publisher and potential authors, has advertised and promoted the book, and has facilitated the collection of articles. In particular, InterPore Office, led by Dr. Karolin C. Weber, and InterPore Communication Committee, chaired by Alberto Guadagnini, have provided very valuable support. Prof. S. Majid Hassanizadeh, Managing Director of InterPore and Editor of the Springer book series "Theory and Applications of Transport in Porous Media", has encouraged the realization of this project from its beginning and advised in many aspects.

InterPore Foundation for Porous Media Science and Technology wishes to sponsor this book and is calling for donations to make it open access. Support InterPore Foundation by making a donation through: `https://www.interpore.org/foundation/foundation-donate` Donations to InterPore Foundation will also directly support researchers from countries with economic difficulties and students.

InterPore

InterPore Foundation

Table of Contents

INTRODUCTION 1

I. A GLIMPSE OF POROUS MEDIA 9

ELECTROSPUN FIBROUS MATS 11
Jeff Gostick, Matthew Kok, and Rhodri Jervis

TAILORED ELECTROSPUN GAS DIFFUSION LAYERS WITH A GRADED PORE STRUCTURE FOR POLYMER ELECTROLYTE MEMBRANE FUEL CELLS 12
Manojkumar Balakrishnan, Pranay Shrestha, Nan Ge, ChungHyuk Lee, Kieran F. Fahy, Roswitha Zeis, Volker P. Schulz, Benjamin D. Hatton, and Aimy Bazylak

DISPERSED NANOSTRUCTURED THIN FILM (dNSTF) CATALYST 13
Karrar Alofari, Anjana Asthana, Tajiri Kazuya, Andrew Haug, and Jeffrey Allen

ELECTROCHEMICAL ACTUATION OF A NANOPOROUS POLYPYRROLE HYBRID MATERIAL ... 14
Manuel Brinker, Tobias Krekeler, and Patrick Huber

LASER-EXCITED ELASTIC GUIDED WAVES REVEAL THE COMPLEX MECHANICS OF NANOPOROUS SILICON 15
Marc Thelen, Nicolas Bochud, Manuel Brinker, Claire Prada, and Patrick Huber

SAMPLE LAYOUT AIDING EFFICIENT SCANS OF HETEROGENEOUS SHEET-LIKE MATERIALS . 16
Michael Stonig, Eduardo Machado Charry, Andre Hilger, Ingo Manke, Matthias Neumann, Volker Schmidt, and Karin Zojer

PROTECTIVE POROUS MEDIA - THE PORE & FIBER STRUCTURE OF FACE MASKS 17
Chven Mitchell and Laura Pyrak-Nolte

INDIVIDUAL FIBERS IN NONWOVEN ARE IDENTIFIED WITH ARTIFICIAL INTELLIGENCE 18
Andreas Griesser

MECHANISM OF IMPERFECT ATOMIC LAYER DEPOSITION IN POROUS MATERIALS 19
Nupur Bihari, Ismo Rauha, Giovanni Marin, Craig Ekstrum, Pierce Mayville, Shane Oberloier, Hele Savin, Maarit Karppinen, and Joshua Pearce

MASS TRANSPORT IN CONFORMALLY COATED POROUS MATERIALS 20
Nupur Bihari, Ismo Rauha, Giovanni Marin, Craig Ekstrum, Pierce Mayville, Yuhuan Fei, Yun Hang Hu, Maarit Karppinen, Hele Savin, and Joshua Pearce

OPTICAL MICROFABRICATION OF POROUS FILMS 21
Natalia Philipp, Nicolas Rapagnani, Paula Caral, Ignacio Gomez Florenciano, and Laura Estrada

BICONTINUOUS POROSITY BY SPINODAL PHASE SEPARATION 22
Andrés Zelcer, Glenna L. Drisko, and Galo Soler-Illia

NUCLEATION AND CRYSTAL GROWTH ON THE SECONDARY SUBSTRATE 23
Mohammad Nooraiepour, Mohammad Masoudi, Nima Shokri, and Helge Hellevang

DOUBLE-POROSITY IN SOIL 24
Su Kong Ngien

BOREAL MOSSES AND LICHENS 25
Laurent Orgogozo, Yohan Davit, Manuel Marcoux, and Michel Quintard

THE GORGONIAN OCTOCORAL *EUNICELLA CAVOLINII* FORMS A POROUS BIOLOGICAL MATERIAL WITH OUTSTANDING MATERIAL PROPERTIES 26
Uwe Schnepf, Raouf Jemmali, and Franz Brümmer

FORAMINIFER 27
Ab Coorn, Fons Marcelis, and Steffen Berg

TROVANTS: THE "LIVING" STONES OF ROMANIA FORMED AS HIGH POROSITY SPHERICAL SANDSTONE CONCRETIONS DEVELOPED AROUND A FOSSIL 28
Vlad C. Manea, Dante Arteaga, Marina Manea, Delia Lazar, Sandra Vega, and Gerardo Carrasco-Núnez

THE COMPLEX PORE SPACES OF OUR BRAIN: A CHALLENGE AND OPPORTUNITY FOR COMPUTER SIMULATIONS 29
Timo Koch, Kent-André Mardal, and Lars Magnus Valnes

II. GEOLOGICAL POROUS MEDIA 31

SCALING AND RESOLUTION EFFECT OF AN ANDESITE SAMPLE 33
Dante Arteaga and Sandra Vega

LENGTH-SCALE DEPENDENCE OF PORE SPACE TOPOLOGY 34
Ruotong Huang, Anna Herring, Adrian Sheppard, Michael Turner, Levi Beeching, Lydia Knuefing, and Robert Middleton

MICROSTRUCTURE OF CRUSHABLE EXPANDED CLAY AGGREGATES 35
Giulia Guida, Francesca Casini, Giulia M.B. Viggiani, Edward Andò, and Gioacchino Viggiani

RETICULATED RUTILE 36
Fons Marcelis and Steffen Berg

MINERAL MAP OF ROTLIEGENDES 37
Fons Marcelis and Steffen Berg

BOISE BROWN 38
Fons Marcelis and Steffen Berg

NORTH NETHERLANDS GAS RESERVOIR 39
Fons Marcelis and Steffen Berg

FIB-SEM IMAGES FROM A CATHODE MATERIAL: UNFILTERED AND FILTERED (CURTAINING FILTER) 40
Andreas Griesser

VESICLES AND FRACTURES IN VOLCANIC ASH 41
Jacopo Taddeucci

A VIEW OF THE INTERNAL PORE STRUCTURE OF A VOLCANIC ROCK: XALTIPAN IGNIMBRITE 42
Jonathan De la Rosa Maldonado and Sandra Vega

IN SITU MICROTOMOGRAPHY OF THE TENSILE FRACTURE AND FRAGMENTATION OF PORPHYRY COPPER ORES 43
Nicolas Francois, Yulai Zhang, Richard Henley, Ron Cruikshank, Ajay Limaye, Michael Turner, Levi Beeching, Andrew Kingston, Mohammad Saadatfar, and Mark Knackstedt

FRACTURED POROUS MEDIA: A CRETACEOUS CARBONATE OUTCROP SAMPLE 44
Sotirios Kokkalas, Syed Nizamuddin, and Sandra Vega

FRACTURE NETWORK IN A SHALE CUBE HYDRAULICALLY FRACTURED IN THE LABORATORY 45
Mei Li, Earl Magsipoc, Aly Abdelaziz, Johnson Ha, Karl Peterson, and Giovanni Grasselli

GEOLOGICAL CARBON STORAGE AT THE UNIVERSITY OF BERGEN, NORWAY 46
Kristoffer Eikehaug, Malin Haugen, Benyamine Benali, Zachary Paul Alcorn, Atle Rotevatn, Jan Martin Nordbotten, and Martin Fernø

III. TWO PHASE FLOW 47

CAPILLARY RISE OF WATER IN NANOPOROUS GLASS AS REVEALED BY X-RAY TOMOGRAPHY-BASED LATTICE-BOLTZMANN SIMULATIONS 49
Guido Dittrich, Juliana Martins de Souza e Silva, Cristine Santos de Oliveira, Sahar Bakshian, Andriy V. Kityk, Martin Steinhart, Dirk Enke, Ralf Wehrspohn, Nima Shokri, and Patrick Huber

CAPILLARY FINGERING PATTERN IN SLOW DRAINAGE FLOWS 50
Marcel Moura, Knut Jørgen Måløy, Eirik Flekkøy, and Renaud Toussaint

DISPERSION PATTERNS WITH TRACER FLOW IN FRACTURES 51
Le Xu, Knut Jørgen Måløy, Renaud Toussaint, Eirik Grude Flekkøy, and Benjy Marks

GPU-ENHANCED PORE-NETWORK SIMULATION OF INSTABILITY DURING FLOW AND TRANSPORT IN POROUS MEDIA 52
Senyou An and Vahid Niasar

VISCOUS FINGERING FRACTAL IN A POROUS MEDIUM 53
Marcel Moura and Knut Jørgen Måløy

STRONG IMBIBITION IN PATTERNED MICROFLUIDICS 54
Benzhong Zhao, Chris MacMinn, and Ruben Juanes

RISING INVADER 55
Joachim Falck Brodin, Knut Jørgen Måløy, Marcel Moura, and Per Arne Rikvold

SIMULATION OF TWO-PHASE FLOW IN BEREA SANDSTONE 56
Christian Hinz

WETTABILITY EFFECTS ON TWO-PHASE FLOW IN HETEROGENEOUS POROUS MEDIA 57
Sahar Bakhshian, Harris Rabbani, and Nima Shokri

OIL DISPLACEMENT BY INJECTION OF COMPLEX DISPERSIONS 58
Raphael Ribeiro and Marcio Carvalho

INVASION PATTERN 59
Joachim Falck Brodin, Per Arne Rikvold, Marcel Moura, and Knut Jørgen Måløy

FORMATION OF A PENDULAR LIQUID RING 60
Navneet Kumar, Jaywant H. Arakeri, and M. S. Bobji

VISUALIZATION OF THE ROLE OF WETTABILITY, PORE SHAPE AND CORNER EFFECTS IN POROUS MEDIA 61
Richard Dawe, Carlos Grattoni, and Kamal Idris

THE INTERPLAY OF WATER SPREADING, EVAPORATION AND IMBIBITION AT NANOPOROUS SILICON SURFACES 62
Laura Gallardo, Juan Sanchez, Hugo Bellezza, Olivier Vincent, and Patrick Huber

IV. PHASE CHANGE AND DIFFUSION 63

EVAPORATION OF WATER IN A GLASS BEAD-PACK 65
Armin Afrough and Karen Feilberg

A FLOATING SALT CRUST OVER EVAPORATING SAND 66
Uri Nachshon

INFLUENCE OF THERMAL GRADIENTS ON INVASION PATTERNS DURING DRYING OF GRANULAR POROUS MEDIUM 67
Divyansh Pandey, Kaustubh Desai, Debashis Panda, Abdolreza Kharaghani, and Vikranth Kumar Surasani

PREFERENTIAL AIR INVASION IN AN EVAPORATING MONOLAYER POROUS MEDIUM 68
Navneet Kumar and Jaywant H. Arakeri

VISUALIZATION OF APPLICATION OF ULTRASONIC DRYING FOR MOIST POROUS PAPER 69
Zahra Noori O'Connor, Jamal Yagoobi, and Burt Tilley

CRACKS IN WET SAND 70
Bjørnar Sandnes, James Campbell, and Deren Ozturk

GRANULAR LABYRINTH 71
Bjørnar Sandnes, Knut Jørgen Måløy, and Eirik Grude Flekkøy

FRICTIONAL FINGER LABYRINTH 72
Kristian Stølevik Olsen, Eirik Grude Flekkøy, Knut Jørgen Måløy, and James Campbell

ASSEMBLY OF PARTICULATE PATTERNED ARRAYS USING LIQUID TEMPLATES 73
Diego Barba Maggi, Román Martino, Marcelo Piva, and Alejandro Boschan

CHARACTERIZATION OF UNSATURATED SOLUTE TRANSPORT WITH X-RAY SYNCHROTRON MICROTOMOGRAPHY 74
Sharul Nizam Hasan, Vahid Niasar, Nikolaos Karadimitriou, Jose R. da Assuncao Godinho, Nghia T. Vo, Senyou An, Arash Rabbani, Dongwon Lee, and Holger Steeb

FLUID STRETCHING AND FOLDING IN 3D POROUS BEAD-PACKS 75
Joris Heyman, Tanguy Le Borgne, Yves Méheust, and Daniel Lester

SPATIO-TEMPORAL PUSH-PULL IMAGING REVEALS THE FRACTAL NATURE OF CHAOTIC SOLUTE LANDSCAPES IN POROUS FLOWS. 76
Joris Heyman, Daniel Lester, and Tanguy Le Borgne

MIXING IN A 3D POROUS MEDIA 77
Mathieu Souzy, Henri Lhuissier, Yves Méheust, Tanguy Le Borgne, Bloen Metzger

DISPERSION OF A BLOB OF TRACERS IN A 3D POROUS MEDIA 78
Mathieu Souzy, Henri Lhuissier, Yves Méheust, Tanguy Le Borgne, Bloen Metzger

MIXING OF A BLOB OF DYE IN A 3D POROUS MEDIA 79
Mathieu Souzy, Henri Lhuissier, Yves Méheust, Tanguy Le Borgne, Bloen Metzger

FINGERPRINTS OF CHAOS IN A TWO-PHASE POROUS MEDIA FLOW 80
Gaute Linga, Joachim Mathiesen, Francois Renard, and Tanguy Le Borgne

V. TRANSPORT IN MICROFLUIDIC DEVICES 81

TRANSITION TO DRYOUT IN DIABATIC MANIFOLD-MICROCHANNELS 83
David Deisenroth

CT IMAGING OF 3D PRINTED STEEL PARTICLE BEDS 84
Sondre Gjengedal, Ole Tore Buset, and Vegard Brøtan

SELF-ASSEMBLY OF DISCOTIC LIQUID CRYSTALS IN NANOPORES AS SEEN BY X-RAY DIFFRACTION AND MONTE CARLO COMPUTER SIMULATION 85
Zhuoqing Li, Kathrin Sentker, Arne Zantop, Milena Lippmann, Oliver Seeck, Andreas Schönhals, Marco G. Mazza, and Patrick Huber

IMBIBITION OF WATER IN NANOPOROUS SILICA (MCM-41) 86
Lars Dammann, Oliver Seeck, Robert Horst Meißner, and Patrick Huber

DIRECT NUMERICAL SIMULATION OF FLOW THROUGH NANOSCALE SHALE PORES IN A MESOSCALE SAMPLE 87
Timothy Kneafsey, David Trebotich, and Terry Ligocki

IN SITU OPTICAL SPECTROSCOPY DURING LIQUID IMBIBITION INTO A NANOPOROUS PHOTONIC CRYSTAL RESOLVES PRECURSOR FILM SPREADING 88
Luisa G. Cencha, Guido Dittrich, Claudio L.A. Berli, Patrick Huber, and Raul Urteaga

STUDY OF STRESS FIELD AND FRACTURE NETWORK DEVELOPMENT WITH ROCK ANALOGS 89
Bolivia Vega and Anthony Kovscek

THERMAL AND OPTICAL IMAGING IN A MICROMODEL 90
Nikolaos Karadimitriou, Philipp Nuske, S. Majid Hassanizadeh, and Rainer Helmig

FLOW-DRIVEN CHANNELIZATION IN A PARTICLE-FILLED POROUS MEDIUM 91
Navid Bizmark and Sujit S. Datta

STREAMLINES IN THE BULK OF A 3D POROUS MEDIA 92
Mathieu Souzy, Henri Lhuissier, Yves Méheust, Tanguy Le Borgne, Bloen Metzger

QUASI-TWO-DIMENSIONAL FOAM FLOW THROUGH AND AROUND A PERMEABLE OBSTACLE 93
Lauren Rose, Natalia Shmakova, Natalya Penkovskaya, Benjamin Dollet, Christophe Raufaste, and Stéphane Santucci

THE FLOW OF SUPERCRITICAL CO_2 FOAM FOR MOBILITY CONTROL 94
Benyamine Benali

MULTIMODAL VISUALIZATION OF CALCIUM CARBONATE ($CaCO_3$) IN THE PORE SPACE OF A MICROMODEL 95
Malin Haugen, Bergit Brattekås, and Martin Fernø

MICROFLUIDIC EXPERIMENTS ON CONSTRAINED OIL REMOBILIZATION INDUCED BY WATER TRANSPORT IN THE OIL PHASE 96
Lifei Yan, Mohammad H. Golestan, Wenyu Zhou, S. Majid Hassanizadeh, Carl F. Berg, Amir Raoof, and Eirik Flekkøy

OIL BLOBS SPIN IN A PORE FULL OF FERROFLUID IN A ROTATING MAGNETIC FIELD 97
Ningyu Wang, Yifei Liu, Luming Cha, Matthew Balhoff, and Masa Prodanovic

BIOFILM ARCHITECTURE IN OPAQUE POROUS MEDIUM 98
Dorthe Wildenschild, Gabriel Iltis, and Sassan Ostvar

VI. TRANSPORT IN POROUS MEDIA 99

PORES IN WATER CONDUCTING CELLS OF PLANTS 101
Steven Jansen

IN SITU VISUALIZATION OF MULTIPHASE FLOW APPLYING PET-MR 102
Bergit Brattekås and Geir Ersland

FLUID FLOW VISUALIZATION BY PET-CT 103
Bergit Brattekås, Martin Fernø, and Heidi Espedal

ADVECTION AND DIFFUSION IN NANOPOROUS SHALE 104
Nirjhor Chakraborty and Zuleima Karpyn

ADSORPTION ENHANCED GAS UPTAKE IN NANODARCY PERMEABILITY SHALE 105
Nirjhor Chakraborty and Zuleima Karpyn

TIME-LAPSE 3D IMAGING BY POSITRON EMISSION TOMOGRAPHY OF COPPER MOBILIZED IN SOIL BY THE HERBICIDE MCPA 106
Michael Kersten and Johannes Kulenkampff

DOES THE INTERFACIAL TRANSITION ZONE INFLUENCE MOISTURE TRANSPORT IN CONCRETE? 107
Laura Dalton, Jacob LaManna, Scott Jones, and Mohammad Pour-Ghaz

GAS TRAPPED IN THE PORE SPACE OF A SANDSTONE 108
Ying Gao, Ab Coorn, Niels Brussee, Hilbert van der Linde, and Steffen Berg

SUBMICRON INSIGHT INTO THE SNAP-OFF PHENOMENON 109
Arjen Mascini, Veerle Cnudde, and Tom Bultreys

DISCONNECTED GAS FLOW IN HYDROPHOBIC POROUS MEDIA 110
Abdulla Alhosani, Branko Bijeljic, and Martin Blunt

FORMATION OF WATER MICRO-DISPERSION IN OIL AS A MECHANISM FOR EFFICIENT DISPLACEMENT IN LOW SALINITY WATERFLOODING 111
Ahmed M. Selem, Nicolas Agenet, Martin J. Blunt, and Branko Bijeljic

PORE-SCALE TRAPPING AND FLOW FIELDS IN CARBONATE ROCKS 112
Kamaljit Singh and Martin Blunt

NANO-IMAGING OF DIATOMITE WITH TRANSMISSION X-RAY MICROSCOPE 113
Bolivia Vega and Anthony Kovscek

PORE SCALE INVESTIGATION OF FREEZE-DRYING 114
Sebastian Gruber, Petra Foerst, Maximilian Thomik, Evangelos Tsotsas, and Nicole Vorhauer-Huget

PLASTICITY DISTRIBUTION OF DIGITAL ROCK FROM ONE POINT OF THE YIELD ENVELOPE 115
Martin Lesueur

IMAGING THE PATH TO FAILURE IN ROCKS 116
François Renard

VISUALIZATION OF THE UNIAXIAL COMPRESSION OF OPEN-CELL FOAMS 117
Matthias Ruf and Holger Steeb

VII. COMPUTATIONAL MODELING 119

OED-BASED METHOD FOR EXTRACTION OF PORE NETWORKS WITH HIGH ASPECT RATIO . 121
Ninghua Zhan, Rui Wu, and Abdolreza Kharaghani

PORE TOPOLOGY METHOD (PTM): A COMPUTATIONALLY EFFICIENT APPROACH FOR PORE-SCALE MODELING AND DESIGN OF POROUS MATERIALS 122
M. Sadegh Riasi, Benjamin Paisley, and Lilit Yeghiazarian

COUPLING OF PORE-NETWORK AND FINITE ELEMENT METHODS FOR RAPID QUANTIFICATION OF DEFORMATION 123
Pejman Tahmasebi

HYBRID RANDOM WALKS FOR RADIATIVE/CONDUCTIVE HEAT TRANSFER IN POROUS MEDIA 124
Gerard Louis Vignoles

FLOW IN BINARY POROUS MEDIA: VISCOUS DISSIPATION 125
Iván Colecchio, Alejandro Boschan, Benoît Noetinger, and Alejandro Otero

A SPHEREPACK AND ITS NAVIER-STOKES SOLUTION APPROXIMATION 126
Javier E. Santos and Masa Prodanovic

FLOW IN TIGHT POROUS MEDIA 127
Pejman Tahmasebi

SPATIAL VARIANCE OF PORE-SCALE FLUID VELOCITIES 128
Timo Koch, Kilian Weishaupt, and Johannes Müller

FLUID FLOW IN SANDSTONE DIGITAL CORE MODEL 129
Kambiz Vafai and Jianchao Cai

FLOW THROUGH THE MICROSTRUCTURE OF NONWOVEN AIR FILTER MEDIA 130
Mehdi Azimian

TORTUOSITY OF THE FLOW IN POROUS MEDIA 131
Maciej Matyka, Zbigniew Koza, and Arzhang Khalili

HIGH RESOLUTION SIMULATION OF FLUID FLOW IN PRESS FELTS USED IN PAPER MANUFACTURING 132
David Trebotich and Terry Ligocki

VISUALIZATION OF WATER ADSORPTION IN CLAY WITH COMPLEX SURFACE WETTABILITY 133
Rui Xu and Masa Prodanovic

CAPROCK TOPOGRAPHY AND INJECTION POSITION INFLUENCE ON THE SOLUBILITY TRAPPING PHENOMENA DURING CO_2 GEOLOGICAL SEQUESTRATION 134
Pradeep Reddy Punnam, Dasika Prabhat Sourya, and Vikranth Kumar Surasani

PORE-SCALE CONTROLS ON CALCITE DISSOLUTION USING DIRECT NUMERICAL SIMULATIONS 135
Sergi Molins and David Trebotich

DROPLET DETACHMENT FROM A GAS DIFFUSION LAYER OF A PEM FUEL CELL 136
Daniel Niblett, Vahid Niasar, Adrian Mularczyk, and Jens Eller

ENGLISH-SPANISH GLOSSARY OF POROUS MEDIA 137
Leyre Alegre-Figuero, Ezequiel F. Médici, and Alejandro D. Otero

CORRECTION TO: ALBUM OF POROUS MEDIA C1
Ezequiel F. Médici and Alejandro D. Otero

INTRODUCTION

The history of human civilization is filled with remarkable narratives of ingenuity that will never cease to evoke in us a sense of wonder regarding our capacity for abstract thinking. Abstract thinking, together with the conceptual communication of that abstract thinking are some of the foundational pillars of human civilization. Among the many concepts that have played a significant role in the evolution of humanity since its early stages and continue to do so is the concept of porous media. While this introduction will serve in part to provide a brief overview of porous media and how they have been portrayed throughout history, the primary motivation for compiling this album is to capture the beauty of porous media in light of the state-of-the-art tools that allow us to explore the world beyond our natural sensorial capabilities.

To define porous media, let us first address the question of what a pore is. Rather than rephrasing its standard scientific definition, which can be found in any textbook in this field, let us revisit the etymology of the word *pore* and explore what can be learned from its origins and evolution through history. The word *pore* as we currently use it in the English language, has its origins in the Greek word πόρος (porous), which has been adopted with minor variations by many languages, including languages with Latin, Germanic, and Slavic roots. The first known written record of the word πόρος has its appearance in the Iliad (Book 14, line 433) [1], with a slightly different meaning from how we use the word today: something more closely resembling a ford than a fully enclosed conduit. Still, the word already captured the essence of fluid flowing through a guided path. With the passage of time, πόρος was used exclusively to identify small, fully enclosed passages or conduits. However, the Greek word, πόρος, has its origins in the Indo-European word *per*, meaning passage or journey[2]. Interestingly, this Indo-European word gave rise to other words commonly used in the field of porous media, such as petroleum (from a combination of the Greek words *petra – (per-tra)* meaning rock, and *elaion* which means oil or olive tree) and permeability. This historical evolution of the word pore already conveys the sense that porous materials have been an intrinsic part of the evolution of human civilization for a long time.

To mention all kinds of porous materials by name would be an almost impossible task to complete due to their large quantity. However, because of the impact some porous media have had throughout history, some human-made porous media (to distinguish them from the naturally occurring geological porous media which has a dedicated section in this album) deserve special attention: ceramics, textiles, filters, and paper. The order in which these items are listed may also reflect their chronological order of appearance, and this order is by no means absolute, as new archaeological evidence is shedding fresh light on our past and the dates and places mentioned subsequently are subject to change. Let us begin with ceramics. The first question a science historian must tackle when embarking on the journey into the fascinating world of porous media is perhaps that of when humanity first realized that one shapeless substance, such as air or water, could intermix with a shaped substance, such as rocks (we now refer to these substances as fluids and solids, respectively), without any apparent modification to the shape of the latter. This property of a solid material is what we now call permeability. Perhaps the answer to the question can be found in the inverse of this property (impermeability, or lack thereof) in one of the oldest human inventions: pottery, and by extension, ceramics. The fact that since ancient times we have made pottery impermeable may be the first indication of our awareness of porous media. Note that the earliest evidence of ceramics used as containers dates back 20,000 years ago[a], predating even the development of agriculture[3] (cover image, Part III). Textiles are also among the oldest human-made porous media. Some of the oldest pieces of human-made twisted flax fiber and tur wool are from 34,000 years ago[5] and they were found at the Dzudzuana Cave, Republic of Georgia. However, one the earliest known evidence of fibers being used to make textiles was found at the Guitarrero Cave in Peru[6] and dates back to 11,000 years ago.

The development of agriculture produced several technological advances, including the manufacturing of purposely designed porous media. We found the earliest evidence of filters in the Brześć Kujawski archaeological site in Poland, dating from 5000 BC; these were used to separate whey from curd while making cheese[b][7]. A painting dated to circa 1450 BC is cited as an early evidence of filters being used in Egypt to strain water or wine[8] (cover image, Part IV). Fast forward several millennia and research conducted by Darcy on the fluid dynamics of filtration can be considered the beginning of the modern scientific framework for characterizing and modeling porous media. The concepts presented in his famous pioneering work, *Les fontaines publiques de la ville de Dijon*[9], has became a

[a] Pieces of pottery found in the Xianrendong Cave, China[3] from 20,000 years ago are the earliest evidence of a culture mastering a pottery technique to make functional containers. The earliest evidence of ceramics is a female figurine from 25,000 years ago found in the Dolni Věstonice archaeological site in the Czech Republic[4]. However, the technique used to make the figure does not appear suitable for making containers.

[b] It is worth mentioning that the words *filter* and *felt* are intimately connected. *filter* comes from the Latin word *filtrum*, which has a shared origin with the word *felt* in the Indo-European word *pel*[2].

E. F. Médici and A. D. Otero (eds.), *Album of Porous Media*,
https://doi.org/10.1007/978-3-031-23800-0_1

law, Darcy's Law, and a standard method for porous media characterization through the measurement of a then new property called permeability (cover image, Part VII). A comprehensive review of the history of porous media theory can be found in the work of R. de Boer [10].

Last but not least in our broad historical categorization of porous media is paper, a porous medium that has incredibly revolutionized the communication and democratization of abstract thought and is perhaps surpassed only by the internet. Paper provides a reliable and practical medium for sharing feelings in the form of art and scientific concepts, for accounting, and even as clothing. We will exclude from this discussion the historical use of other non-cellulose-based materials for similar purposes, including leather, tree bark, and shells. The word *paper* has its origins in the Greek word πάπυρος (papyrus), probably of ancient Egyptian origin, which was used to describe a type of smooth and thin surface made from *Cyperus Papyrus* fiber with sufficiently good quality for ink or pigment deposition. Papyrus does not factor into the contemporary definition of paper, which is produced through the process of breaking, separating, and drying cellulose fibers and then pressing them into a thin sheet. This paper-making technique was conceived, and produced in large scale, at two locations in the world: China[11] and Mexico[12, 13]. However, all paper manufacturing techniques used today can be traced almost exclusively to Chinese origins. The colonization of the Americas brought an abrupt end to what was once a thriving paper-making culture. Today, only a few remains of that industry can be found in the form of only few surviving codices and the accounts of the early conquistadors. In fact, paper was so important in Mesoamerica that it was offered as tribute to the rulers[c]. The paper industry reemerge again in the Americas with the first paper mill and press granted by the Spanish crown in 1539, which used the Chinese paper-making tradition[14].

Depiction of abstract thoughts has played a pivotal role in the development of written communication. The earliest depictions of abstract thoughts can be found in petroglyphs and pictographs from around the world, featuring a variety of themes and styles. Interestingly, some of these earlier depictions eventually evolved into many of the symbols used in written languages. This perdurable method of communication is considered one of the most critical achievements in human evolution: a major leap that allows us to share valuable information effectively and efficiently through time and space, at least with those in cultural groups in agreement on its meaning. Despite the common origins shared by the art of depiction and written language, the art of depiction was, and still is, a necessary tool for technical communication that complements the content described in written language. Some of the earliest depictions of porous media, capturing its modern conceptualization of a solid matrix with voids, are in the 1665 drawings by R. Hooke presented in *Micrographia*[15] (cover image, Part I). It is worth mentioning that R. Hooke took advantage of a recently developed scientific invention that would become the standard instrument for visualizing and characterizing porous media: the microscope. Another important aspect of *Micrographia* is that it was printed—not handwritten—on paper. The invention of the printing press is another major development in human history in which porous media played a key role as we mentioned before.

The invention of photography in the 19th century greatly improved the art of depiction, including the field of porous media. In fact, scientific communication currently relies heavily on photography to depict observations of the natural world, to the point that an observation unaccompanied by a photograph lacks a certain degree of credibility. In addition to photography as a depiction tool, recent advances in visualization techniques have catalyzed an enormous wealth of insight in every field of science and engineering by extending our senses far beyond what we are capable of perceiving naturally. These fascinating and rapid developments in visualization techniques have motivated several editions of atlases summarizing these advances in various specialized fields. The field of porous media has also benefited enormously from these developments in visualization techniques. Three modern advances in visualization techniques that have uniquely improved our perception of porous media are X-rays[16] (cover image, Part II), electron microscopy[17] (cover image, Part V), and positron emission tomography (cover image, Part VI). Indeed, developments in these techniques have led to the enhanced morphological characterization of porous media, expanding our understanding of the various physical processes that occur at the pore level, such as mass transport, capillarity, swelling, and fracturing. These observations have, in turn, led to superior usage practices for existing porous materials and the design of new products involving porous materials. However, to date, there is no book compilation or atlas with visualization in porous media as its primary focus. Therefore, this album provides a collection of state-of-the-art visualizations covering various aspects of the science and engineering behind porous media and is intended to serve as a reference for future research and education.

This collection of scientific works includes high-resolution pore morphology, new porous materials, visualization of various transport processes that occur inside porous materials, and interpretations of numerical modeling of porous materials. These works were submitted by members of the porous media community and scrutinized via a peer-review process. Within the album, these scientific works are organized into

[c] It is worth noting that the American paper-making tradition was not only interrupted by the colonization of the Americas, but was also forgotten in mainstream scientific literature until very recently when new archaeological discoveries and reexamination of old anecdotal accounts revealed a sense of the magnitude of the production scale, cultural importance, and technique sophistication of the Mesoamerican paper.

seven themed parts covering different applications involving porous media and different transport mechanisms active at the pore level. Each part includes a collection of images with captions briefly describing the novelty of the observations and the visualization technique used. The seven parts are as follows:

I. A glimpse of porous media
II. Geological porous media
III. Two phase flow
IV. Phase change and diffusion
V. Transport in microfluidic devices
VI. Transport in porous media
VII. Computational modeling

The intent of Part I is to present the reader with visualizations from a broad range of fields that have porous media as essential element. Part II is a special section dedicated to geological porous media addressing the different phenomena manifesting in the solid matrix and in the pore space. Part III is dedicated to the visualization of the remarkable and sometimes counterintuitive dynamics resulting from the interaction between two or more fluids at the pore level, where surface tension is the dominant force. Part IV covers other commonly occurring transport processes in porous media that are dominated by the law of diffusion, such as heat and mass transport and phase change. In Part V, the focus is on observing these transport processes using microfluidic devices, which have become an important analog analysis tool for studying porous media either by isolating the effect of the stochastic nature of the pore size or permitting visual access to what would otherwise be an opaque media. However, significant effort has been dedicated to the *in situ* visualization of the dynamics of the solid matrix and the transport processes in pore spaces. This is the subject of Part VI. Finally, Part VII is dedicated to the visualization of transport processes in porous media through computational modeling.

Reflecting the cultural background of both editors and to reach a larger audience, we took the challenging initiative of complementing the album with a translation to the Spanish language. This album is also accompanied with a list of the terminology equivalency (whenever it was possible) between English and Spanish of the most relevant technical terms used by the porous media community that was organically emerging during the translation. By encouraging bilingual publications, we hope this effort will serve scientists, science communicators, engineers, and educators in bridging science communication and education between both languages.

Our hope in putting this collection of works together is to help disseminate information on the various visualization and experimental techniques, as well as the challenges faced by the different fields that utilize porous media and might otherwise be unaware of efforts in the other fields. We also hope to engage readers with the fascinating microscopic world and hopefully spark their curiosity—especially that of the next generation of scientists and engineers—about the vast field of porous media.

Ezequiel F. Médici – Alejandro D. Otero

1. Murray A.T. and Wyatt W.F. (1925). *Homer. Iliad, Volume II. Books 13-24. Loeb Classical Library 171.* Harvard University Press.
2. Pokorny J. (1959). *Indogermanisches etymologisches wörterbuch.* Francke.
3. Wu X. *et al.* (2012). *Science*, 336(6089):1696.
4. Vandiver P.B. *et al.* (1989). *Science*, 246(4933):1002.
5. Kvavadze E. *et al.* (2009). *Science*, 325(5946):1359.
6. Jolie E.A. *et al.* (2011). *Current Anthropology*, 52(2):285.
7. Salque M. *et al.* (2013). *Nature*, 493(7433):522.
8. Wilkinson J.G. and Birch S. (1878). *The manners and customs of the ancient Egyptians*, volume 2. John Murray.
9. Darcy H. (1856). *Les fontaines publiques de la ville de Dijon exposition et application.* Victor Dalmont.
10. De Boer R. (2000). *Theory of porous media: highlights in historical development and current state.* Springer Science & Business Media.
11. Tsuen-Hsuin T. (1985). *Science and Civilisation in China. Chemistry and Chemical Technology. Part I. Paper and Printing.*, volume 5. Cambridge University Press.
12. Neumann F.J. (1973). *History of Religions*, 13(2):149.
13. Benz B.F. *et al.* (2006). *Ancient Mesoamerica*, 17(2):283.
14. Fernández Í.F. (2010). *Documentación de las Ciencias de la Información*, 33:69.
15. Hooke R. (1665). *Micrographia: or Some Physiological Description of Minute Bodies Made by Magnifying Glasses with Observations and Enquires Threupon.* Royal Society.
16. Röntgen W.C. (1896). *Science*, 3(59):227.
17. Ruska E. (1987). *Reviews of modern physics*, 59(3):627.

La historia de la civilización humana está llena de notables narraciones de ingenio que nunca dejarán de evocar en nosotros una sensación de asombro con respecto a nuestra capacidad para el pensamiento abstracto. El pensamiento abstracto, junto con la comunicación conceptual de ese pensamiento abstracto, son algunos de los pilares fundacionales de la civilización humana. Entre los muchos conceptos que han desempeñado un papel importante en la evolución de la humanidad desde sus primeras etapas, y continúan haciéndolo, se encuentra el concepto de medios porosos. Si bien esta introducción servirá en parte para brindar una breve descripción general de los medios porosos y cómo han sido retratados a lo largo de la historia, la principal motivación para compilar este álbum es capturar la belleza de los medios porosos a la luz del estado del arte de las herramientas que nos permiten explorar el mundo más allá de nuestras capacidades sensoriales naturales.

Para definir los medios porosos, primero debemos abordar la cuestión de qué es un poro. En lugar de reformular su definición científica estándar, que se puede encontrar en cualquier libro de texto en este campo, revisemos la etimología de la palabra *poro* y exploremos lo que se puede aprender de sus orígenes y evolución a lo largo de la historia. La palabra *poro*, tal como se usa actualmente en el idioma español, tiene su origen en la palabra griega πόρος (poros), que ha sido adoptada con variaciones menores por muchos idiomas, incluidos idiomas con raíces latinas, germánicas y eslavas. El primer registro escrito conocido de la palabra πόρος tiene su aparición en la Ilíada (Libro 14, línea 433) [1], con un significado ligeramente diferente de cómo usamos la palabra hoy: algo más parecido a un vado que a un conducto completamente cerrado. Aún así, la palabra ya capturaba la esencia de un fluido que fluye a través de un camino guiado. Con el paso del tiempo, πόρος se utilizó exclusivamente para identificar pasajes o conductos pequeños y completamente cerrados. Sin embargo, la palabra griega, πόρος, tiene su origen en la palabra indoeuropea *per*, que significa conducir o llevar[2]. Curiosamente, esta palabra indoeuropea dio lugar a otras palabras de uso común en el campo de los medios porosos, como petróleo (de una combinación de las palabras griegas *petra – (per-tra)* que significa roca y *elaion* que significa aceite u olivo) y permeabilidad. Esta evolución histórica de la palabra poro ya transmite la sensación de que los materiales porosos han sido una parte intrínseca de la evolución de la civilización humana durante mucho tiempo.

Mencionar todos los tipos de materiales porosos por su nombre sería una tarea casi imposible de completar debido a su gran cantidad. Sin embargo, por el impacto que han tenido a lo largo de la historia, algunos medios porosos hechos por el hombre (para distinguirlos de los medios porosos geológicos que a su vez tienen una sección exclusiva en este álbum) merecen una atención especial: cerámica, textiles, filtros y papel. El orden en el que se enumeran estos elementos también puede interpretarse como su orden cronológico de aparición, sin embargo, este orden no es absoluto, ya que nuevas evidencias arqueológicas arrojan una visión actualizada sobre nuestro pasado y las fechas y lugares mencionados posteriormente están sujetos a cambios. Comencemos con la cerámica. La primera pregunta que debe abordar un historiador de la ciencia cuando se embarca en la travesía del fascinante mundo de los medios porosos es quizás la de cuándo la humanidad se dio cuenta por primera vez de que una sustancia sin forma, como el aire o el agua, podía entremezclarse con una sustancia con forma, como las rocas (ahora nos referimos a estas sustancias como fluidos y sólidos, respectivamente), sin ninguna modificación aparente en la forma de este último. Esta propiedad de los materiales sólidos es lo que ahora llamamos permeabilidad. Quizás la respuesta a la pregunta se encuentre en la inversa de esta propiedad (impermeabilidad, o falta de ella) en uno de los inventos humanos más antiguos: la alfarería, y por extensión, la cerámica. El hecho de que desde la antigüedad hayamos impermeabilizado la cerámica puede ser el primer indicio de nuestra conciencia de los medios porosos. Hay que tener en cuenta que la evidencia más antigua del uso de la cerámica como recipiente data de hace 20000 años[a], anterior incluso al desarrollo de la agricultura[3] (imagen en la portada de la Parte III). Los textiles también se encuentran entre los medios porosos más antiguos creados por el hombre. Algunos de los pedazos más antiguos de hilados de fibra de lino y lana de tur hecha por los humanos data de hace 34000 años[5] y se encontró en la cueva de Dzudzuana, en la República de Georgia. Sin embargo, la evidencia más temprana hasta ahora conocida del uso de fibras para fabricar textiles se encontró en la Cueva de Guitarrero en Perú[6] y data de hace 11000 años.

El desarrollo de la agricultura produjo varios avances tecnológicos, incluido la fabricación de materiales porosos diseñados para algún propósito especifico. Encontramos la evidencia más temprana de filtros en el sitio arqueológico Brześć Kujawski en Polonia, que data del año 5000 a. C.; estos se usaron para separar el suero de la cuajada para hacer queso[b][7]. Una pintura que data de alrededor de 1450 a. C. se cita como una evidencia temprana del uso de filtros en Egipto para colar agua o vino[8] (imagen de portada, Parte IV). Si avanzamos rápido varios milenios nos encontramos con la investigación realizada por Darcy sobre la dinámica de fluidos en la filtración que puede considerarse el comienzo del marco científico moderno para caracterizar y modelar los medios porosos. Su famoso trabajo pionero, *Les fontaines publiques de la ville de Dijon*[9], se convirtió en una ley, la Ley de Darcy, y un método estándar para la caracterización de medios porosos a través de la medición de una nueva, en aquel entonces, propiedad llamada permeabilidad (imagen en la portada de la Parte VII). Una revisión exhaustiva de la historia de la teoría de los medios porosos se encuentra en el trabajo de R. de Boer [10].

Por último, pero no menos importante, en nuestra amplia categorización histórica de los medios porosos está el

[a] Piezas de cerámica encontradas en la cueva de Xianrendong, China[3] de hace 20000 años son la evidencia más temprana de una cultura que dominó la técnica de la alfarería para hacer recipientes funcionales. Sin embargo, la evidencia más antigua de cerámica es una estatuilla femenina de hace 25000 años encontrada en el sitio arqueológico de Dolni Věstonice en la República Checa[4]. Sin embargo, la técnica utilizada para realizar la figura no parecía adecuada para la realización de recipientes.

[b] Vale la pena mencionar que las palabras *filtro* y *fieltro* están íntimamente conectadas. *filtro* proviene de la palabra latina *filtrum*, que tiene un origen compartido con la palabra *fieltro* en la palabra indoeuropea *pel*[2].

papel, un medio poroso que ha revolucionado increíblemente la comunicación y la democratización del pensamiento abstracto y quizás solo sea superado por Internet. El papel proporciona un medio fiable y práctico para compartir sentimientos en forma de conceptos artísticos y científicos, para la contabilidad e incluso como ropa. Excluiremos de esta discusión el uso histórico de otros materiales que no están basados en fibras de celulosa para fines similares, incluidos el cuero, la corteza de los árboles y las conchas. La palabra *papel* tiene su origen en la palabra griega πάπυρος (papiro), probablemente de origen egipcio antiguo, que fue usada para describir un tipo de superficie lisa y delgada hecha de fibras de *Cyperus Papyrus* con una calidad suficientemente buena para la deposición de tinta o pigmento. Lo que hoy en día se denomina Papyrus no incluye la definición contemporánea de papel, que se produce a través del proceso de romper, separar y secar las fibras de celulosa y luego presionarlas en una hoja delgada. Esta técnica de fabricación de papel fue concebida y producida a gran escala en dos lugares del mundo: China[11] y México[12, 13]. Sin embargo, todas las técnicas de fabricación de papel utilizadas actualmente se pueden rastrear casi exclusivamente a orígenes chinos. La colonización de América trajo un final abrupto a lo que alguna vez fue una próspera cultura de fabricación de papel. Hoy en día, los únicos restos que pueden encontrarse de esa industria son los pocos códices sobrevivientes y los relatos dejados por los primeros conquistadores. De hecho, el papel era tan importante en Mesoamérica que se ofrecía como tributo a los gobernantes[c]. La industria del papel resurge nuevamente en América con la primera fábrica de papel y prensa concedida por la corona española en 1539, que utilizó la tradición china de fabricación de papel[14].

La representación de pensamientos abstractos ha desempeñado un papel fundamental en el desarrollo de la comunicación escrita. Las primeras representaciones de pensamientos abstractos se pueden encontrar en petroglifos y pictografías alrededor de todo el mundo, con una variedad de temas y estilos. Curiosamente, algunas de estas representaciones eventualmente evolucionaron dando origen a muchos de los símbolos utilizados por los lenguajes escritos. Este método perdurable de comunicación se considera uno de los logros más críticos en la evolución humana: un gran salto que nos permite compartir información valiosa de manera efectiva y eficiente a través del tiempo y el espacio, al menos dentro de ciertos grupos culturales que coinciden en su sentido. A pesar de los orígenes comunes que comparten el arte de la representación y el lenguaje escrito, el arte de la representación fue, y sigue siendo, una herramienta necesaria para la comunicación técnica que complementa al lenguaje escrito con contenido descriptivo. Algunas de las primeras representaciones de medios porosos, que capturan su conceptualización moderna de una matriz sólida con espacios vacíos, se encuentran en los dibujos de 1665 de R. Hooke presentados en *Micrographia*[15] (imagen en la portada de la Parte I). Cabe mencionar que R. Hooke aprovechó un invento científico desarrollado recientemente que se convertiría en el instrumento estándar para visualizar y caracterizar medios porosos: el microscopio. Otro aspecto importante de *Micrographia* es que fue impresa, no escrita a mano, en papel. La invención de la imprenta es otro gran avance en la historia humana en el que los medios porosos jugaron un papel clave, como se dijo anteriormente.

La invención de la fotografía en el siglo XIX mejoró enormemente el arte de la representación, incluido el campo de los medios porosos. De hecho, la comunicación científica actual se basa en gran medida en la fotografía para representar las observaciones del mundo natural, hasta el punto de que una observación que no vaya acompañada de una fotografía carece de cierto grado de credibilidad. Además de la fotografía como herramienta de representación, los avances recientes en las técnicas de visualización han catalizado una enorme riqueza de conocimientos en todos los campos de la ciencia y la ingeniería al extender nuestros sentidos mucho más allá de lo que somos capaces de percibir de forma natural. Estos fascinantes y rápidos avances en las técnicas de visualización han motivado varias ediciones de atlas que resumen estos avances en varios campos especializados. El campo de los medios porosos también se ha beneficiado enormemente de estos avances en las técnicas de visualización. Tres avances modernos en las técnicas de visualización que han mejorado de manera única nuestra percepción de los medios porosos son los rayos X[16] (imagen en la portada de la Parte II), la microscopía electrónica[17] (imagen en la portada de la Parte V) y la tomografía por emisión de positrones (imagen en la portada de la Parte VI). De hecho, los avances en estas técnicas han llevado a una mejor caracterización morfológica de los medios porosos, expandiendo nuestra comprensión de los diversos procesos físicos que ocurren a nivel de los poros, como el transporte de masa, la capilaridad, la hinchazón y la fractura. Estas observaciones, a su vez, han llevado a mejores prácticas de uso para los materiales porosos existentes y al diseño de nuevos productos que involucran materiales porosos. Sin embargo, hasta la fecha, no existe una compilación de libros o atlas con visualización en medios porosos como su enfoque principal. Por lo tanto, este álbum proporciona una colección de visualizaciones de vanguardia que cubren varios aspectos de la ciencia y la ingeniería detrás de los medios porosos y pretende servir como referencia para futuras investigaciones y para la educación.

Esta colección de trabajos científicos incluye morfología de poros de alta resolución, nuevos materiales porosos, visualización de varios procesos de transporte que ocurren dentro de materiales porosos e interpretaciones del modelado numérico de materiales porosos. Estos trabajos fueron enviados por miembros de la comunidad de medios porosos y examinados a través de un proceso de revisión por pares. Dentro del álbum, estos trabajos científicos están organizados en siete partes temáticas que cubren diferentes aplicaciones que involucran medios porosos y diferentes mecanismos de transporte activos a nivel de poro. Cada parte incluye una colección de imágenes con leyendas que describen brevemente la novedad de las observaciones y

[c] Vale la pena señalar que la tradición americana de fabricación de papel no solo fue interrumpida por la colonización de América, sino que también fue olvidada en la literatura científica dominante hasta hace muy poco, cuando nuevos descubrimientos arqueológicos y la revisión de relatos anecdóticos antiguos revelaron un sentido de la magnitud de la escala de producción, importancia cultural y sofisticación técnica del papel mesoamericano.

la técnica de visualización utilizada. Las siete partes son las siguientes:

I. Un vistazo a los medios porosos
II. Medios porosos geológicos
III. Flujo bifásico
IV. Cambio de fase y difusión
V. Transporte en dispositivos de microfluidos
VI. Transporte en medios porosos
VII. Modelado computacional

La intención de la Parte I es presentar al lector visualizaciones de una amplia gama de campos que tienen a los medios porosos como elementos centrales. La Parte II es una sección especial dedicada a los medios porosos geológicos abordando los diferentes fenómenos que se manifiestan en una matriz sólida y en el espacio poroso. La Parte III está dedicada a la visualización de la destacable, y a veces contraintuitiva, dinámica que resulta de las interacciones entre dos o más fluidos al nivel de los poros, donde la tensión superficial es la fuerza dominante. La Parte IV cubre otros procesos de transporte que ocurren comúnmente en medios porosos que están dominados por la ley de la difusión, como el transporte de calor y masa y el cambio de fase. En la Parte V, la atención se centra en la observación de estos procesos de transporte utilizando dispositivos de microfluidos, que se han convertido en una importante herramienta de análisis analógico para estudiar medios porosos, ya sea aislando el efecto de la naturaleza estocástica del tamaño del poro o permitiendo el acceso visual a lo que de otro modo sería un medio opaco. Sin embargo, se ha dedicado un esfuerzo significativo a la visualización *in situ* de la dinámica de la matriz sólida y los procesos de transporte en los espacios porosos. Este es el tema de la Parte VI. Finalmente, la Parte VII está dedicada a la visualización de los procesos de transporte en medios porosos a través del modelado computacional.

Reflejando el trasfondo cultural de ambos editores y con el fin de alcanzar a una mayor audiencia, tomamos la desafiante iniciativa de complementar el álbum con una traducción al idioma español. Este álbum también es acompañado con una lista de equivalencias terminológicas (siempre que fuese posible) entre inglés y español de los términos técnicos más relevantes utilizados por la comunidad de medios porosos que fue surgiendo orgánicamente durante la traducción. Al alentar las publicaciones bilingües, esperamos que este esfuerzo sirva a los científicos, comunicadores científicos, ingenieros y educadores para fomentar la comunicación científica y la educación en ambos idiomas.

Nuestra expectativa al reunir esta colección de trabajos es ayudar a difundir información sobre las diversas técnicas experimentales y de visualización, así como los desafíos que enfrentan los diferentes campos que utilizan medios porosos y, de lo contrario, podrían no estar al tanto de los esfuerzos en los otros campos. También esperamos involucrar a los lectores con el fascinante mundo microscópico y despertar su curiosidad, especialmente la de la próxima generación de científicos e ingenieros, sobre el vasto campo de los medios porosos.

Ezequiel F. Médici – Alejandro D. Otero

1. Murray A.T. and Wyatt W.F. (1925). *Homer. Iliad, Volume II. Books 13-24. Loeb Classical Library 171.* Harvard University Press.
2. Pokorny J. (1959). *Indogermanisches etymologisches wörterbuch.* Francke.
3. Wu X. *et al.* (2012). *Science*, 336(6089):1696.
4. Vandiver P.B. *et al.* (1989). *Science*, 246(4933):1002.
5. Kvavadze E. *et al.* (2009). *Science*, 325(5946):1359.
6. Jolie E.A. *et al.* (2011). *Current Anthropology*, 52(2):285.
7. Salque M. *et al.* (2013). *Nature*, 493(7433):522.
8. Wilkinson J.G. and Birch S. (1878). *The manners and customs of the ancient Egyptians*, volume 2. John Murray.
9. Darcy H. (1856). *Les fontaines publiques de la ville de Dijon exposition et application.* Victor Dalmont.
10. De Boer R. (2000). *Theory of porous media: highlights in historical development and current state.* Springer Science & Business Media.
11. Tsuen-Hsuin T. (1985). *Science and Civilisation in China. Chemistry and Chemical Technology. Part I. Paper and Printing.*, volume 5. Cambridge University Press.
12. Neumann F.J. (1973). *History of Religions*, 13(2):149.
13. Benz B.F. *et al.* (2006). *Ancient Mesoamerica*, 17(2):283.
14. Fernández Í.F. (2010). *Documentación de las Ciencias de la Información*, 33:69.
15. Hooke R. (1665). *Micrographia: or Some Physiological Description of Minute Bodies Made by Magnifying Glasses with Observations and Enquires Thereupon.* Royal Society.
16. Röntgen W.C. (1896). *Science*, 3(59):227.
17. Ruska E. (1987). *Reviews of modern physics*, 59(3):627.

I. A GLIMPSE OF POROUS MEDIA

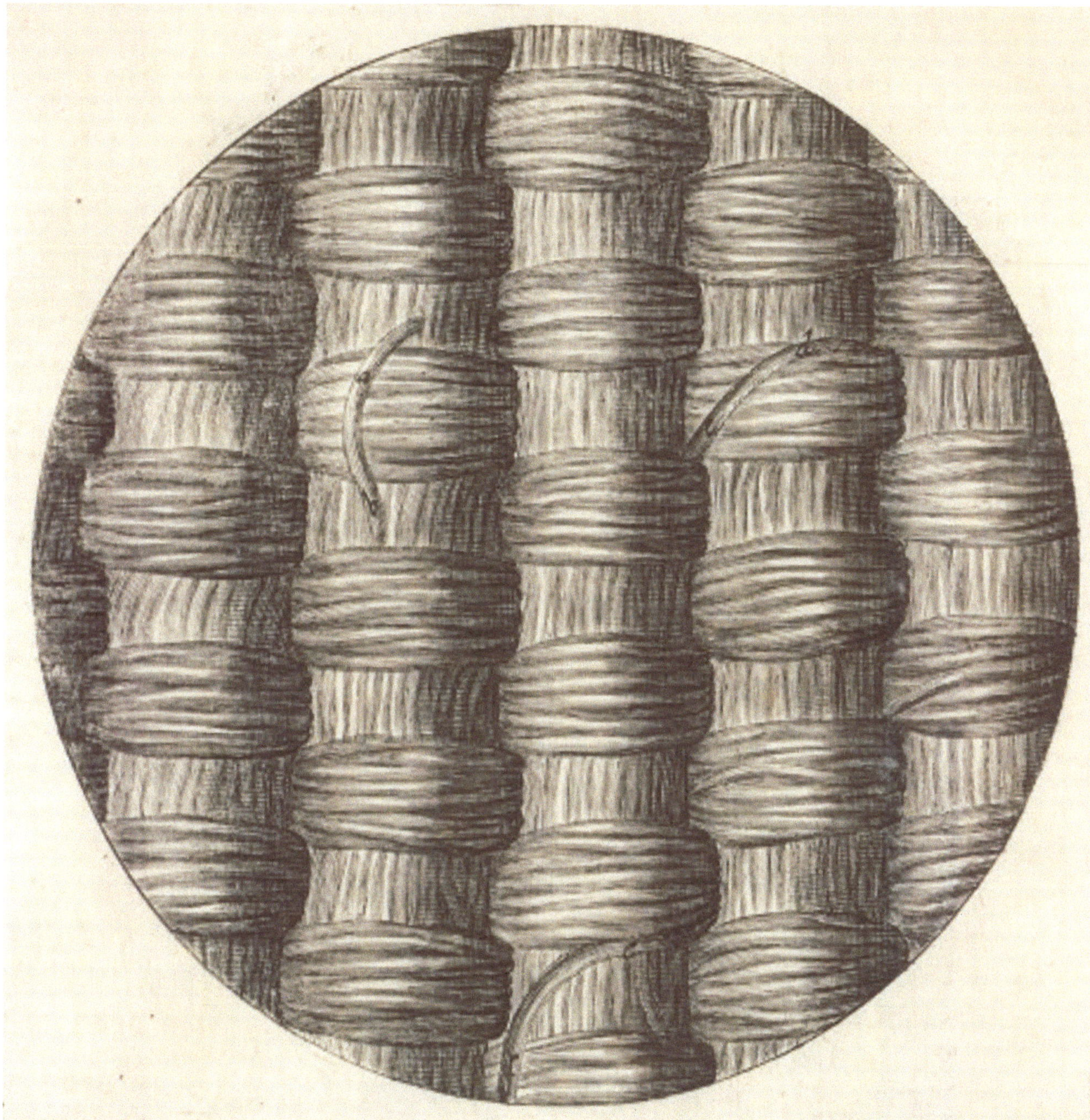

Earliest modern depiction and first known micrograph image of a porous media showing high fidelity details of the solid matrix of a silk fabric using an optical microscope by R. Hooke from 1665[1]. Courtesy Library of Congress, Rare Book and Special Collections Division.

Representación moderna más antigua y primera imagen micrográfica conocida de un medio poroso que muestra detalles de alta fidelidad de la matriz sólida de una tela de seda usando un microscopio óptico por R. Hooke de 1665[1]. Cortesía Library of Congress, Rare Book and Special Collections Division.

1. Hooke R. (1665). *Micrographia: or Some Physiological Description of Minute Bodies Made by Magnifying Glasses with Observations and Enquires Threupon.* Royal Society.

ELECTROSPUN FIBROUS MATS

Jeff Gostick[1], *Matthew Kok*[2], *and Rhodri Jervis*[2]

[1] University of Waterloo
[2] University College London

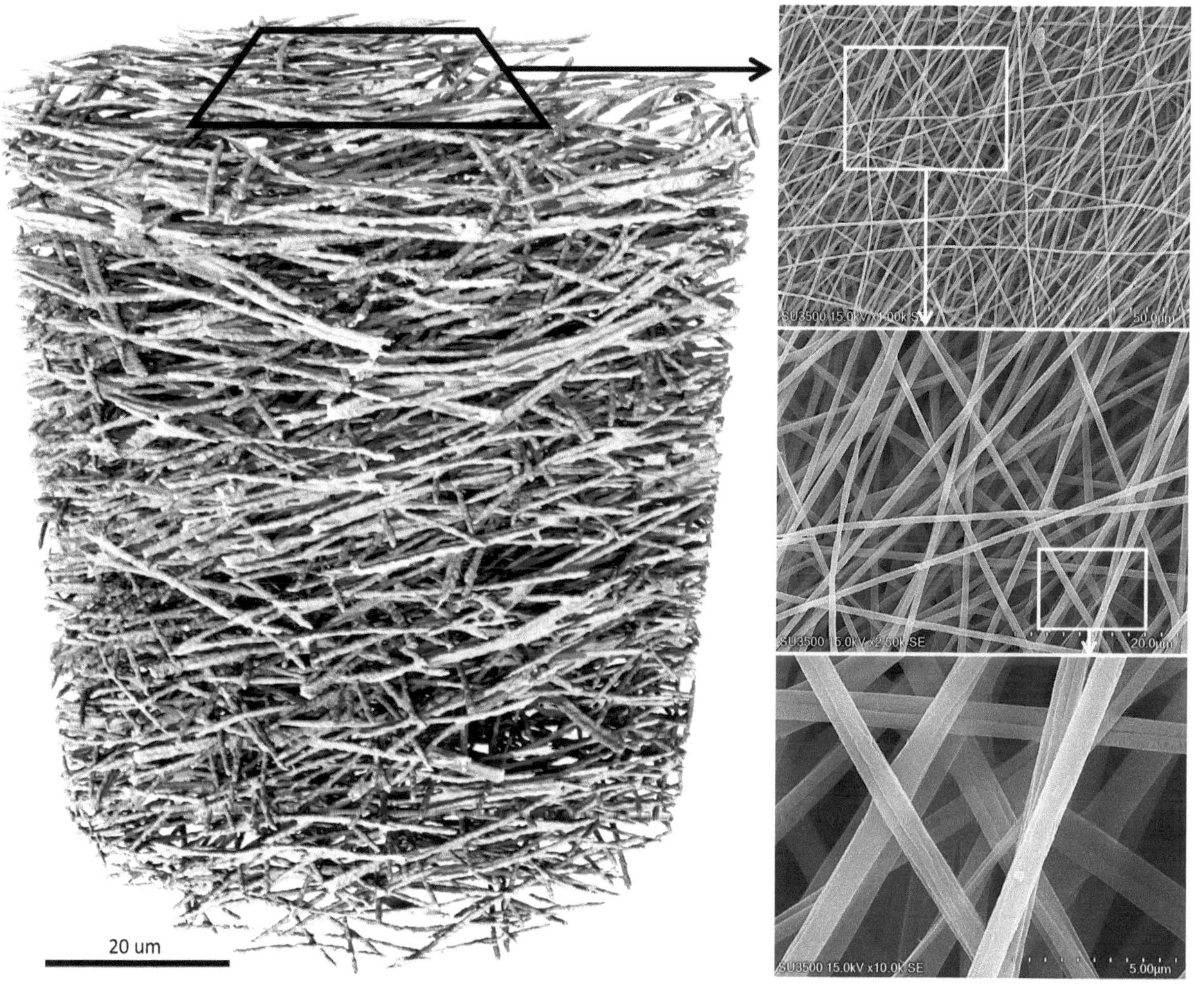

Electrospinning is a technique whereby a polymer-solvent solution is ejected from a syringe, and by the application of a high DC voltage (>10 kV) is drawn into very small fibers (~100-500 nm). The fibers land on an electrically grounded collector, and the process results in a fibrous mat structure with high porosity (>90 %) with a thickness below 1 mm. The end uses of electrospun mats are very diverse, including flow battery electrodes (once carbonized)[1], tissue scaffolds and wound dressings (if a biodegradable polymer was used), filtration membranes[2], and catalyst supports[3], to name a few. The imaging of such materials can easily be done using SEM, but nano-resolution X-ray tomography has also been applied to yield 3D images[4].

1. Liu S. *et al.* (2017). *J. Electrochem. Soc.*, 164(9):A2038.
2. Gopal R. *et al.* (2006). *J. Membrane Sci.*, 281(1-2):581.
3. Chan S. *et al.* (2018). *J. Mater Sci.*, 53(16):11633.
4. Jervis R. *et al.* (2018). *Energy Tech.*, 6(12):2488.

El electrohilado es una técnica en la que una solución de polímero solvente se extrude de una jeringa, la que al aplicarle un alto voltaje de CC (>10 kV) forma fibras muy pequeñas (~100-500 nm). Estas fibras se depositan en un colector conectado a tierra, resultando en un tejido fibroso de alta porosidad (>90 %) y grosor menor de 1 mm. Las aplicaciones de los tejidos electrohilados son muy variadas e incluyen electrodos de las baterías de flujo (una vez carbonizados)[1], soportes de tejidos y apósitos para heridas (si se utiliza un polímero biodegradable), membranas de filtración[2], y soportes para catalizadores[3], por mencionar algunos. La visualización de estos materiales puede hacerse fácilmente con MEB. No obstante, las tomografías de rayos X computarizadas de nanoresolución también se han aplicado con éxito para obtener imágenes en 3D[4].

Contact: Jeff Gostick <jgostick@uwaterloo.ca>

E. F. Médici and A. D. Otero (eds.), *Album of Porous Media*,
https://doi.org/10.1007/978-3-031-23800-0_2

TAILORED ELECTROSPUN GAS DIFFUSION LAYERS WITH A GRADED PORE STRUCTURE FOR POLYMER ELECTROLYTE MEMBRANE FUEL CELLS

Manojkumar Balakrishnan[1], *Pranay Shrestha*[1], *Nan Ge*[1], *ChungHyuk Lee*[1], *Kieran F. Fahy*[1], *Roswitha Zeis*[2], *Volker P. Schulz*[3], *Benjamin D. Hatton*[4], *and Aimy Bazylak*[1]

[1] Thermofluids for Energy and Advanced Materials Laboratory, Department of Mechanical and Industrial Engineering, University of Toronto
[2] Karlsruhe Institute of Technology, Helmholtz Institute Ulm
[3] Department of Mechanical Engineering, Cooperative State University Mannheim
[4] Functional and Adaptive Surfaces Group, Department of Materials Science and Engineering, University of Toronto

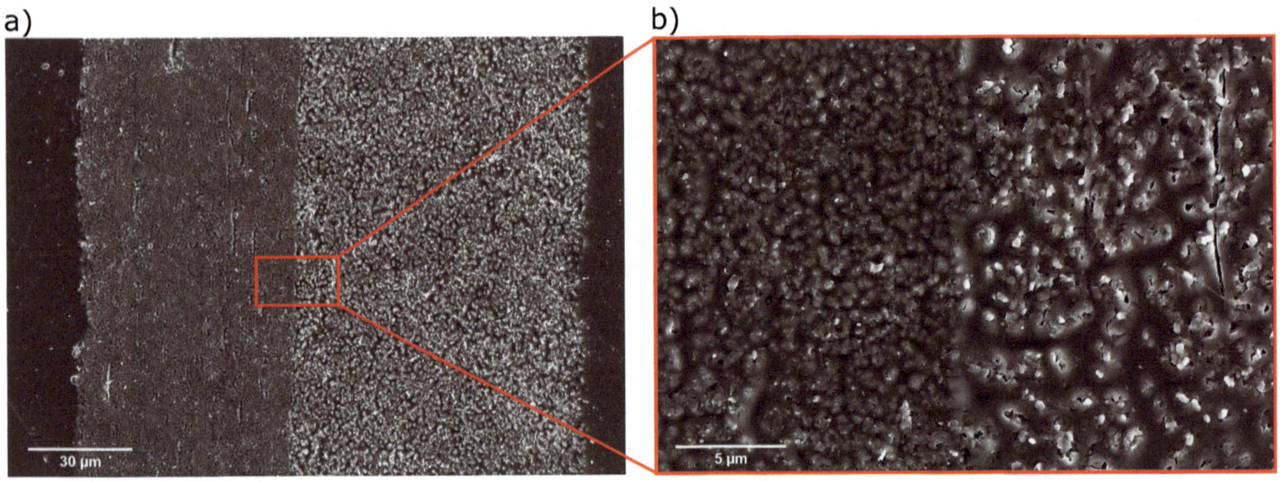

Polymer electrolyte membrane (PEM) fuel cells are sustainable and versatile alternatives for internal combustion engines in the transportation sector. However, high manufacturing costs prohibit the technology from being widely utilized. Here we showcase a novel, tailored electrospun gas diffusion layer (eGDL) with a graded pore structure (from the catalyst layer to the flow field) to improve the high current density performance of PEM fuel cells and reduce the fuel cell stack size and cost. The images feature the cross-section of the eGDL to highlight the unique pore structure. Specifically, a) showcases the complete cross-section of the eGDL with the graded electrospun layers, and b) highlights the variation in pore size achieved via electrospinning at the interface between the two electrospun layers. The sample was prepped by embedding the porous eGDL in resin, and the images were obtained via scanning electron microscopy. Details on the novel eGDL, the manufacturing technique, fuel cell performance, and imaging can be found in our publication[1].

1. Balakrishnan M. *et al.* (2020). *ACS Appl. Energy Mater.*, 3(3):2695.

Contact: Aimy Bazylak <aimy.bazylak@utoronto.ca>

Las pilas de combustible de membrana de electrolito polimérico (PEM, por sus siglas en inglés) son alternativas sostenibles y versátiles para los motores de combustión interna en el sector del transporte. Sin embargo, los elevados costos de producción impiden su uso generalizado. Aquí exponemos una novedosa capa de difusión de gas electrohilada (eGDL, por sus siglas en inglés) con una estructura porosa graduada (desde la capa del catalizador hasta los canales de flujo) para mejorar el rendimiento a alta densidad de corriente de las pilas de combustible PEM y reducir el tamaño del conjunto de las pilas de combustible y su costo. Las imágenes destacan una visión transversal de la eGDL para resaltar la singular estructura porosa. Específicamente, a) muestra una sección transversal completa de la eGDL con capas graduadas por electrohilado, y b) resalta la variación del tamaño del poro lograda mediante el electrohilado en la interfaz entre las dos capas electrohiladas. La muestra fue preparada al impregnar la eGDL porosa con resina, y las imágenes se obtuvieron con un microscopio electrónico de barrido. Los detalles de la novedosa eGDL, la técnica de producción, el rendimiento de la pila de combustible y la imagen pueden encontrarse en nuestra publicación[1].

E. F. Médici and A. D. Otero (eds.), *Album of Porous Media*,
https://doi.org/10.1007/978-3-031-23800-0_3

DISPERSED NANOSTRUCTURED THIN FILM (dNSTF) CATALYST

Karrar Alofari[1], *Anjana Asthana*[1], *Tajiri Kazuya*[1], *Andrew Haug*[2], *and Jeffrey Allen*[1]

[1] Michigan Technological University
[2] 3M Corp

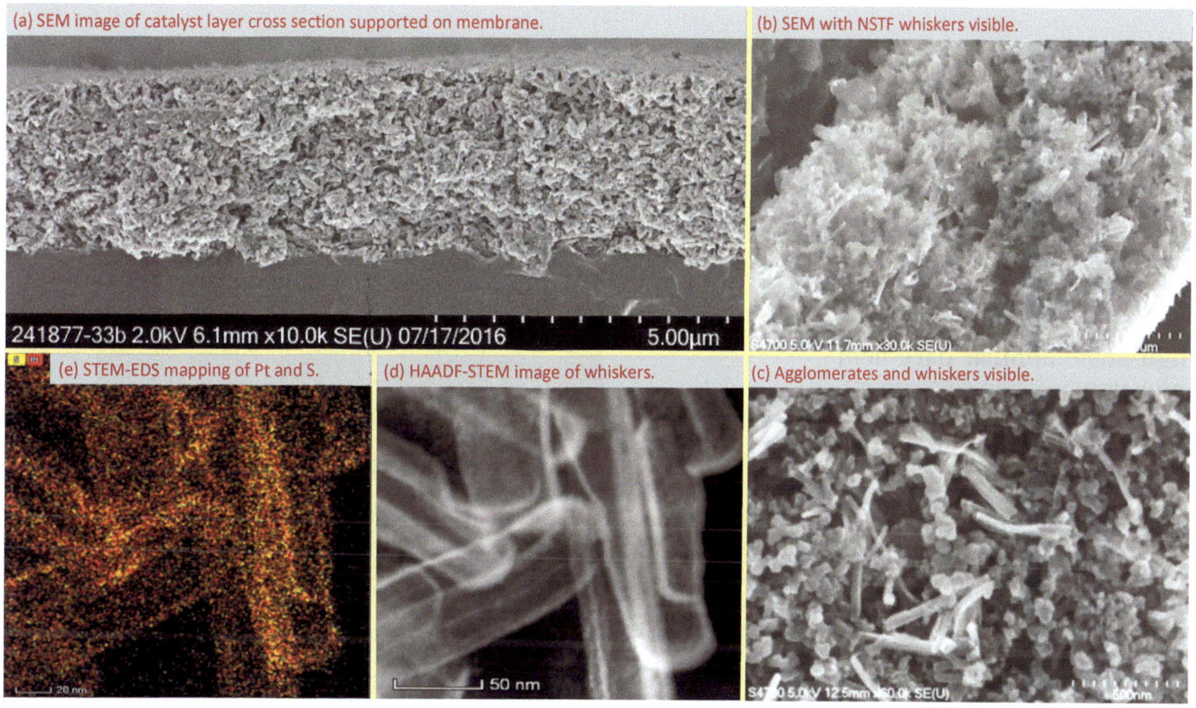

Typical proton exchange membrane fuel cell (PEMFC) electrodes are multi-micron thick, porous films comprised of platinum dispersed over nm-scale carbon particles bound by ion conducting polymer, (ionomer). Electrodes are generally made as an ink that is deposited and dried on a substrate. Nanostructured thin film (NSTF) whiskers (40 nm×500 nm) are an alternative electrode comprised of an organic perylene red core coated with platinum. Porous cathode electrodes (shown here) comprised of Pt NSTF whiskers dispersed in a matrix of carbon nanoparticles and ionomer have achieved an improved oxygen reduction activity and reduced electrochemical corrosion. These dispersed nanostructured thin films (dNSTF) are less than 10 nm thick. As a result, traditional porous media characterization techniques cannot be applied. Imaging technology, however, can be used to quantify structural and morphological distributions in the electrode. Microstructural characterization of the PEMFC samples was carried out by SEM and Aberration-corrected STEM/TEM (FEI Titan Themis 200kV). These tools, in conjunction with EDS, provide high-resolution imaging from which pore size distribution, chemical composition, as well as ionomer distribution can be determined.

Contact: Karrar Alofari <ktalofar@mtu.edu>

Los electrodos típicos de las pilas de combustible de membrana de intercambio protónico (PEMFC, por sus siglas en inglés) son capas porosas con un grosor de varios micrones compuestas por platino disperso sobre partículas de carbono de dimensiones nanométricas unidas por un polímero conductor de iones (ionómero). En general, los electrodos se fabrican como una tinta que se deposita y se seca sobre un sustrato. Los filamentos de las láminas delgadas nanoestructurada (NSTF, por sus siglas en inglés) (40 nm×500 nm) son electrodos alternativos compuestos por un núcleo rojo de perileno orgánico recubierto con platino. Los electrodos de cátodo poroso (que se muestran aquí) compuestos por filamentos de Pt NSTF dispersos en una matriz de nanopartículas de carbono e ionómero han logrado un mejoramiento de la actividad de la reducción del oxígeno y una disminución de la corrosión electroquímica. Estas láminas delgadas nanoestructuradas dispersas (dNSTF) tienen un grosor de menos de 10 nm. Como resultado, las técnicas tradicionales de caracterización de los medios porosos no pueden aplicarse. Sin embargo, técnicas de visualización pueden utilizarse para cuantificar las distribuciones estructurales y morfológicas en el electrodo. La caracterización microestructural de las muestras de PEMFC se realizó mediante MEB y MEBT/MET corregidos de aberración (FEI Titan Themis 200kV). Estas herramientas, junto con EDS, proporcionan visualizaciones de alta resolución a partir de las cuales se puede determinar la distribución de tamaño de poro y la composición química, así como la distribución de ionómero.

E. F. Médici and A. D. Otero (eds.), *Album of Porous Media*,
https://doi.org/10.1007/978-3-031-23800-0_4

ELECTROCHEMICAL ACTUATION OF A NANOPOROUS POLYPYRROLE HYBRID MATERIAL

Manuel Brinker[1], *Tobias Krekeler*[1], *and Patrick Huber*[2]

[1] Hamburg University of Technology
[2] Hamburg University of Technology and Deutsches Elektronen-Synchrotron DESY

(a)

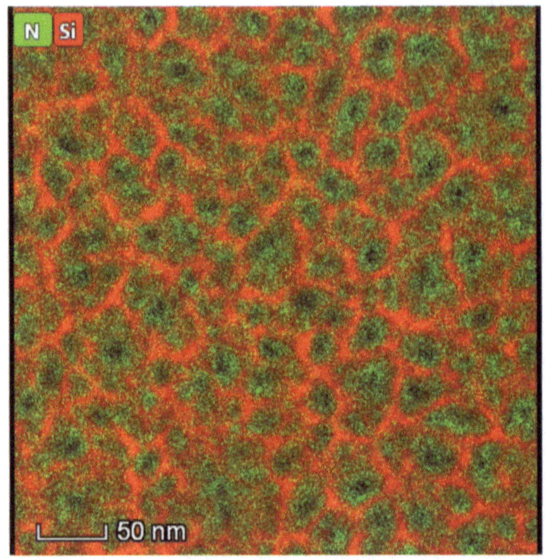

(b)

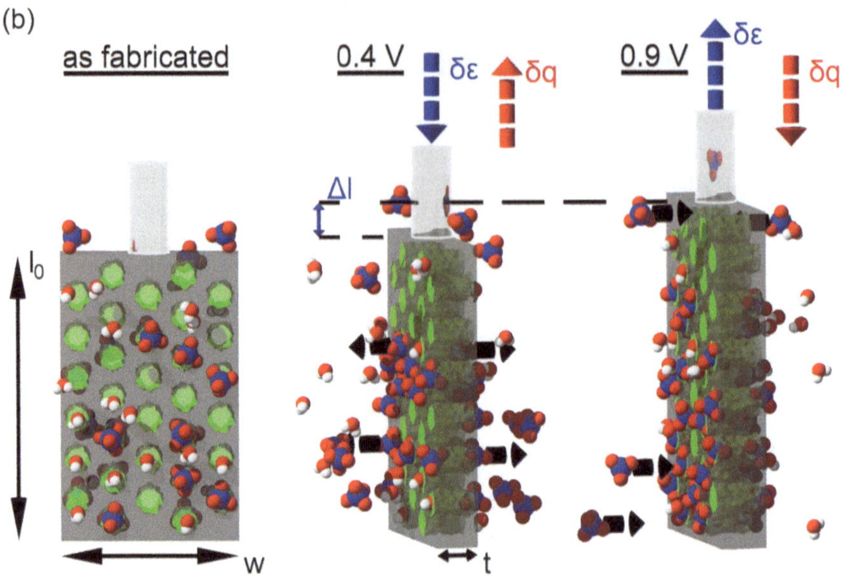

(a) High-angle annular dark-field scanning transmission electron micrograph (TEM) top-view of an as-fabricated nanoporous silicon membrane functionalized with the electrically conductive polymer polypyrrole by electropolymerization. Green and red color code indicates nitrogen (N) and silicon (Si) concentration obtained by energy-dispersive X-ray detection (EDX). (b) Schematics of the electroactuation experiment on a nanoporous silicon membrane (gray) filled with polypyrrole (green). The sample is immersed in an electrolyte consisting of aqueous (red and white molecules) perchloric acid ($HClO_4$, blue and red). If a respective potential, 0.9 V, is applied to the material, anions from the electrolyte can be incorporated into the polymer, which results in a swelling of the polymer and thus of the whole membrane. Vice versa, decreasing the potential to 0.4 V leads to an expulsion of anions, followed by a contraction of the sample. Δl denotes the change in length[1].

(a) Micrografía de campo oscuro anular con microscopio electrónico de transmisión (MET) de la vista superior de una membrana de silicio nanoporosa, tal como fue fabricada, funcionalizada con polímero conductor polipirrol por electropolimerización. Los colores verde y rojo indican la concentración de nitrógeno (N) y silicio (Si) obtenida a través de la fluorescencia de rayos X por energía dispersiva (EDX). (b) Esquema del experimento de electroactuación en una membrana nanoporosa de silicio (gris) llena de polipirrol (verde). La muestra está sumergida en un electrolito formado por ácido perclórico ($HClO_4$, azul y rojo) acuoso (moléculas rojas y blancas). Si se aplica una tensión de 0,9 V al material, aniones del electrolito pueden incorporarse al polímero, lo que resulta en una expansión del polímero, y por lo tanto, de toda la membrana. Y viceversa, disminuir el voltaje a 0,4 V conlleva la expulsión de los aniones, seguida de una contracción de la muestra. Δl denota el cambio en la longitud[1].

1. Brinker M. *et al.* (2020). *Sci. Adv.*, 6(40).

Contact: Manuel Brinker <manuel.brinker@tuhh.de>

E. F. Médici and A. D. Otero (eds.), *Album of Porous Media*,
https://doi.org/10.1007/978-3-031-23800-0_5

LASER-EXCITED ELASTIC GUIDED WAVES REVEAL THE COMPLEX MECHANICS OF NANOPOROUS SILICON

Marc Thelen[1], *Nicolas Bochud*[2], *Manuel Brinker*[1], *Claire Prada*[3], *and Patrick Huber*[1,4]

[1] Institute for Materials and X-Ray Physics, Hamburg University of Technology, Germany
[2] Univ Paris Est Creteil, Univ Gustave Eiffel, CNRS, UMR 8208, MSME, Créteil, France
[3] Institut Langevin, ESPCI Paris, Université Paris Sciences et Lettres, CNRS, Paris, France
[4] Deutsches Elektronen-Synchrotron DESY

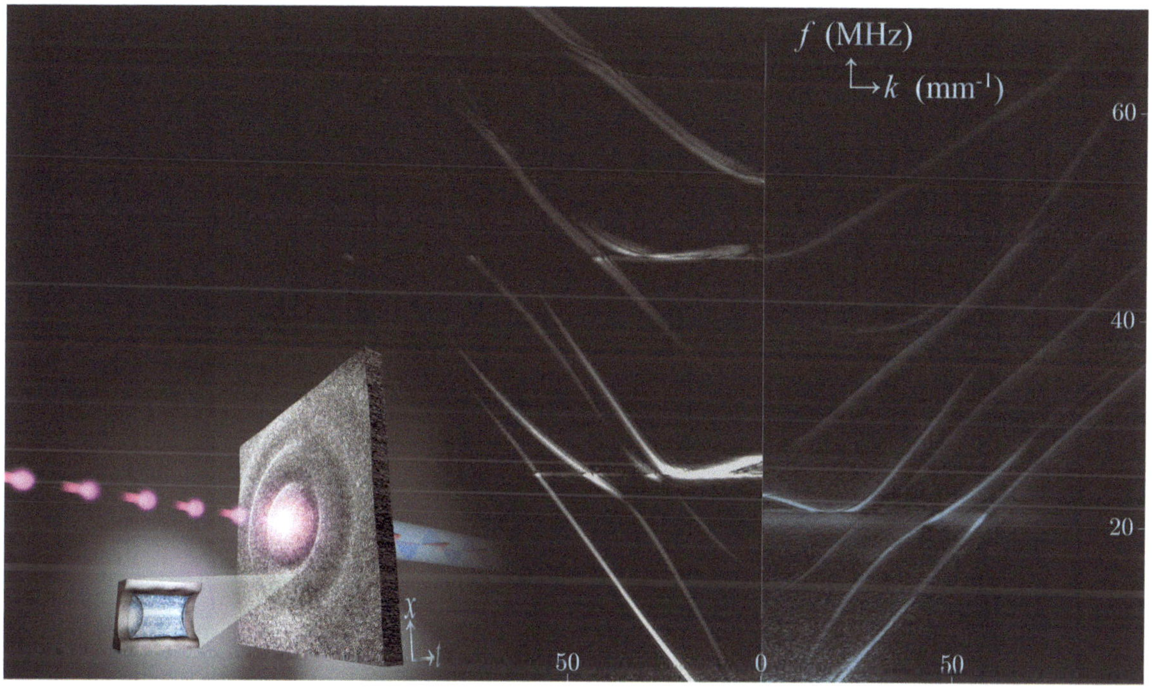

Left: Schematic representation of the laser-ultrasonic measurement principle for non-contact mechanical characterization of hybrid nanomaterials. The 100 μm thick membrane is based on electron micrographs of nanoporous silicon with a pore diameter of 7 nm. UV laser pulses generate broadband ultrasonic vibrations in the membrane by thermo-elastic conversion. The out-of-plane displacements of the elastic waves guided along the membrane with empty or liquid-infused nanopores are then detected by a laser interferometer scanning the surface in x-direction.
Right: The dispersion curves of the guided modes are observed in the wavenumber-frequency (k – f) plane by applying a 2D Fourier transform to the measured spatio-temporal (x – t) displacements. These curves provide a signature of the mechanical properties of the membrane for empty (gray, left) and liquid-infused (blue, right) nanopores. For instance, the decrease of the effective bulk wave velocities of the material due to the absorption of the liquid can be traced back to the cut-off frequencies (at k = 0) and the slope of the guided modes. In addition, applying a suitable waveguide model further allows recovering quantitative information on the effective stiffness and even the pore's conicity of the material [1].

1. Thelen M. *et al.* (2021). *Nat. Commun.*, 12(1).

Izquierda: Representación esquemática del principio de medición por láser ultrasónico para la caracterización mecánica sin contacto de nanomateriales híbridos. La membrana de 100 μm de espesor está basada en micrografías electrónicas de silicio nanoporoso con poros de 7 nm de diámetro. Los pulsos de láser UV generan vibraciones ultrasónicas de banda ancha en la membrana debido a la conversión termoelástica. Los desplazamientos fuera del plano de las ondas elásticas guiadas a lo largo de la membrana con nanoporos vacíos o impregnados de líquido son detectados por un láser interferométrico que escanea la superficie en la dirección x.
Derecha: Las curvas de dispersión de los modos guiados son observadas en el plano número de onda-frecuencia (k – f) al aplicar la transformada de Fourier en 2D a los desplazamientos espacio temporales (x – t). Estas curvas proporcionan una descripción de las propiedades mecánicas de la membrana con nanoporos vacíos (gris, izquierda) e impregnados de líquido (azul, derecha). Por ejemplo, la disminución de las velocidades efectivas de onda en el material debido a la absorción del líquido puede ser atribuida a las frecuencias límites (a k = 0) y a la inclinación de los modos guiados. Asimismo, la aplicación de un modelo de guía de onda apropiado permite además la recuperación de información cuantitativa sobre la rigidez efectiva e incluso la conicidad de los poros del material[1].

Contact: Marc Thelen <marc.thelen@tuhh.de>

E. F. Médici and A. D. Otero (eds.), *Album of Porous Media*,
https://doi.org/10.1007/978-3-031-23800-0_6

SAMPLE LAYOUT AIDING EFFICIENT SCANS OF HETEROGENEOUS SHEET-LIKE MATERIALS

Michael Stonig[1], Eduardo Machado Charry[1], Andre Hilger[2], Ingo Manke[2], Matthias Neumann[3], Volker Schmidt[3], and Karin Zojer[4]

[1] Graz University of Technology
[2] Institute of Applied Materials, Helmholtz-Zentrum Berlin für Materialien und Energie
[3] Institute of Stochastics, Ulm University
[4] Christian Doppler Laboratory for Mass Transport through Paper and Institute of Solid State Physics, Graz University of Technology

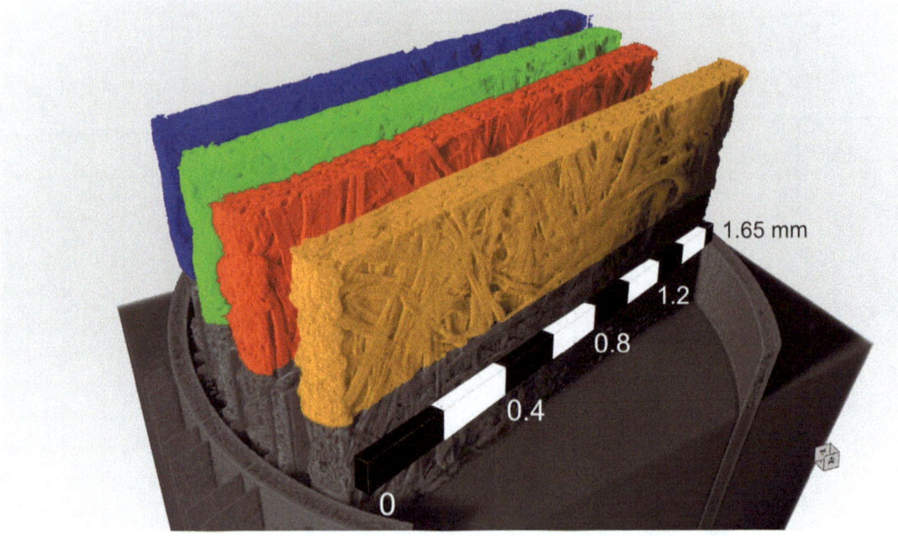

3D image of a sample layout that enables a more efficient structure acquisition of heterogeneous sheet-like materials like paper with X-ray computed microscopy[1]. Sheets are very thin compared to their lateral extension, e.g., paper sheets consisting of cellulose-based fibers. Local variations in the pore space might be considerably further apart than the sheet is thick. Hence, air or liquids transported through such sheets might experience locally different bottlenecks, for example, due to a reduced porosity or more torturous pathways. To quantify the number and properties of such bottlenecks it is necessary to probe many locations in the sheets while keeping a resolution large enough to resolve the pore space. Probing the sheet location by location using X-ray computed microscopy is highly time-consuming. The key idea of our sample layout is to permit the scan of multiple cutouts from the sheet at once. From the bottom to the top, the artistically enhanced image reveals the essential arrangement of the cutouts alike a peeled-off banana. Four paper strips, highlighted in color, are arranged into a stack. In this stack of paper strips, Kapton strips are intercalated as spacers (shown as upright grey sheets of alternating thickness in the image center). The resulting stack sits tightly fitted in a Kapton tube (grey cylinder at the bottom). All such acquired cutouts combined inform on local sheet variations on the centimeter scale.

1. Neumann M. *et al.* (2021). *Microsc Microanal*, 27(6):1305.

Contact: Karin Zojer <karin.zojer@tugraz.at>

Imagen en 3D de una disposición de muestras que permite una adquisición eficiente de la microestructura de materiales laminares heterogéneos por medio de tomografía computarizada basada en rayos X[1]. En estos materiales el grosor de las láminas es muy pequeño en comparación con la extensión lateral. Un ejemplo de estos materiales son las hojas de papel que consisten en fibras a base de celulosa. El papel puede presentar variaciones locales del espacio poroso considerablemente más separadas en comparación con el grosor de las láminas. Por lo tanto, el desplazamiento de aire o líquidos a lo largo de estas láminas puede presentar diferentes cuellos de botella distribuidos espacialmente, debido a reducciones de la porosidad o a la presencia de caminos más tortuosos. Para cuantificar el número y las propiedades de tales cuellos de botella, es necesario sondear múltiples ubicaciones de las láminas manteniendo al mismo tiempo una resolución que permita una buena caracterización del espacio poroso. Sondear la lámina ubicación por ubicación por medio de tomografía computarizada es muy lento. La idea clave de la disposición de nuestra muestra es permitir el escaneo de varios cortes de la lámina al mismo tiempo. Al observar la disposición, en la imagen mejorada artísticamente, desde la parte inferior hasta la superior, esta luce como una banana pelada. Cuatro tiras de papel, resaltadas en color, están agrupadas en posición vertical. En esta disposición, tiras de Kapton son utilizadas como separadores (láminas verticales grises en el centro de la imagen). El grupo resultante queda firmemente colocado en un tubo de Kapton (cilindro gris de la parte inferior). La exploración de múltiples recortes al mismo tiempo nos permite determinar variaciones locales en la escala del centímetro.

E. F. Médici and A. D. Otero (eds.), *Album of Porous Media*,
https://doi.org/10.1007/978-3-031-23800-0_7

PROTECTIVE POROUS MEDIA - THE PORE & FIBER STRUCTURE OF FACE MASKS

Chven Mitchell and Laura Pyrak-Nolte
Purdue University

Face masks were an essential aspect of the COVID-19 pandemic safety protocols. Understanding the structure of the mask provides insight into the best protection to reduce the transmission of viruses and other particulate matter. A Zeiss Xradia 510 versa X-Ray Microscope was used to image the fiber network, porosity, and overall structure of (a) an n95 mask, (b) a medical mask, and (c) a gaiter with pixel edge length resolutions of $\sim$46 μm & $\sim$3 μm, $\sim$50.2 μm & $\sim$3 μm, and $\sim$50.2 μm & $\sim$5.1 μm, respectively. The gaiter contains fibers ranging 5.54 - 33.26 μm in diameter. The number of layers in the gaiter depends on how the cloth is folded, and multiple layers may improve the filtering efficacy. The medical mask consists of 3 layers with fiber diameters in the top and bottom layers between 8.1 to 32.4 μm. The fibers in the central layer are <8 μm. The n95 mask is mostly comprised of fine fibers that are 4 μm or less. However, the fibers located in the top outer layer of the n95 mask vary between 12 – 19 μm. Due to its fine-scale woven structure and the absence of a respirator, the n95 mask would in all likelihood hinder dual scale flow and offer the most protection against aerosols and large droplets.

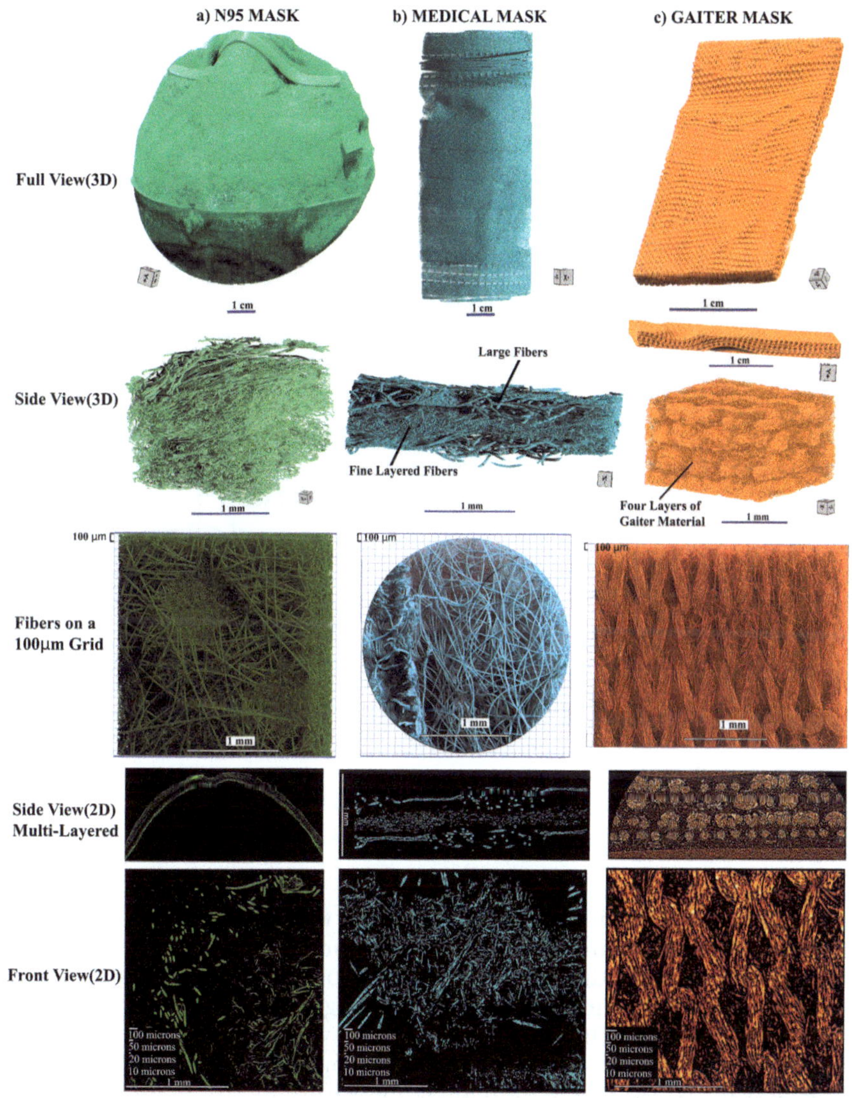

Los cubrebocas, barbijos o las máscaras fueron un aspecto esencial de los protocolos de seguridad de la pandemia de COVID-19. Comprender la estructura del cubrebocas proporciona conocimiento sobre la mejor protección para reducir la transmisión de virus y otros materiales particulados. Se utilizó un microscopio Zeiss Xradia 510 versa de rayos X para obtener imágenes del entramado de la tela, porosidad y estructura general de (a) una máscara n95, (b) una máscara médica y (c) una polaina para el cuello con resoluciones de tamaño de pixel de $\sim$46 μm y $\sim$3 μm, $\sim$50,2 μm y $\sim$3 μm, y $\sim$50,2 μm y $\sim$5,1 μm, respectivamente. La polaina contiene fibras que van de los 5,54 a los 33,26 μm de diámetro. El número de capas en la polaina depende de cómo se doble la tela, y varias capas podrían mejorar la eficacia de la filtración. La máscara médica tiene 3 capas y un diámetro de fibras en las capas superior e inferior entre 8,1 y 32,4 μm. Las fibras en la capa central tienen <8 μm. La máscara n95 está compuesta principalmente de fibras finas de 4 μm o menos. Sin embargo, las fibras de la capa superior externa de la máscara n95 varían entre 12 y 19 μm. Debido a la estructura tejida de entramado fino y a la ausencia de respirador, la máscara n95 impediría con toda seguridad el flujo de doble escala, ofreciendo mayor protección frente a los aerosoles y gotas grandes.

Contact: Chven Mitchell <mitch240@purdue.edu>

E. F. Médici and A. D. Otero (eds.), *Album of Porous Media*,
https://doi.org/10.1007/978-3-031-23800-0_8

INDIVIDUAL FIBERS IN NONWOVEN ARE IDENTIFIED WITH ARTIFICIAL INTELLIGENCE

Andreas Griesser
Math2Market GmbH

A nonwoven acquisition layer of a diaper was scanned and segmented into fibers (gray) and pores (transparent). Four individual scans were taken by Bruker Micro CT and stitched together into a single gigantic data set of 15,600×4,000×1,800 voxels with side length 2.7 μm. Using a neural network / AI approach in the GeoDict software, the individual fibers were identified in the scan. This is illustrated in the front half of the image, where each fiber shows in its own color. Those colors are also available in the gray portion, but not shown to illustrate the work done by the algorithms. Once the individual fibers are known, many types of further analysis become very easy: statistics of fiber lengths, fiber diameters, fiber curvatures, etc. The identification of fibers in this type of porous material is important and necessary for the accurate and analytically correct modeling of this type of material on the computer. The models can be then digitally modified and ultimately, this process will lead to the design of new, improved, real porous materials.

Una capa de absorción de pañal no tejida fue escaneada y segmentada en fibras (gris) y poros (transparente). Se realizaron cuatro escaneos individuales con una micro TC Bruker y se unieron en un único conjunto gigante de datos de 15600×4000×1800 vóxels con una distancia lateral de 2,7 μm. Mediante el uso de una red neuronal e IA en el software GeoDict, se identificaron individualmente las fibras en el escaneado. Esto se ilustra en la mitad frontal de la imagen, en la que cada fibra muestra su propio color. Estos colores también están disponibles en la porción gris, pero no se muestran para ilustrar el trabajo hecho por los algoritmos. Una vez que se conocen las fibras individuales, se pueden realizar fácilmente otros análisis: estadísticas de largos, diámetros y curvaturas de las fibras, etc. La identificación de las fibras en este tipo de material poroso es importante y necesaria para el preciso y analíticamente correcto modelado de este tipo de material en la computadora. Los modelos pueden modificarse digitalmente y, finalmente, este proceso conducirá al diseño de nuevos y mejores materiales porosos reales.

Contact: Barbara Planas <barbara.planas@math2market.de>

E. F. Médici and A. D. Otero (eds.), *Album of Porous Media*,
https://doi.org/10.1007/978-3-031-23800-0_9

MECHANISM OF IMPERFECT ATOMIC LAYER DEPOSITION IN POROUS MATERIALS

Nupur Bihari[1], *Ismo Rauha*[2], *Giovanni Marin*[2], *Craig Ekstrum*[1], *Pierce Mayville*[1], *Shane Oberloier*[1], *Hele Savin*[2], *Maarit Karppinen*[2], *and Joshua Pearce*[3]

[1] Michigan Technological University
[2] Aalto University
[3] Western University, Canada

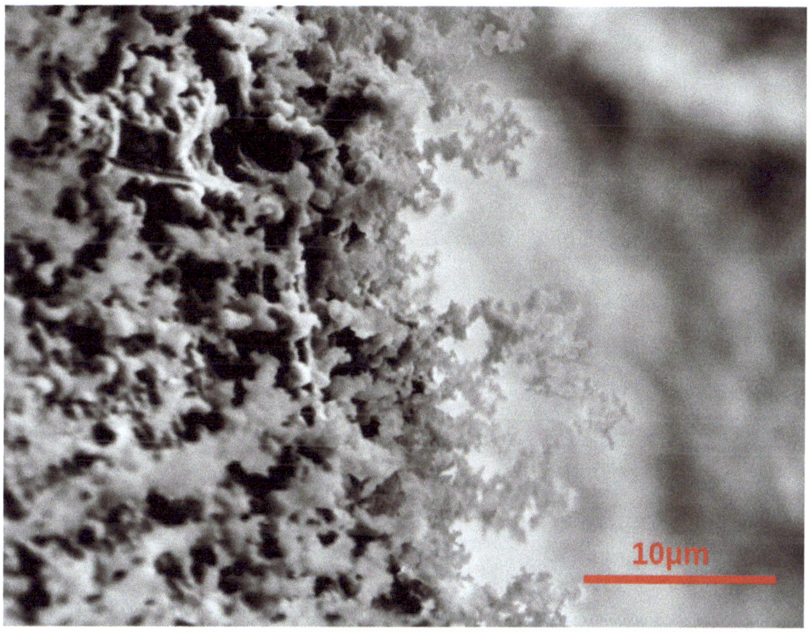

Polymer-based materials used for 3D printing are often porous[1]. While these materials are coated with an adsorption based method for conformal coatings, atomic layer deposition (ALD), the incoming precursor gets exposed to a very large surface area. Once the surface is saturated, a purge cycle is initiated, which takes away the un-adsorbed, gas-phase excess precursor. However due to the highly porous nature of 3D printed plastic materials, the degree in tortuosity in the purge gas path is very high. When this purge gas is introduced, it is unable to reach all the gas phase precursor molecules, often leading to incomplete removal of un-adsorbed precursor species. When the oxidizer is introduced, it not only reacts with the adsorbed precursor as desired, but also with the precursor trapped in the voids. This leads to the ALD growth having a significant parasitic Chemical Vapor Deposition (CVD) component to it. This is seen as the fractal-like AlO_x growth emerging out of the plastic polypropylene substrate as imaged here. The resultant coating has a high degree of surface roughness due to the CVD growth and may break off to form particle based contaminants. A Philips XL40 ESEM was used to image this polypropylene surface coated with AlO_x fractals.

1. Bihari N. *et al.* (2020). *J. Vacuum Sci. & Tech. A*, 38(5):053204.

Contact: Nupur Bihari <nbihari@mtu.edu>

Los materiales basados en polímeros para la impresión en 3D suelen ser porosos[1]. Mientras estos materiales son recubiertos con un método para recubrimientos conformados basado en la adsorción, deposición de capas atómicas (DCA), el precursor entrante queda expuesto a una superficie muy grande. Cuando la superficie queda saturada, se inicia un ciclo de purga, que remueve el exceso de precursor no adsorbido en fase gaseosa. Sin embargo, debido a la naturaleza altamente porosa de los materiales plásticos impresos en 3D, el grado de tortuosidad en el camino de la purga del gas es muy elevado. Cuando se introduce este gas purgador, no puede alcanzar todas las moléculas de la fase gaseosa del precursor, por lo que a menudo la eliminación de las especies precursoras no adsorbidas es incompleta. Cuando se introduce el oxidante, no solo reacciona con el precursor adsorbido tal y como se desea, sino también con el precursor atrapado en los huecos. Esto conlleva que el crecimiento de la DCA tenga una deposición química de vapor (DQV) parasitaria significativa. Esto se observa en la imagen como el crecimiento en forma de fractal del AlO_x que emerge del sustrato plástico de polipropileno. El recubrimiento resultante tiene un alto grado de rugosidad en la superficie debido al crecimiento de la DQV y podría romperse para formar partículas de contaminantes. Para obtener la imagen de esta superficie de polipropileno recubierto de AlO_x fractales se utilizó un MEBA Philips XL40.

E. F. Médici and A. D. Otero (eds.), *Album of Porous Media*,
https://doi.org/10.1007/978-3-031-23800-0_10

MASS TRANSPORT IN CONFORMALLY COATED POROUS MATERIALS

Nupur Bihari[1], *Ismo Rauha*[2], *Giovanni Marin*[2], *Craig Ekstrum*[1], *Pierce Mayville*[1], *Yuhuan Fei*[1], *Yun Hang Hu*[1], *Maarit Karppinen*[2], *Hele Savin*[2], *and Joshua Pearce*[3]

[1] Michigan Technological University
[2] Aalto University
[3] Western University, Canada

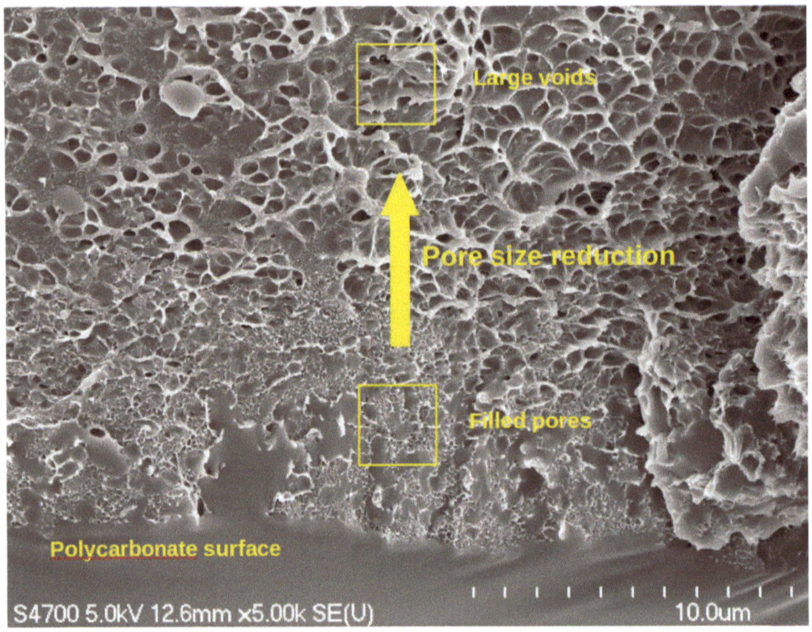

Coating 3D printed plastic materials has been shown to be an effective method of reducing outgassing[1]. However, the sealing of these materials is not well understood because of the large variation in pore sizes, long path lengths for incoming and outgoing gas species and high degree of tortuosity. The problem is further exacerbated when this coating is done using different gas species at elevated temperatures or reduced pressures. The coating species are trapped in the pores, plugging them. This reduces outgassing, as volatile species from the material do not have a viable path outward. This also prevents incoming coating species from going deeper into the plastic and filling voids. This creates a material where voids close to the surface are more effectively plugged than the pores deeper in the bulk. When this coated plastic is used in extreme temperatures or lowered pressures, this can cause significant robustness issues leading to part failure. Understanding mass transport mechanisms in coated porous materials can have various applications in vacuum environments, including making semiconductor processing tools and hardware used for space-based applications. Image shows polycarbonate coated with Atomic Layer Deposited AlO_x.

1. Heikkinen I.T. *et al.* (2020). *Surface and Coatings Tech.*, 386:125459.

Contact: Nupur Bihari <nbihari@mtu.edu>

Se ha mostrado que recubrir los materiales plásticos impresos en 3D es un método efectivo para reducir la desgasificación[1]. Sin embargo, no se comprende bien el sellado de estos materiales debido a la gran variación en los tamaños de los poros, la gran longitud de los caminos para las especies de gases entrantes y salientes, y el alto grado de tortuosidad. El problema se exacerba más cuando el recubrimiento se hace utilizando diferentes especies de gas a temperaturas elevadas o presiones reducidas. Las especies de recubrimiento quedan atrapadas en los poros, obstruyéndolos. Esto reduce la desgasificación, a medida que las especies volátiles del material no encuentran un camino viable al exterior. Esto también evita que especies de recubrimiento entrante vayan a zonas más profundas del plástico y llenen los huecos. Esto crea un material en el que los huecos cercanos a la superficie se obstruyen con más eficacia que los poros ubicados en la profundidad del medio. Cuando este plástico recubierto se utiliza en temperaturas extremas o presiones bajas, puede causar significativos problemas de robustez que pueden conducir al fallo de la pieza. Comprender los mecanismos del transporte de masa en materiales porosos recubiertos puede tener varias aplicaciones en ambientes con vacío, incluyendo la fabricación de herramientas para el procesamiento de semiconductores y equipos utilizados en aplicaciones espaciales. La imagen muestra un policarbonato recubierto con una capa atómica depositada de AlO_x.

E. F. Médici and A. D. Otero (eds.), *Album of Porous Media*,
https://doi.org/10.1007/978-3-031-23800-0_11

OPTICAL MICROFABRICATION OF POROUS FILMS

Natalia Philipp[1,2], *Nicolas Rapagnani*[1], *Paula Caral*[1], *Ignacio Gomez Florenciano*[1], *and Laura Estrada*[1,2]

[1] Universidad de Buenos Aires, Facultad de Ciencias Exactas y Naturales, Departamento de Física. Argentina.
[2] CONICET - Universidad de Buenos Aires, Instituto de Física de Buenos Aires (IFIBA). Argentina.

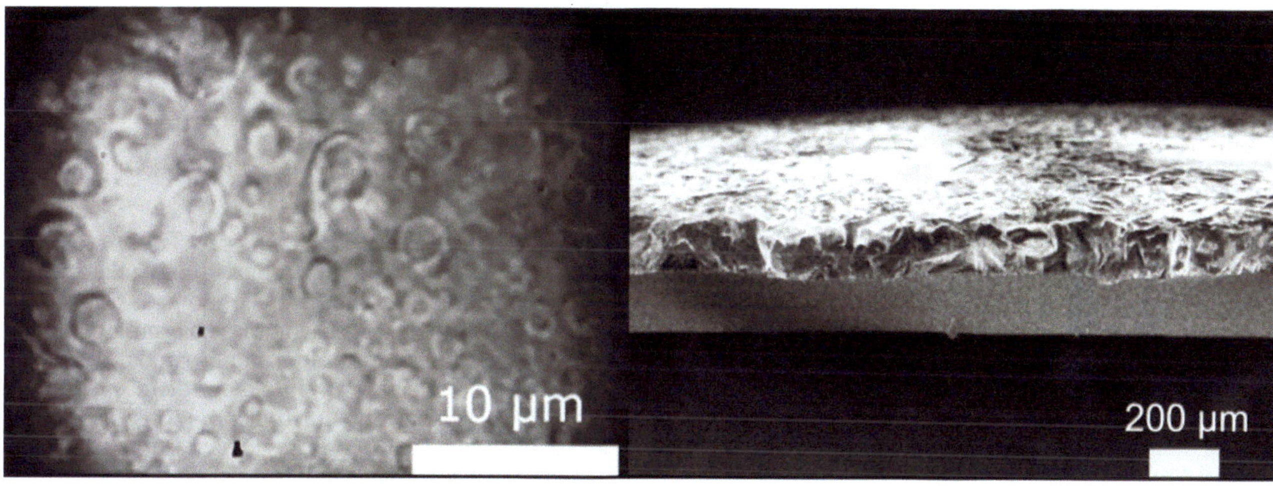

In the last years, efforts have been directed towards the development and fabrication of porous materials with a wide range of applications, from pharmaceutical research to drug delivery matrices. In this work porous films and 3D structures were fabricated by exposing a commercial UV optical adhesive as a photoresist to the irradiation of a UV lamp or, for 3D microfabrication, the irradiation with an IR femtosecond laser from a two-photon microscope (2PM)[1, 2]. Using the software-controlled scanning mirrors available in any 2PM, arbitrary structures both 2D and 3D can be prepared in a few minutes for a fraction of the cost of a traditional lithography technique. The reproducibility and quality of the obtained patterns, provide an enormous advantage for cheap and straightforward design of porous materials allowing to scale up the manufacturing processes. The possibility of incorporating into the uncured photoresist materials such as drugs, metallic or magnetic nanoparticles, and fluorescent agents, have been demonstrated[3, 4] and opens new opportunities for the fabrication of low-cost porous-active devices. Left: optical image (top view), Right: SEM image (lateral view).

1. Zheng L. *et al.* (2019). *Nanoscale Res Lett*, 14(1).
2. Costantino S. *et al.* (2005). *Microsc. Res. Tech.*, 68(5):272.
3. Infuehr R. *et al.* (2007). *Applied Surface Science*, 254(4):836.
4. Kawata S. *et al.* (2001). *Nature*, 412(6848):697.

En los últimos años, se han dirigido esfuerzos hacia el desarrollo y la fabricación de materiales porosos con un amplio abanico de aplicaciones que van desde la investigación farmacéutica hasta la fabricación de matrices para la administración de fármacos. En este trabajo, láminas porosas y estructuras 3D fueron fabricadas exponiendo un adhesivo óptico comercial, curable en la region UV del espectro, a la irradiación de una lámpara UV, o para la microfabricación en 3D, a la irradiación de un láser IR de femtosegundos proveniente de un microscopio por absorción de dos fotones (2PM)[1, 2]. Utilizando los espejos de barrido controlados por software disponibles en cualquier 2PM, pueden prepararse en pocos minutos estructuras arbitrarias tanto en 2D como en 3D, por una fracción del costo de una litografía tradicional. La reproducibilidad y calidad de los patrones obtenidos proporcionan una enorme ventaja al momento de obtener un diseño barato y sencillo de materiales porosos que permiten escalar el proceso de fabricación. La posibilidad de incorporar en la fotoresina no curada materiales tales como fármacos, partículas metálicas magnéticas y fluorescentes ha sido demostrada[3, 4] abriendo nuevas oportunidades para la fabricación de dispositivos porosos activos de bajo costo. Izquierda: imagen óptica (vista superior), Derecha: imagen MEB (vista lateral).

Contact: Laura Estrada <lestrada@df.uba.ar>

E. F. Médici and A. D. Otero (eds.), *Album of Porous Media*,
https://doi.org/10.1007/978-3-031-23800-0_12

BICONTINUOUS POROSITY BY SPINODAL PHASE SEPARATION

Andrés Zelcer[1], *Glenna L. Drisko*[2], *and Galo Soler-Illia*[3]

[1] Centro de Investigaciones en Bionanociencias (CIBION), CONICET, Argentina
[2] Institut de Chimie de la Matière Condensée de Bordeaux, Université de Bordeaux, CNRS, France
[3] Instituto de Nanosistemas, Universidad Nacional de San Martín, Argentina

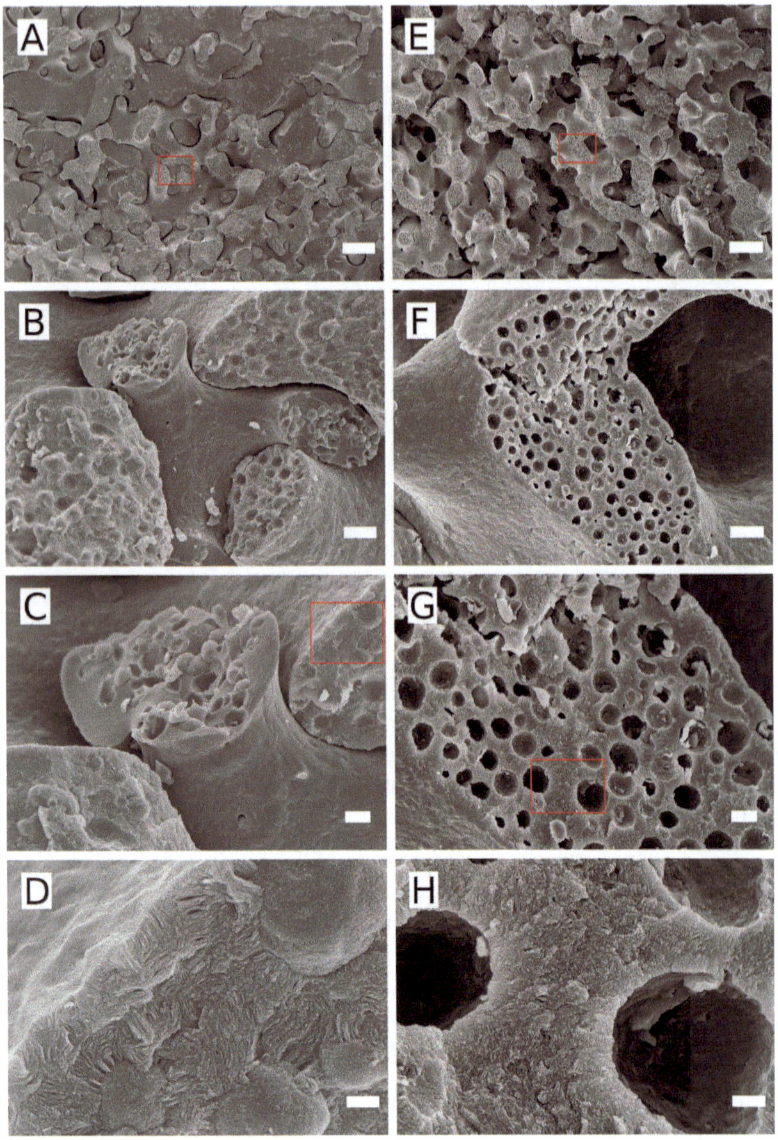

La separación de fases por descomposición espinodal es una técnica utilizada para preparar materiales cerámicos porosos para su uso como fases estacionarias en cromatografía[1]. Estas muestras fueron preparadas mediante polimerización simultánea de reactivos orgánicos (alcohol furfurílico) e inorgánicos (tetraetoxisilano) en presencia de un surfactante y solventes (agua, etanol)[2]. Después de la descomposición de fases espinodal, solidifican una fase rica en material orgánico y una rica en material inorgánico, reteniendo pequeñas gotas de la otra fase y, en el caso de la fase inorgánica, micelas de surfactantes ordenadas. La extracción de la parte orgánica mediante combustión conlleva un material que muestra una porosidad multiescala. Los poros dejados por las micelas surfactantes no son visibles debido al recubrimiento metálico, pero su organización es evidente debido a la disposición del material pulverizado. Paneles A-D: material antes de extraer la fase orgánica. Paneles E-H: material puramente inorgánico. Los rectángulos rojos marcan la zona aumentada en la siguiente imagen. Escala: A-E 50 μm, B-F 5 μm, C-G 2 μm, D-H 500 nm. Todas las imágenes se obtuvieron mediante FE-SEM (Car-Zeiss SUPRA 40).

Spinodal phase separation is a technique used to prepare porous ceramic materials for its use as chromatographic stationary phases[1]. These samples were prepared by simultaneous polymerization of organic (furfuryl alcohol) and inorganic (tetraethoxysilane) reactants in the presence of a surfactant and solvents (water, ethanol)[2]. After the spinodal phase separation, an organic-rich and an inorganic-rich phase solidify, trapping small droplets of the other phase and, in the case of the inorganic phase, ordered surfactant micelles. Removal of the organic part by combustion leads to a material showing multiscale porosity. The pores left by the surfactant micelles are not visible due to the metal coating, but its arrangement is evidenced by disposition of the sputtered material. Panels A-D: material before removal of the organics. Panels E-H: pure inorganic material. The red boxes mark the area zoomed into the next image. Scale bars: A-E 50 μm, B-F 5 μm, C-G 2 μm, D-H 500 nm. All images were obtained by FE-SEM (Carl-Zeiss SUPRA 40).

1. Nakanishi K. and Tanaka N. (2007). *Acc. Chem. Res.*, 40(9):863.
2. Drisko G.L. *et al.* (2012). *Microporous and Mesoporous Materials*, 148(1):137.

Contact: Andrés Zelcer <andres.zelcer@cibion.conicet.gov.ar>

E. F. Médici and A. D. Otero (eds.), *Album of Porous Media*,
https://doi.org/10.1007/978-3-031-23800-0_13

NUCLEATION AND CRYSTAL GROWTH ON THE SECONDARY SUBSTRATE

Mohammad Nooraiepour[1], *Mohammad Masoudi*[1], *Nima Shokri*[2], *and Helge Hellevang*[1]

[1] Department of Geosciences, University of Oslo, Norway
[2] Institute of Geo-Hydroinformatics, Hamburg University of Technology, Germany

Substrate surface properties govern the stochastic dynamics of solid precipitation during reactive transport in the porous media. Probability and affinity for nucleation and subsequent growth are higher on or adjacent to the newly formed crystals than the foreign substrate surface [1, 2]. Detailed high-resolution scanning electron microscopy (SEM) with secondary electrons (SE) together with energy-dispersive X-ray spectroscopy (EDS) provided direct evidence of this selective phenomenon. For the synthesization of calcium carbonates, stock solutions of well-defined weight of crystalline salts of calcium chloride ($CaCl_2$) and sodium bicarbonate ($NaHCO_3$) were added to the DI-water (Milli-Q water)[1]. The subfigures show the formation of early phase crystal, transformation stage, progressive growth, and interconnection of calcium carbonate crystalline geometries on or adjacent to the secondary substrates (previously precipitated crystals). The selection of secondary substrates over the sandstone surface is evident in nanometer to micron scales in the SEM micrographs. The preferential nucleation sites provide favorable surface characteristics and will be selected first against the available unreacted foreign surface[2, 3].

Las propiedades de la superficie del sustrato gobiernan las dinámicas estocásticas de la precipitación sólida durante el transporte reactivo en los medios porosos. La probabilidad y la afinidad para la nucleación y el consiguiente crecimiento son más elevados en los nuevos cristales formados o a su alrededor, que en la superficie foránea del sustrato [1, 2]. Microscopías electrónicas de barrido (MEB) de alta resolución con electrones secundarios (ES) junto con espectroscopías de rayos X por energía dispersiva (EED) proporcionaron evidencia directa de este fenómeno selectivo. Para la sintetización de los carbonatos de calcio, se añadieron al agua desionizada (agua Milli-Q) soluciones madre con peso bien definido de sales cristalinas de cloruro de calcio ($CaCl_2$) y de bicarbonato de sodio ($NaHCO_3$)[1]. Los recuadros muestran la formación de un cristal de fase temprana, la etapa de transformación, el crecimiento progresivo y la interconexión de las geometrías cristalinas del carbonato de calcio en los sustratos secundarios o adyacentes (cristales previamente precipitados). La selección de sustratos secundarios sobre la superficie de arenisca es evidente en las escalas de nanómetro a micrones en las micrografías MEB. Los sitios preferentes de nucleación proporcionan características favorables de la superficie y se seleccionarán primero frente a las superficies foráneas disponibles que no hayan reaccionado[2, 3].

1. Nooraiepour M. *et al.* (2021). *ACS Omega*, 6(42):28072.
2. Nooraiepour M. *et al.* (2021). *Sci. Rep.*, 11(1):16397.
3. Hellevang H. *et al.* (2019). *E3S Web Conf.*, 98:04006.

Contact: Mohammad Nooraiepour <monoo@uio.no>

E. F. Médici and A. D. Otero (eds.), *Album of Porous Media*,
https://doi.org/10.1007/978-3-031-23800-0_14

DOUBLE-POROSITY IN SOIL

Su Kong Ngien
Department of Civil Engineering, College of Engineering, Universiti Malaysia Pahang

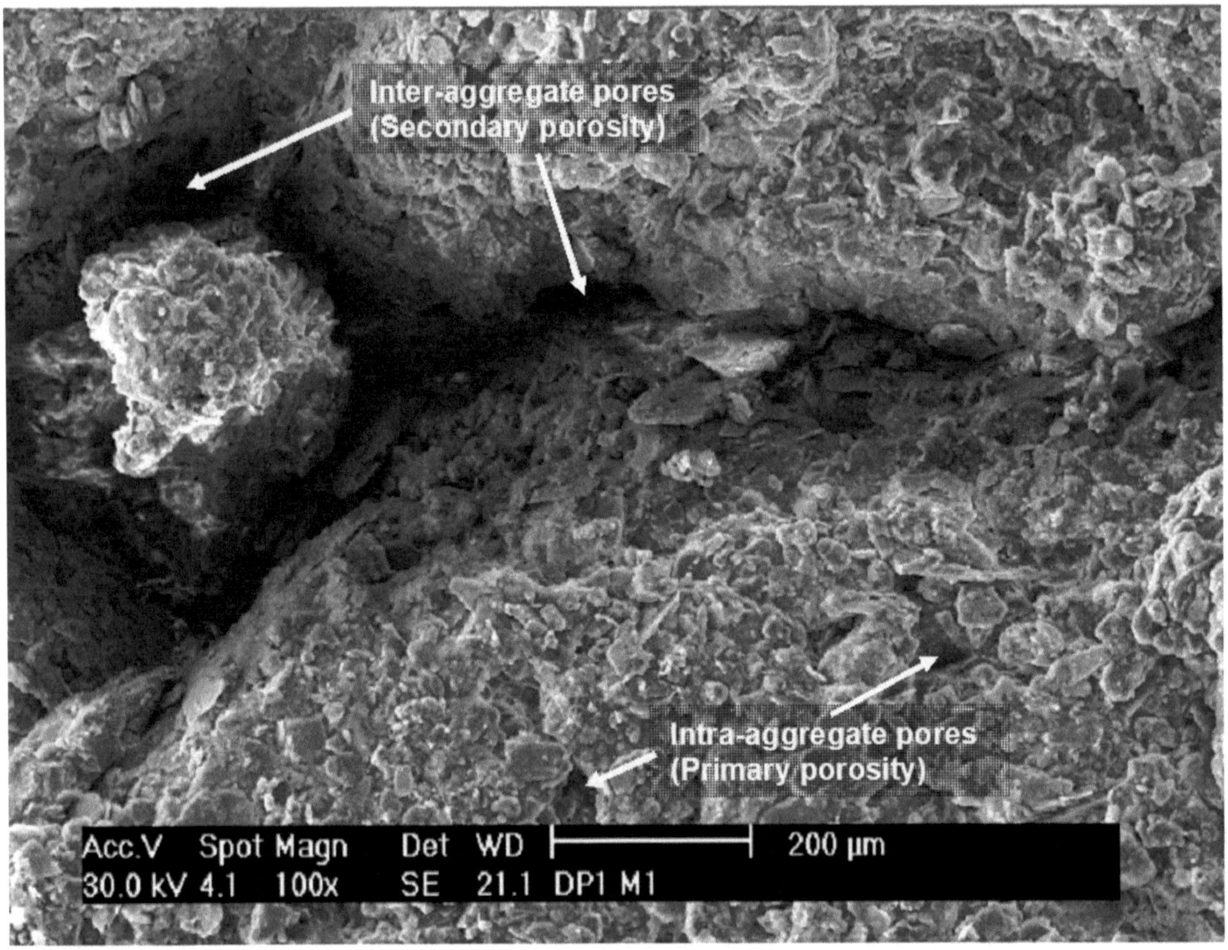

This image shows inter-aggregate and intra-aggregate pores, one of the double-porosity phenomena that may exist in soil. Double-porosity occurs in soil when two sets of porosities having highly contrasting hydraulic properties co-exists in the fabric of the soil. The occurrence of double-porosity in soil can be due to natural formation, activities of soil fauna and plant roots, as well as soil aggregation that results from tillage. These factors create cracks, fissures, soil pipes and inter-aggregate pores that are also known as secondary porosity features in the soil. When non-aqueous phase liquid pollutants are introduced to double-porosity soil, they are found to travel further within a shorter amount of time through the secondary porosity features compared to migration in single-porosity soil. Therefore, secondary porosity features can be said to be paths of least resistance for immiscible fluids and will become preferential pathways for the flow of immiscible pollutants in the subsurface. This image was captured via Scanning Electron Microscopy at a magnification of 100 times.

Contact: Su Kong Ngien <bejita__2@hotmail.com>

Esta imagen muestra poros interagregados e intraagregados, uno de los fenómenos de la doble porosidad que pueden existir en suelos. La doble porosidad puede ocurrir en suelos cuando dos tipos de porosidad con propiedades hidráulicas altamente contrastantes coexisten en el entramado del suelo. La ocurrencia de suelo con doble porosidad puede deberse a una formación natural, actividades de la fauna y de las raíces de las plantas, así como a la agregación de suelo debida a la labranza. Estos factores crean grietas, fisuras, conductos en la tierra y poros interagregados que también son conocidos como características de porosidad secundaria en el suelo. Cuando contaminantes líquidos de fase no acuosa son introducidos en suelos con doble porosidad, se observa que se desplazan más lejos en poco tiempo a través de las estructuras particulares de porosidad secundaria en comparación con los desplazamientos en suelos con una única porosidad. Por lo tanto, se puede afirmar que estas estructuras particulares de porosidad secundaria son caminos de mínima resistencia para los fluidos inmiscibles y serán el camino preferido por el flujo de contaminantes inmiscibles en el subsuelo. Esta imagen fue obtenida con un microscopio electrónico de barrido, a una magnificación de 100 veces.

E. F. Médici and A. D. Otero (eds.), *Album of Porous Media*,
https://doi.org/10.1007/978-3-031-23800-0_15

BOREAL MOSSES AND LICHENS

Laurent Orgogozo[1], *Yohan Davit*[2], *Manuel Marcoux*[2], *and Michel Quintard*[2]

[1] University of Toulouse - Observatory Midi-Pyrénées - Geosciences Environment Toulouse laboratory
[2] Institut de Mécanique des Fluides de Toulouse

Permafrost covers a quarter of northern hemisphere lands and is highly sensitive to climate warming. The role of hydrological drains and thermal insulators of the moss and lichen layers, which are the most widespread vegetation covers on tundra and taïga ecosystems, is largely acknowledged to be of primary importance for permafrost dynamics but is poorly quantified. Tomographic images of samples of these peculiar porous media may be used to digitaly study the heat and water transfers and thus contribute to this quantification. Three examples of Tomographic images of samples of sphagnum and lichens from the INTERACT environmental monitoring station of Khanymey (Western Siberia) are shown in relation with their plot scale distribution on the peatlands of this region. These tomographic images are studied in the framework of the HiPerBorea project[1].

El permafrost recubre una cuarta parte de las tierras del hemisferio norte y es extremadamente sensible al calentamiento global. El papel de los drenajes hídricos y de los aislantes térmicos de las capas de musgo y liquen, cubierta vegetal más extendida en los ecosistemas de la tundra y de la taiga, es altamente reconocido en las dinámicas del permafrost, sin embargo están escasamente cuantificados. Las imágenes tomográficas de muestras de este peculiar medio poroso pueden utilizarse para estudiar digitalmente las transferencias de calor y agua y, por lo tanto, contribuir a su cuantificación. Observamos tres ejemplos de imágenes tomográficas de muestras de musgo esfagno y líquenes de la estación de monitoreo medioambiental INTERACT de Khanymey (al oeste de Siberia) en relación con su distribución en la parcela en las turberas de esta región. Estas imágenes tomográficas se estudian en el marco del proyecto HiPerBorea[1].

1. URL `https://hiperborea.omp.eu/`.

Contact: Laurent Orgogozo <laurent.orgogozo@get.omp.eu>

E. F. Médici and A. D. Otero (eds.), *Album of Porous Media*,
https://doi.org/10.1007/978-3-031-23800-0_16

THE GORGONIAN OCTOCORAL *EUNICELLA CAVOLINII* FORMS A POROUS BIOLOGICAL MATERIAL WITH OUTSTANDING MATERIAL PROPERTIES

Uwe Schnepf[1], Raouf Jemmali[2], and Franz Brümmer[1]

[1] University of Stuttgart, Institute of Biomaterials and Biomolecular Systems, Research Unit Biodiversity and Scientific Diving, Germany
[2] German Aerospace Center (DLR), Institute of Structures and Design, Ceramic Composites and Structures, Germany

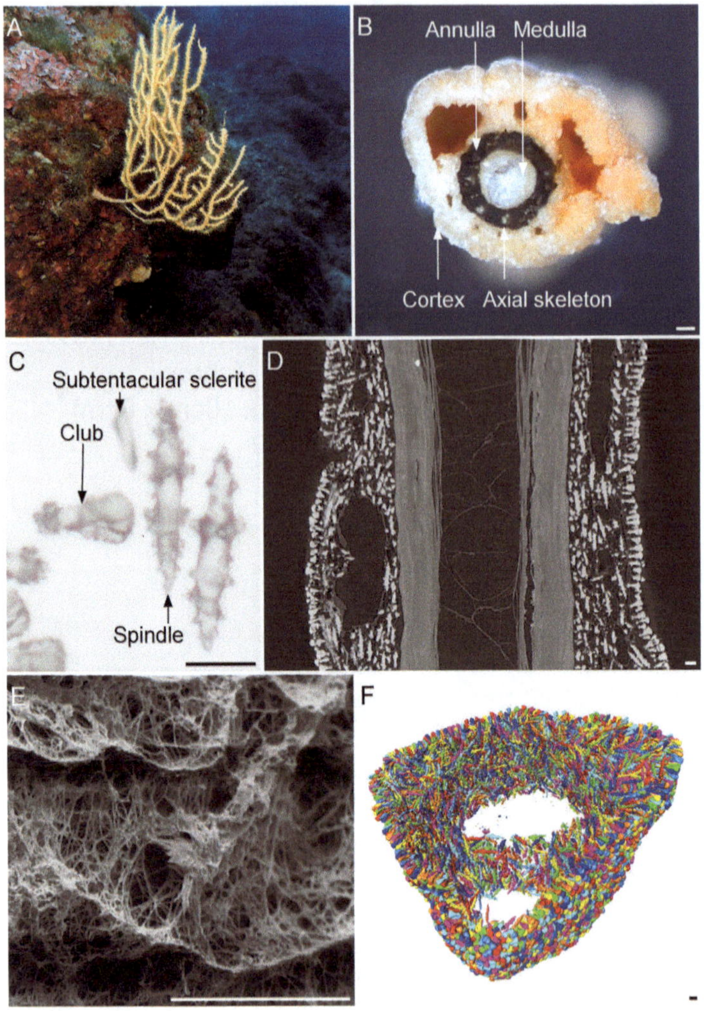

Los gorgónidos (*Cnidaria: Octocorallia*) son corales marinos, formados por pólipos en un material poroso que se organizan en colonias con forma de árbol[1]. Las ramas se orientan en ángulo recto a la corriente del agua. Para evitar daños, los gorgónidos se enredan y doblan fácilmente, manteniendo una rigidez relativa[2]. Así que, ¿cuáles son las estructuras que median estas propiedades mecánicas? Para contestar esto, estudiamos *Eunicella cavolinii*, una especie del Mediterráneo. **A** Los corales fueron documentados *in situ* con un equipo de buceo y fotografía submarina. **B** Un microscopio óptico mostró que las colonias consistían de un esqueleto axial y de un córtex anaranjado. El esqueleto axial está formado por dos estructuras hechas de un símil de colágeno, la gorgonina[3]: el anillo negro, un anillo esclerótico con pequeñas cantidades de minerales y la médula blanca. La corteza está formada por un cenénquime poroso lleno de **C** tres tipos de escleritos de $CaCO_2$[4]: escleritos en forma de garrotes, ejes y subtentaculares. **D** En las secciones de μRayos X, las de forma de garrotes estaban orientadas perpendicularmente al esqueleto axial, mientras que las de forma de eje estaban orientadas en paralelo. **E** En la imagen MEB, la médula mostró una fina red de fibras entrelazadas en paralelo al esqueleto axial. **F** Reconstrucción pseudocolorida de μRayos X en 3D. Escala: 50 μm.

Gorgonians (Cnidaria: Octocorallia) are marine corals, formed by polyps in a porous material that are organized in a tree-like colony[1]. Branches are oriented at right angle to the water current. To prevent damage, gorgonians can be easily twisted and bent while being relatively stiff[2]. So which structures mediate those mechanical properties? To answer this, we studied *Eunicella cavolinii*, a Mediterranean species. **A** Corals were documented *in situ* by scuba diving and underwater photography. **B** Light microscopy showed that colonies consist of the axial skeleton and the orange-colored cortex. The axial skeleton is built of two structures made of collagen-like gorgonin[3]: the black annulla, a sclerotic ring with minor amounts of minerals, and the white medulla. The cortex consists of a porous coenenchym filled with **C** three types of $CaCO_2$ sclerites[4]: clubs, spindles, and subtentacular sclerites. **D** In μX-ray slices, clubs were oriented perpendicularly to the axial skeleton, while spindles were parallelly oriented. **E** In SEM, the medulla showed a fine web of cross-linked fibers parallel to the axial skeleton. **F** Pseudo-colored μX-ray 3D reconstruction. Scale bars: 50 μm.

1. Sánchez J.A. *et al.* (2019). In *Coral Reefs of the World*, (729–747). Springer International Publishing.
2. Etnier S.A. (2003). *The Biological Bulletin*, 205(1):36.
3. Ehrlich H. (2019). *Marine Biological Materials of Invertebrate Origin*. Springer International Publishing.
4. Vielzeuf D. *et al.* (2017). *Crystal Growth & Design*, 17(10):5080.

Contact: Uwe Schnepf <uwe.schnepf@bio.uni-stuttgart.de>

E. F. Médici and A. D. Otero (eds.), *Album of Porous Media*,
https://doi.org/10.1007/978-3-031-23800-0_17

FORAMINIFER

Ab Coorn, Fons Marcelis, and Steffen Berg
Shell Global Solutions International B.V.

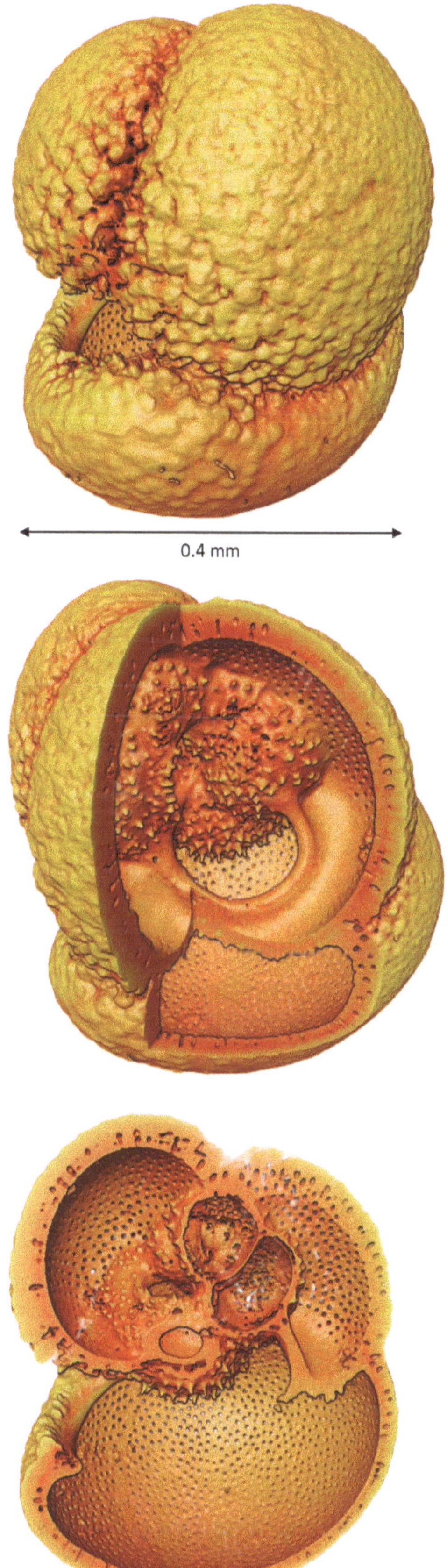

3 different cross sectional views (3D rendering) of a Foraminifera[1] imaged in 3D by X-ray computed micro-tomography. Foraminifera are single-celled organisms that form shells with multiple chambers reflecting the different growth stages, which are visualized here by the three different cross-sectional views of the same organism. Foraminifera belong to the group of micro-fossils. By determining the exact species based on shape etc. they are used to determine the geological age of subsurface formations of sedimentary, porous rock. At the same time, the wall of their shell has small pores of micrometer diameter or less, i.e. they are (micro-)porous themselves. The Foraminifera shown here has been imaged by X-ray computed micro-tomography at a resolution of 0.6 micrometer/voxel which is close to the maximum resolution of conventional micro-CT instruments.

3 vistas transversales distintas (visualizadas en 3D) de una imagen de *Foraminifera*[1] en 3D mediante microtomografía computarizada de rayos X. Los foraminíferos son organismos unicelulares que forman conchas con distintas cámaras que reflejan los distintos estados de crecimiento; visualizados aquí mediante tres vistas transversales distintas del mismo organismo. Los foraminíferos pertenecen al grupo de microfósiles. Al determinar la especie exacta según la forma, etc., se puede determinar la edad geológica de las formaciones de rocas porosas de origen sedimentario en el subsuelo. Así mismo, la pared de su caparazón tiene pequeños poros de micrómetros de diámetro o aun menores, es decir, ellos mismos son (micro)porosos. El foraminífero que se muestra aquí ha sido visualizado con una microtomografía computarizada de rayos X con una resolución de 0,6 micrómetros/vóxel, que es casi la resolución máxima de los dispositivos convencionales micro-TC.

1. Boardman R. *et al.* (1987). *Fossil Invertebrates*. Wiley.

Contact: Steffen Berg <steffen.berg@shell.com>

E. F. Médici and A. D. Otero (eds.), *Album of Porous Media*,
https://doi.org/10.1007/978-3-031-23800-0_18

TROVANTS: THE "LIVING" STONES OF ROMANIA FORMED AS HIGH POROSITY SPHERICAL SANDSTONE CONCRETIONS DEVELOPED AROUND A FOSSIL

Vlad C. Manea[1], Dante Arteaga[1], Marina Manea[2], Delia Lazar[3], Sandra Vega[1], and Gerardo Carrasco-Núñez[1]

[1] Laboratorio Universitario de Microtomografía de Rayos X (LUMIR), Centro de Geociencias, Universidad Nacional Autónoma de México
[2] Computational Geodynamics Laboratory, Centro de Geociencias, Universidad Nacional Autónoma de México
[3] Hunt Oil Company, Bucharest, Romania.

The term *trovants* was introduced for the first time by the Romanian geologist Gheorghe Murgoci more than one hundred years ago[1], and refers to concretionary sandstone formations. They are believed to form as a result of the fusion of layers of sand in the presence of high-concentrated calcareous water. The *trovant* geological reservation at Costești, Romania, hold a large and diverse in size and shape collection of *trovants* embedded in alluvial strata. A new 3D images of a small-size cm size *trovant* (a) were acquired using a Carl Zeiss Xradia Versa-510 X-ray μCT at LUMIR Laboratory. In the images (28.5 μm/voxel resolution) featured in our main graphic, we revealed the presence of subspherical structures built around a central nucleus represented by a full shell (b). We apply a binarization filtering technique for a cubic $600\times600\times200\mu$m voxel sub volume (c), and show the presence of an interconnected porous (ϕ = 5-9 %) assembly, which provides main permeability fluid-paths that control *trovants* formation through mineral precipitation around the inner shell structures (d). All images and simulations are acquired/performed using Avizo 2020.2 software.

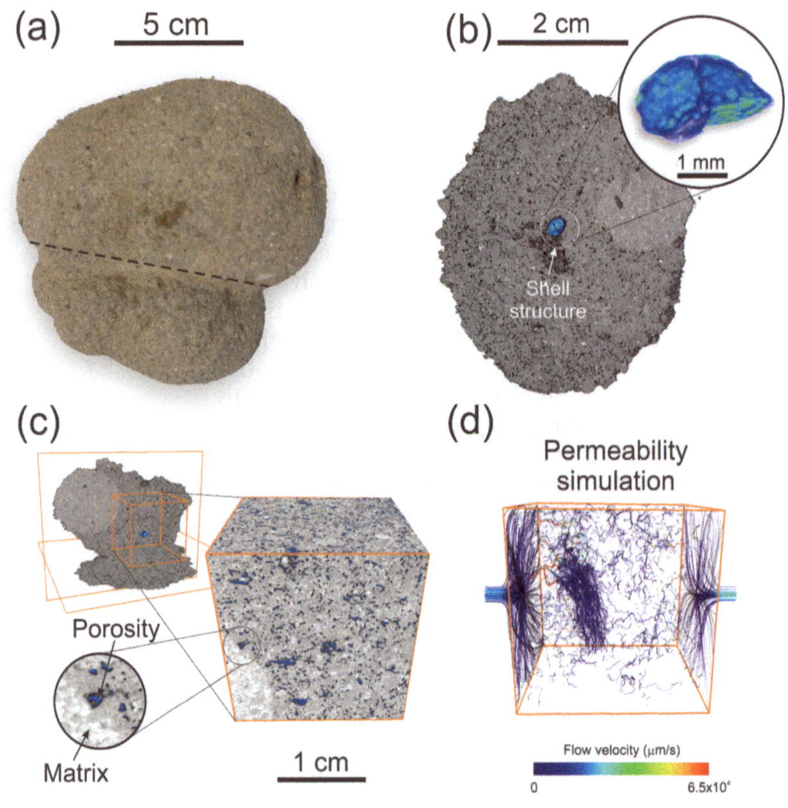

El término *trovant* fue introducido por primera vez por el geólogo rumano Gheorghe Murgoci hace más de cien años[1] y se refiere a formaciones de concreciones de arenisca. Se cree que estas formaciones son el resultado de la fusión de capas de arena ante la presencia de agua con una alta concentración calcárea. La reserva geológica de *trovants* en Costești, Rumanía, contiene una colección grande y diversa en tamaño y forma de *trovants* incrustados en estratos aluvionales. Se obtuvieron nuevas imágenes en 3D de un pequeño *trovant* del orden del cm (a) utilizando un μTC de rayos X Carl Zeiss Xradia Versa-510 en el laboratorio LUMIR. En las imágenes (28,5 μm/vóxel de resolución) presentadas en nuestro gráfico principal, revelamos la presencia de estructuras subesféricas creadas alrededor de un núcleo central representado por una concha completa (b). Hemos aplicado la técnica del filtrado binario usando subvolúmenes de vóxeles cúbicos de $600\times600\times200$ μm (c), y mostramos la presencia de un agrupamiento de poros (ϕ = 5-9 %) interconectados, que proporciona los caminos principales para la permeabilidad de fluidos que controlan la formación de *trovants* mediante precipitación mineral alrededor de las estructuras internas de la concha (d). Todas la imágenes y simulaciones se obtuvieron y realizaron mediante el software Avizo 2020.2.

1. Murgoci G. (1907). *Analele Institutului Geologic al României.*

Contact: Vlad Constantin Manea <vlad@geociencias.unam.mx>

E. F. Médici and A. D. Otero (eds.), *Album of Porous Media*,
https://doi.org/10.1007/978-3-031-23800-0_19

THE COMPLEX PORE SPACES OF OUR BRAIN: A CHALLENGE AND OPPORTUNITY FOR COMPUTER SIMULATIONS

Timo Koch[1,2], *Kent-André Mardal*[1,2], *and Lars Magnus Valnes*[3]

[1] University of Oslo
[2] Simula Research Laboratory
[3] Oslo University Hospital - Rikshospitalet

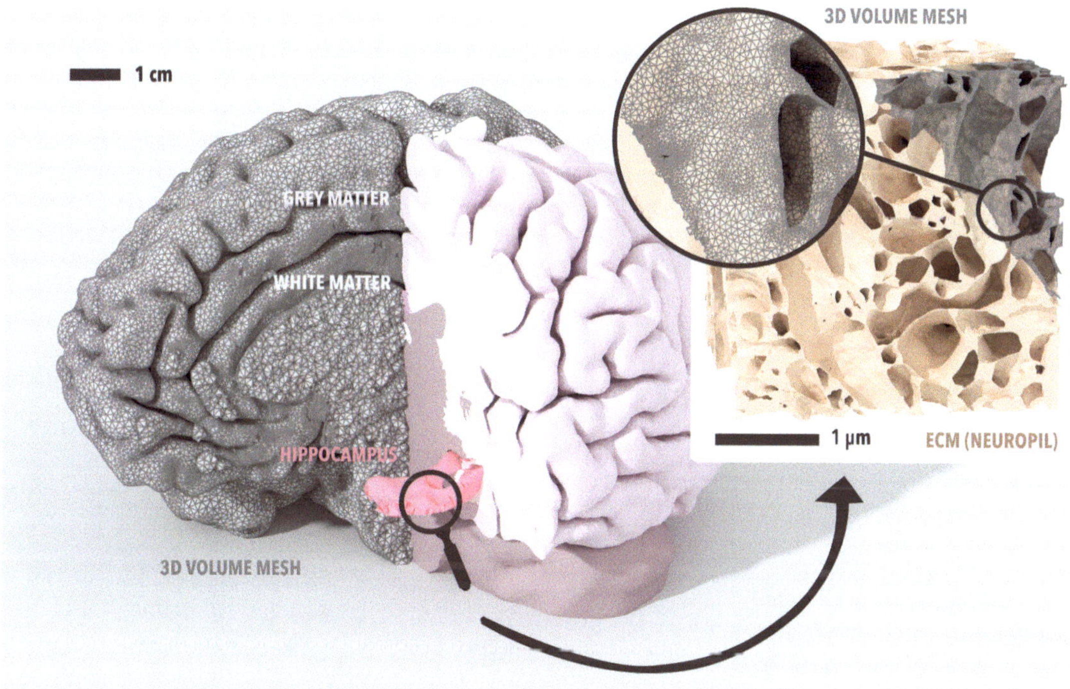

How waste is cleared from the brain is a mystery as it lacks lymphatic vessels. Accumulation of metabolic waste such as amyloid-β and CSF-tau is associated with Alzheimer's and Parkinson's disease. The "glymphatic" theory[1] proposes passive transport of these proteins in the tortuous pore space of the extra-cellular matrix (ECM) shown as sand-colored solid (cells are void). Fluid in the ECM carries waste products. Research shows that our brain clears metabolic waste during sleep faster than when we are awake[2]. The ECM is characterized by tube- and sheet-like structures of 10-100 nm width[3, 4], and increases in volume by 50 % during sleep which enables faster fluid transport. Next to clinical and biological methods, computer simulations allow investigating the mechanisms behind waste transport[4, 5]. A key aspect is an accurate representation of the complex geometries at different scales by 3D computational meshes. The shown volume meshes were generated from T1 MRI data (brain)[5] and electron microscopy data (ECM)[3].

1. Iliff J.J. *et al.* (2012). *Sci. Transl. Med.*, 4(147).
2. Eide P.K. *et al.* (2021). *Brain*, 144(3):863.
3. Kinney J.P. *et al.* (2012). *J. Comp. Neurol.*, 521(2):448.
4. Holter K.E. *et al.* (2017). *Proc. Natl. Acad. Sci. U.S.A.*, 114(37):9894.
5. Valnes L.M. *et al.* (2020). *Sci Rep*, 10(1).

Al carecer de vasos linfáticos, es un misterio cómo se expulsan los desechos del cerebro. La acumulación de desechos metabólicos como β-amiloide y LCR-tau está asociada a las enfermedades de Alzheimer y Parkinson. La teoría «glinfática»[1] propone el transporte pasivo de estas proteínas en los espacios porosos y tortuosos de la matriz extracelular (MEC) mostrada como un sólido de color arena (las celdas son huecas). El fluido en la MEC lleva los desechos. La investigación señala que nuestro cerebro elimina mucho más rápidamente los desechos metabólicos cuando estamos dormidos que cuando estamos despiertos[2]. La MEC se caracteriza por unas estructuras similares a tubos y láminas de 10-100 nm de ancho[3, 4], y aumenta en volumen hasta un 50 % mientras dormimos, lo que permite un transporte más rápido. Junto con los métodos clínicos y biológicos, las simulaciones numéricas permiten investigar los mecanismos detrás del transporte de desechos[4, 5]. Un aspecto clave es una representación clara de las complejas geometrías en diferentes escalas por mallas 3D. Los volúmenes de las mallas mostradas fueron generadas con datos de IRM T1 (cerebro)[5] y datos de microscopio electrónico de barrido (MEB)[3].

Contact: Timo Koch <timokoch@math.uio.no>

E. F. Médici and A. D. Otero (eds.), *Album of Porous Media*,
https://doi.org/10.1007/978-3-031-23800-0_20

II. GEOLOGICAL POROUS MEDIA

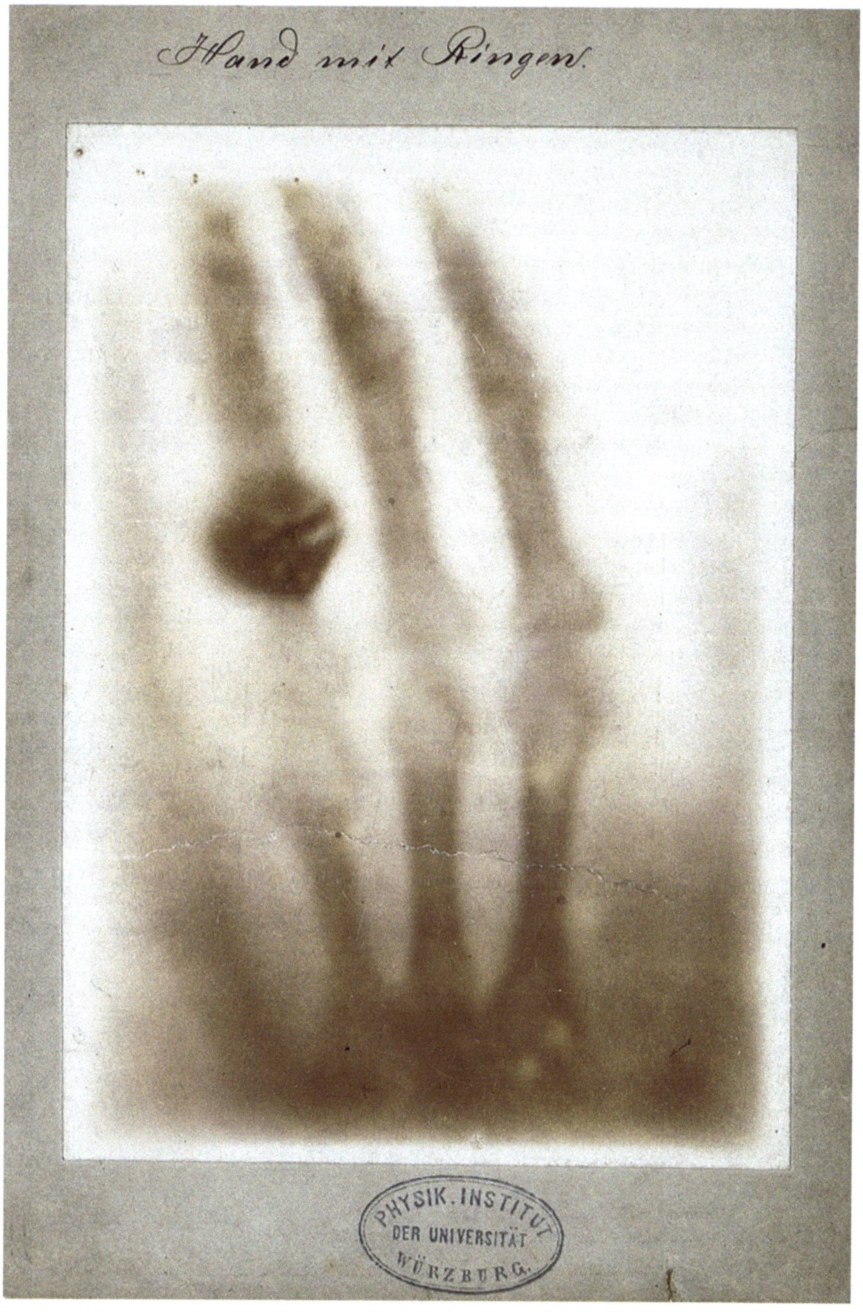

Hand mit Ringen (Hand with Rings): a print of one of the first X-rays by Wilhelm Röntgen (1845–1923) of the left hand of his wife Anna Bertha Ludwig. It was presented to Professor Ludwig Zehnder of the Physik Institut, University of Freiburg, on 1 January 1896.

Hand mit Ringen (Mano con anillos): impresión de una de las primeras radiografías de Wilhelm Röntgen (1845–1923) de la mano izquierda de su esposa Anna Bertha Ludwig. Fue presentada al profesor Ludwig Zehnder del Physik Institut de la Universidad de Friburgo el 1 de enero de 1896.

SCALING AND RESOLUTION EFFECT OF AN ANDESITE SAMPLE

Dante Arteaga and Sandra Vega
Laboratorio Universitario de Microtomografía de Rayos X (LUMIR), Centro de Geociencias, Universidad Nacional Autónoma de México

The present figure shows an example of how the pore space is captured at different sizes and their corresponding resolutions using X-ray microtomography. The sample shown here is an Andesite from an outcrop located at 664431 (x) 2184836 (y) coordinates UTM in Los Humeros Caldera in Mexico[1, 2]. The images come from a physical plug (a) of 2.54 cm of diameter at 29 micron/voxel of resolution; the (b) and (c) images are digital plugs of 1.54 and 0.54 cm of diameter from the center of (a), at 15 and 5 micron/voxel of resolution, respectively. The variation in the plug sizes is a consequence of the used technique, for which the resolution is inversely proportional to the size of the samples. (I), (II) and (III) show different views of the three sample sizes together: (I) from outside, (II) as concatenated images; and (III) a 2D view of the red section displayed on (II). The captured pore space is highlighted in blue, and the solid space is at grey scale. The porosity measured in the images (a), (b) and (c) are 7, 23 and 34 %, respectively. All this illustrates how the pore space is revealed differently at various images resolutions.

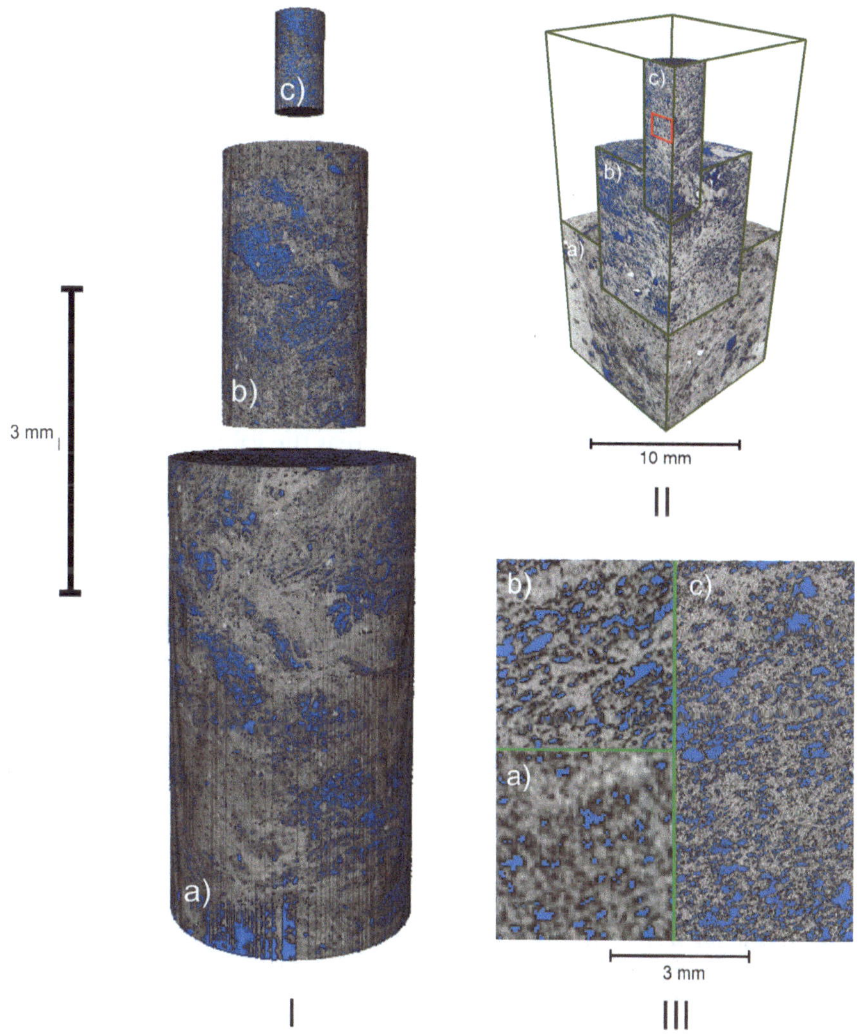

Las presentes figuras muestran un ejemplo de cómo el espacio poroso es captado en diferentes tamaños y sus correspondientes resoluciones utilizando microtomografías de rayos X. La muestra que se observa aquí es de Andesita de un afloramiento rocoso ubicado en las coordenadas (UTM) 664431 (x) 2184836 (y) en la caldera de Los Humeros en México[1, 2]. Las imágenes provienen de un cilindro (a) de 2,54 cm de diámetro con una resolución de 29 micrones/vóxel; las imágenes (b) y (c) son cilindros digitales de 1,54 y 0,54 cm de diámetro respectivamente desde el centro de (a), con una resolución de 15 y 5 micrones/vóxel respectivamente. La variación en el tamaño de los cilindros es una consecuencia de la técnica empleada, para la cual la resolución es inversamente proporcional al tamaño de las muestras. (I), (II) y (III) muestran simultáneamente diferentes vistas de tres muestras de distinto tamaño: (I) desde fuera, (II) como imágenes concatenadas y (III) vista en 2D de la sección roja de (II). El espacio poroso captado está resaltado en azul y el espacio sólido en una escala de grises. La porosidad medida en las imágenes (a), (b) y (c) es de 7, 23 y 34 %, respectivamente. Todo esto ilustra cómo el espacio poroso se revela de manera distinta a diferente resolución de imagen.

1. Ferriz H. and Mahood G.A. (1984). *J. Geophys. Res.*, 89(B10):8511.
2. URL `https://doi.org/10.22201/igg.terradigitalis.2017.2.24.78`.

Contact: Dante Arteaga <darteaga@geociencias.unam.mx>

E. F. Médici and A. D. Otero (eds.), *Album of Porous Media*,
https://doi.org/10.1007/978-3-031-23800-0_21

LENGTH-SCALE DEPENDENCE OF PORE SPACE TOPOLOGY

Ruotong Huang[1,2,3], *Anna Herring*[1,2,4], *Adrian Sheppard*[1,2], *Michael Turner*[1,2,3], *Levi Beeching*[1,2,3], *Lydia Knuefing*[1,2], *and Robert Middleton*[1]

[1] Department of Materials Physics, Research School of Physics, The Australian National University, Australia
[2] ARC Training Centre for M3D Innovation, Research School of Physics, The Australian National University, Australia
[3] CTLab, Research School of Physics, The Australian National University, Australia
[4] Department of Civil and Environmental Engineering, Tickle College of Engineering, The University of Tennessee, USA

3D-visualisation of a Bentheimer sandstone sample ($2\times2\times8.3$ mm^3) using Paraview:
1. The high-resolution micro-CT data (1.66 μm voxel size, acquired: Australian National University, CTLab [1, 2, 3]),
2. Two phase segmentation (black and yellow represents void and grain, respectively),
3. Network model of the connected macropores (generated with OPENPNM),
4. Disconnected microporosity defined as isolated voids smaller than 3.15×10^{-6} mm^3.
In addition to the high-quality tomographic data (1), the visualization also demonstrates the connected and disconnected pore space components in the sample (3 and 4, respectively). The volume-normalized Euler characteristic ($\bar{\chi}$) for the sample considering macropores (3) is -2854 mm^{-3}, whereas the value including the disconnected micropores (4) is -560 mm^{-3}. The deviation in $\bar{\chi}$ shows that the presence of small isolated voids in sandstone samples significantly impact topological value. Depending on the porous medium and the flow processes under investigation, microporosity must be considered and potentially removed to obtain relevant topological values for applications such as immiscible fluid displacements [4].

Visualización en 3D de una muestra de arenisca de Bentheimer ($2\times2\times8{,}3$ mm^3) utilizando Paraview:
1. Datos de micro-TC de alta resolución (tamaño de vóxel de 1,66 μm, adquirido de: Australian National University, CTLab [1, 2, 3]);
2. Segmentación en dos fases (el negro y el amarillo representan los huecos y los gránulos respectivamente);
3. Modelo de red de macroporos conectados (generado con OPENPNM);
4. Microporosidad desconectada definida como huecos aislados de tamaño inferior a $3{,}15\times10^{-6}$ mm^3.
Además de los datos de la tomografía de alta calidad (1), la visualización también demuestra los componentes conectados y desconectados del espacio poroso en la muestra (3 y 4, respectivamente). La función característica de Euler normalizada por el volumen ($\bar{\chi}$) de la muestra teniendo en consideración los macroporos (3) es de -2854 mm^{-3}, mientras que el valor incluyendo los microporos desconectados (4) es de -560 mm^{-3}. La desviación en $\bar{\chi}$ muestra que la presencia de pequeños huecos aislados en las muestras de arenisca impacta significativamente el valor topológico. Según el medio poroso y los procesos de flujo bajo investigación, la microporosidad debe tenerse en consideración y potencialmente eliminarse para obtener valores topológicos relevantes para aplicaciones como los desplazamientos de fluidos inmiscibles [4].

1. Kingston A.M. *et al.* (2018). *IEEE Trans. Comput. Imaging*, 4(3):447.
2. Myers G.R. *et al.* (2016). In S.R. Stock, B. Müller and G. Wang, editors, *SPIE Proceedings*. SPIE.
3. Latham S.J. *et al.* (2018). *IEEE Trans. Comput. Imaging*, 4(2):271.
4. Huang R. *et al.* (2021). *Transp Porous Med*, 137(3):489.

Contact: Ruotong Huang <u5942786@anu.edu.au>

E. F. Médici and A. D. Otero (eds.), *Album of Porous Media*,
https://doi.org/10.1007/978-3-031-23800-0_22

MICROSTRUCTURE OF CRUSHABLE EXPANDED CLAY AGGREGATES

Giulia Guida[1], *Francesca Casini*[1], *Giulia M.B. Viggiani*[2], *Edward Andò*[3], *and Gioacchino Viggiani*[4]

[1] Università degli Studi di Roma Tor Vergata
[2] Cambridge University
[3] EPFL-école polytechnique fédérale de Lausanne
[4] Laboratoire 3SR, Grenoble

The images show two different and complementary views of a mixture of Light Expanded Clay Aggregate (LECA) particles with size ranging between 0.25-1.0 mm. The upper-left is a SEM image showing a projected view of the particles, while the lower-left is a slice of a micro Computed Tomography (μCT) 3D reconstruction. The SEM image shows clearly the roughness and the asperities of the surface of the particles, while the X-ray tomography reveals the heterogeneity existing between different particles in terms of density (greyscale proportional to X-ray attenuation), grain shape and internal porosity. The μCT images on the right show the state of the material scanned at different levels of imposed one-dimensional deformation, indicating significant grain crushing. Particle scale phenomena can thus be tracked throughout compression giving an objective measurement of each particle's life expectancy. A comparative analysis of the microstructural features of the material in its initial and degraded state was used to characterise the overall material compressibility and attempt to link it to the properties of individual grains[1].

1. Guida G. *et al.* (2018). *Géotechnique Letters*, 8(2):155.

Contact: Giulia Guida <giulia.guida@uniroma2.it>

Las imágenes muestran dos vistas distintas y complementarias de una mezcla de partículas de arlita (leca, o arcilla expandida) de un tamaño que oscila entre los 0,25 y 1,0 mm. La imagen superior izquierda es una imagen tomada con microscopio electrónico de barrido (MEB) que muestra una vista proyectada de las partículas, mientras que la imagen inferior izquierda es un corte transversal de una reconstrucción 3D de una microtomografía computarizada (μTC). La imagen MEB muestra claramente la rugosidad y las asperezas de la superficie de las partículas, mientras que la tomografía de rayos X revela la heterogeneidad que existe entre las distintas partículas en términos de densidad (escala de grises proporcional a la atenuación de rayos X), forma del grano y porosidad interna. Las imágenes de μTC de la derecha muestran el estado del material escaneado para distintos niveles de deformación unidimensional impuesta, que indican un aplastamiento significativo del grano. Por lo tanto, los fenómenos de escala en las partículas pueden rastrearse mediante compresión, dando una medida de la expectativa de vida de cada partícula. Se utilizó un análisis comparativo de las características microestructurales del material en su estado inicial y degradado para caracterizar la compresibilidad general del material e intentar vincularlo a las propiedades de los granos individuales[1].

E. F. Médici and A. D. Otero (eds.), *Album of Porous Media*,
https://doi.org/10.1007/978-3-031-23800-0_23

RETICULATED RUTILE

Fons Marcelis and Steffen Berg
Shell Global Solutions International B.V.

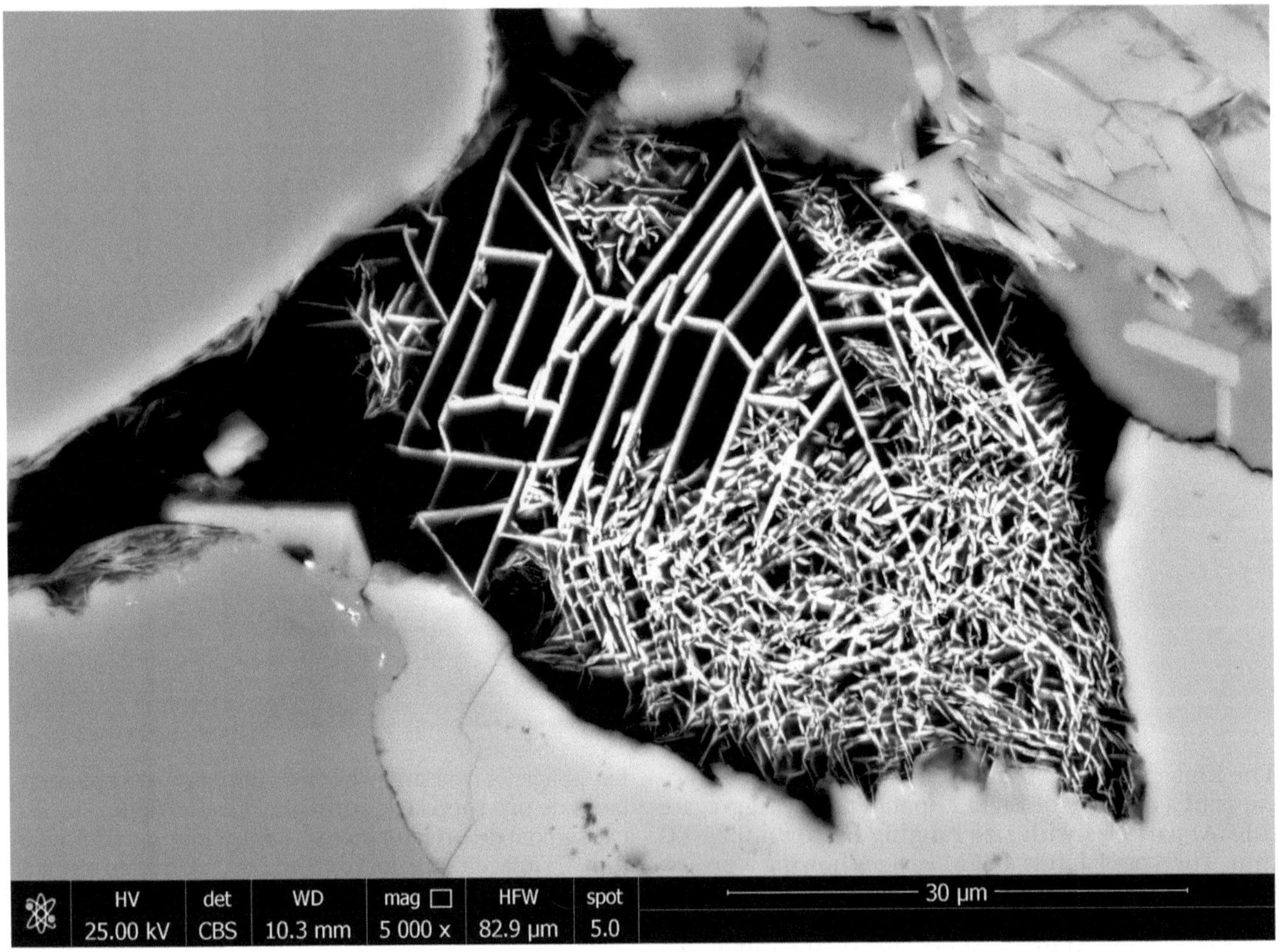

Reticulated Rutile (TiO_2) framework in a pore of a tight Middle East sandstone sample. A very fine network of rutile (titanium dioxide) has grown inside the pore space between quartz grains. Such pore bridging structures can reduce permeability of rock by orders of magnitude. Acquisition by Quanta FEG 650F electron microscope operated at 25 keV, spotsize 5 and column mounted ring BSE detector. The sample is high grade polished and its surface Carbon sputter coated.

Estructura de rutilo (TiO_2) reticulado en un poro de una muestra de arenisca de Oriente Medio. Una red muy fina de rutilo (dióxido de titanio) ha crecido dentro del espacio poroso entre los granos de cuarzo. Estas estructuras porosas intersticiales tienden a reducir la permeabilidad de la roca por varios órdenes de magnitud. La imagen fue obtenida mediante un microscopio electrónico de barrido Quanta FEG 650F operado a 25 kV, spotsize de 5 y anillo detector BSE montado en la columna. La muestra tiene un pulido de alta calidad y su superficie recubierta con una pulverización con carbono.

Contact: Steffen Berg <steffen.berg@shell.com>

E. F. Médici and A. D. Otero (eds.), *Album of Porous Media*,
https://doi.org/10.1007/978-3-031-23800-0_24

MINERAL MAP OF ROTLIEGENDES

Fons Marcelis and Steffen Berg
Shell Global Solutions International B.V.

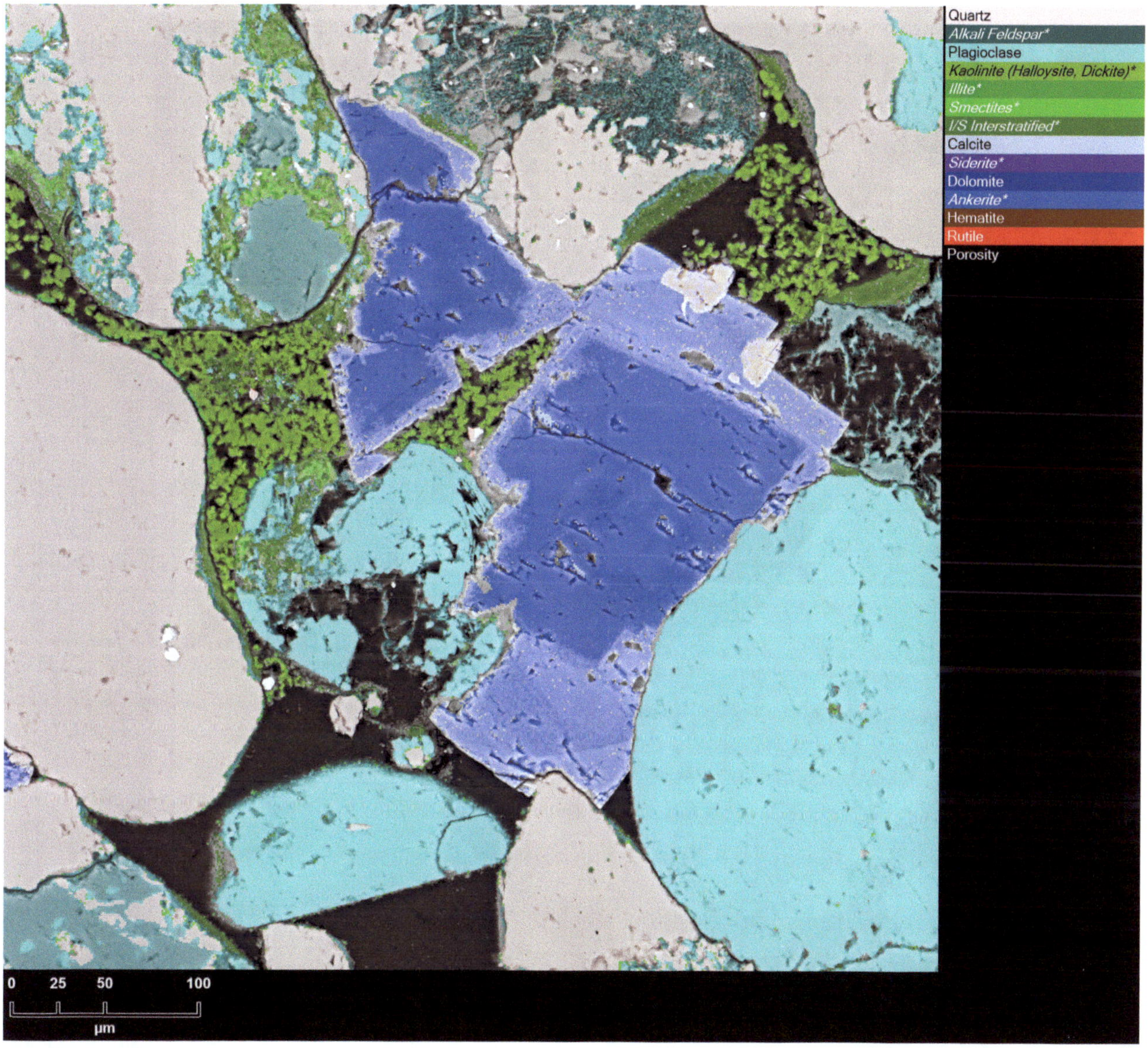

Mineral map of a Rotliegendes-aged sandstone sample from a North Netherlands gas reservoir. The sandstone is composed of eolian quartz and feldspar grains partially cemented by diagenetic zonal dolomite with Iron-rich Ankerite outer rim. Pore space partly occluded by secondary kaolinite, likely the product of replacive K-feldspar grains. Acquisition by Thermo Fisher Mineral Mapping program using a Quanta FEG 650F electron microscope operated at 25 kV, spotsize 5 and column mounted ring BSE detector combined by dual Bruker Xflash 6 EDX detectors. Image processing by Nanomin software. EDX pixels resolution 1 micron. The sample is epoxy resin impregnated and high grade polished; final surface Carbon sputter coated.

Contact: Steffen Berg <steffen.berg@shell.com>

Mapa mineral de una muestra de arenisca del periodo Rotliegendes de un yacimiento de gas al norte de los Países Bajos. La arenisca está compuesta por cuarzo eólico y granos de feldespato parcialmente cementados con dolomita diagenética zonal con un banda externa de ankerita rica en hierro. Espacio poroso parcialmente ocluido por caolinita secundaria, probablemente resultado del reemplazo con granos de feldespato K. La imagen fue obtenida mediante el programa de mapeado mineral de Thermo Fisher utilizando un microscopio electrónico de barrido Quanta FEG 650F operado a 25 kV, spotsize de 5 y anillo detector BSE montado en la columna y combinado con doble detectores Bruker Xflash 6 EDX. El procesamiento de la imagen es realizado con el software Nanomin. La resolución del EDX es de 1 micrón. La muestra está impregnada de resina epoxídica y pulida en alta calidad, con un recubrimiento superficial final de pulverización con carbono.

E. F. Médici and A. D. Otero (eds.), *Album of Porous Media*,
https://doi.org/10.1007/978-3-031-23800-0_25

BOISE BROWN

Fons Marcelis and Steffen Berg
Shell Global Solutions International B.V.

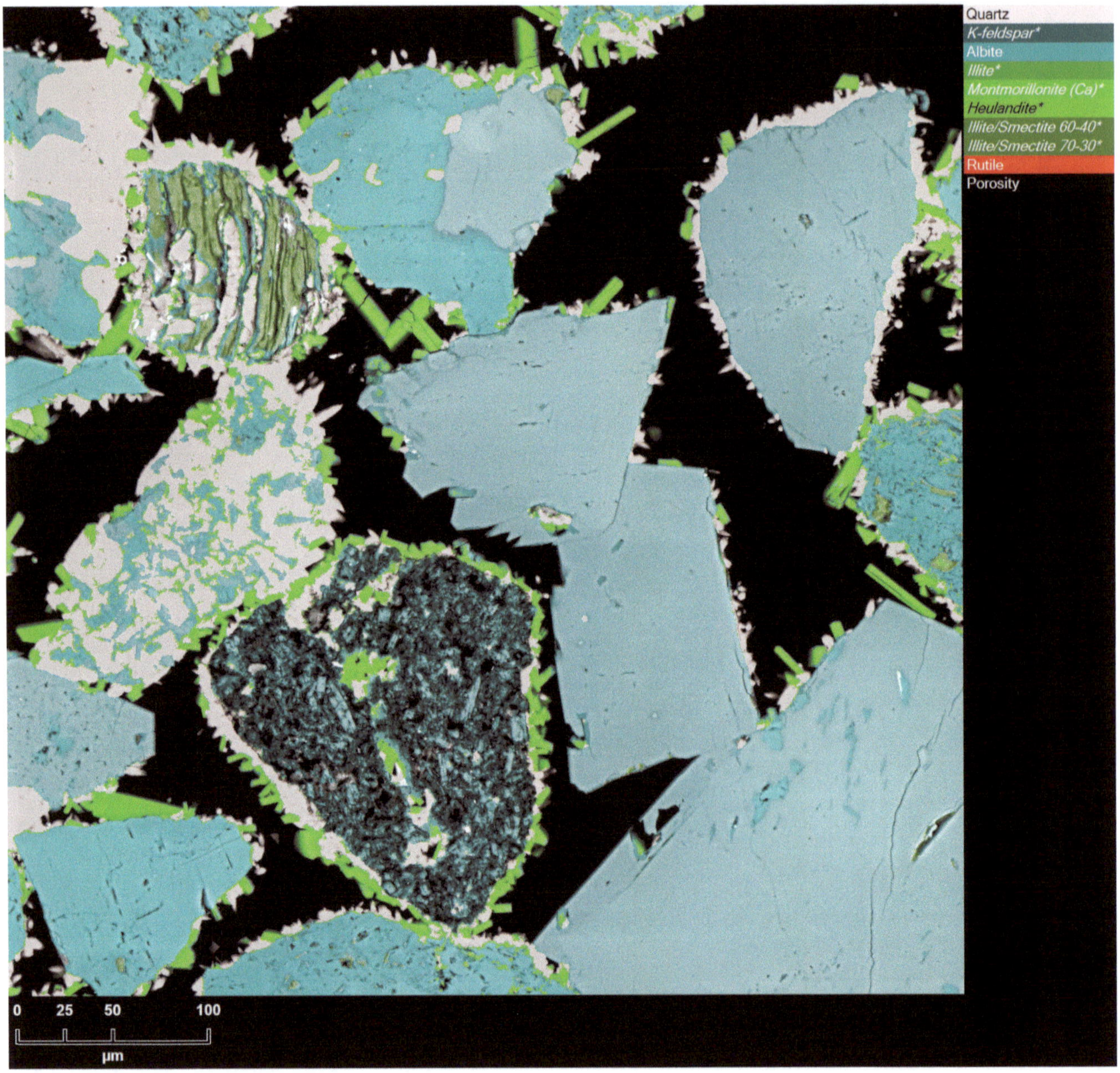

Mineral map of outcrop sandstone Boise Brown. The quartz and feldspar detrital grains are overgrain with secondary quartz which again is overgrown by a late phase of zeolite – in this case Heulandite which has a blocky crystal appearance. Acquisition by Thermo Fisher Mineral Mapping program using a Quanta FEG 650F electron microscope operated at 25 kV, spotsize 5 and column mounted ring BSE detector combined by dual Bruker Xflash 6 EDX detectors. Image processing by Nanomin software. EDX mapping resolution 1 micron.

Mapa mineral de arenisca de afloramiento rocoso en Boise Brown. Los granos detríticos de cuarzo y feldespato se encuentran desbordados por cuarzo secundario que a su vez está desbordado por una fase tardía de zeolita, en este caso heulanditas que tienen apariencia de bloque de cristal. La imagen fue obtenida mediante el programa de mapeado mineral de Thermo Fisher utilizando un microscopio electrónico de barrido Quanta FEG 650F operado a 25 kV, spotsize de 5 y anillo detector BSE montado en la columna y combinado con doble detectores Bruker Xflash 6 EDX. El procesamiento de la imagen es realizado con el software Nanomin. La resolución de mapeado con EDX es de 1 micrón.

Contact: Steffen Berg <steffen.berg@shell.com>

E. F. Médici and A. D. Otero (eds.), *Album of Porous Media*,
https://doi.org/10.1007/978-3-031-23800-0_26

NORTH NETHERLANDS GAS RESERVOIR

Fons Marcelis and Steffen Berg
Shell Global Solutions International B.V.

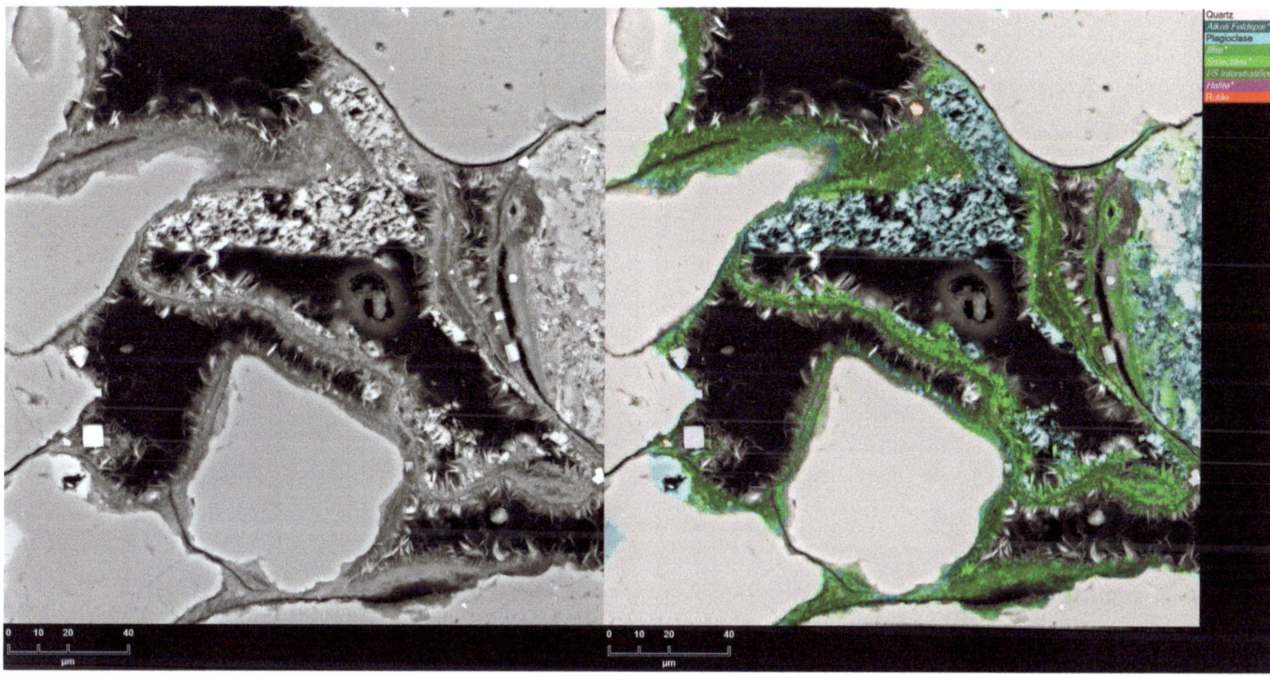

Very high magnification and resolution backscatter electron (BSE) map and the corresponding Mineral Map of a subsurface sandstone sample from a North Netherlands gas reservoir. The sample shows authigenic grain coating illite/smectite and diagenetic events of radial illite growth, a later phase of chlorite (white crystals). The phases have reduced the pore space but especially the permeability of the rock. Acquisition by Thermo Fisher Mineral Mapping program using a Quanta FEG 650F electron microscope operated at 25 kV, spotsize 5 and column mounted ring BSE detector combined by dual Bruker Xflash 6 EDX detectors. Image processing by Nanomin software. The BSE image was acquired with a pixel size resolution of 0.1 micron; the overlain EDX map has a resolution 0.5 micron. The sample is epoxy resin impregnated and the high-grade polished surface is Carbon sputter coated.

Mapa de gran aumento y resolución mediante retrodispersión de electrones (BSE, por sus siglas en inglés) y su correspondiente mapa mineral de una muestra de arenisca del subsuelo de un yacimiento de gas al norte de los Países Bajos. En la muestra se observa un recubrimiento granular autigénico de illita y esmectita y eventos diagenéticos de crecimiento radial de illita, una fase tardía del clorito (cristales blancos). Las fases han reducido el espacio del poro, en particular la permeabilidad de la roca. La imagen fue obtenida mediante el programa de mapeado mineral de Thermo Fisher utilizando un microscopio electrónico de barrido Quanta FEG 650F operado a 25 kV, spotsize de 5 y anillo detector BSE montado en la columna y combinado con detectores dobles Bruker Xflash 6 EDX. El procesamiento de la imagen es realizado con el software Nanomin. La imagen de BSE se realizó con una resolución de 0,1 micrones, el mapa sobrepuesto de EDX tiene una resolución de 0,5 micrones. La muestra está impregnada de resina epoxídica y el pulido de alta calidad de la superficie tiene un recubrimiento de pulverización con carbono.

Contact: Steffen Berg <steffen.berg@shell.com>

E. F. Médici and A. D. Otero (eds.), *Album of Porous Media*,
https://doi.org/10.1007/978-3-031-23800-0_27

FIB-SEM IMAGES FROM A CATHODE MATERIAL: UNFILTERED AND FILTERED (CURTAINING FILTER)

Andreas Griesser
Math2Market GmbH

FIB-SEM images of a High-power $LiNiCoAlO_2$-$LiCoO_2$ cathode were scanned and are provided by the Karlsruhe Institute of Technology (KIT-IAM-ET, left). Two different types of active materials can be distinguished, one homogeneous type and one with an internal granular structure. The latter may break into pieces, and the dark zones are pores. These pores are filled with Carbon black (lighter gray), which is an appropriate infiltration material, but also the real issue in these images. A curtaining filter from the GeoDict software package, applied during image processing and segmentation, removed the vertical streak artefacts that are very common in FIB-SEM images, and thus facilitates and allows further analysis of the sample. The image shows the unfiltered image (left) and the filtered image (right).

Se escanearon imágenes de MEB con haz iones enfocados del cátodo de alta potencia de $LiNiCoAlO_2$-$LiCoO_2$ provistas por el Karlsruhe Institute of Technology (KIT-IAM-ET, izquierda). Se pueden distinguir dos tipos de materiales activos: uno homogéneo y otro con una estructura interna granular. Este último puede romperse en trozos y las zonas oscuras son poros. Estos poros están llenos de carbono negro (gris claro), un material apropiado para la infiltración, aunque también es un verdadero problema en estas imágenes. Un filtro de cortina del paquete de software GeoDict, aplicado durante el procesamiento y segmentación de la imagen, eliminó las rayas verticales comunes en las imágenes de MEB con haz iones enfocados, y así facilitó análisis adicionales de la muestra. La visualización exhibe la imagen sin filtros (izquierda) y la imagen con filtro (derecha).

Contact: Barbara Planas <barbara.planas@math2market.de>

E. F. Médici and A. D. Otero (eds.), *Album of Porous Media*,
https://doi.org/10.1007/978-3-031-23800-0_28

VESICLES AND FRACTURES IN VOLCANIC ASH

Jacopo Taddeucci
Istituto Nazionale di Geofisica e Vulcanologia

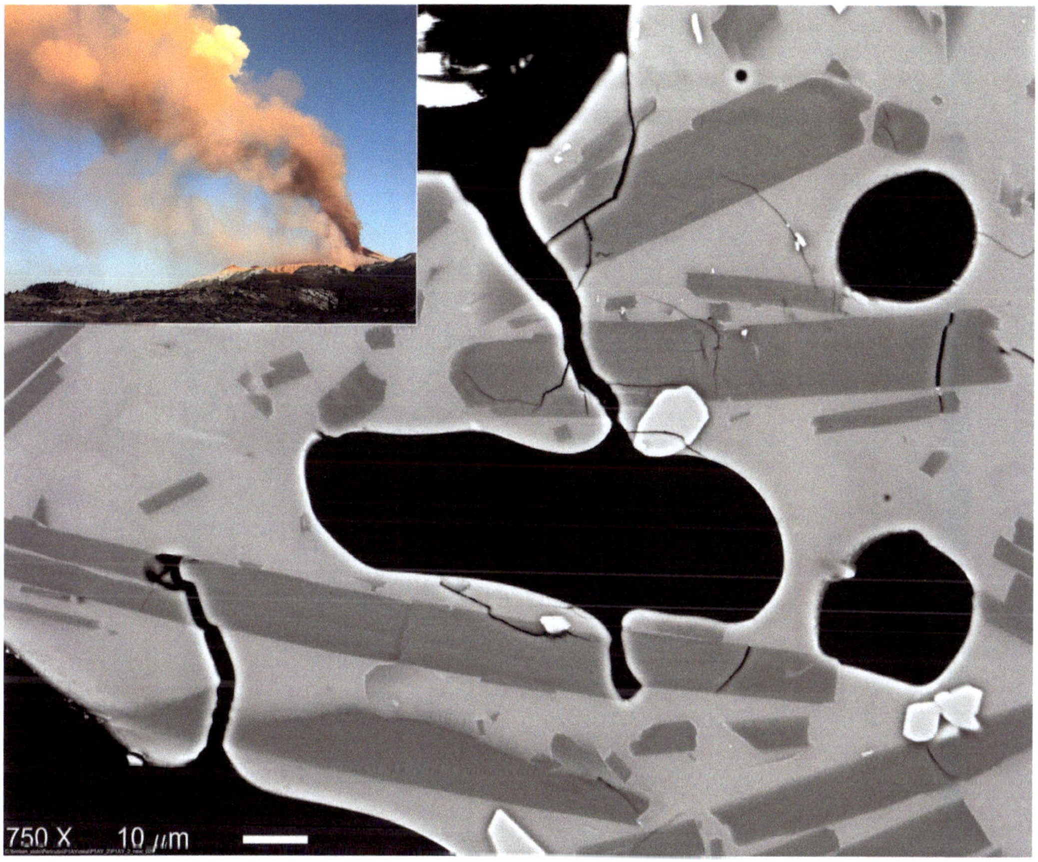

Explosive volcanic eruptions eject fragmental magma particles that range in size from microns to meters (in the inset, an eruption plume from Etna volcano is rich in mm-sized ash particles). Scanning electron micrographs of the particle's interior reveal the nature of magma at the moment of the eruption, with solid crystals (in dark and light grey tones) and gas bubbles and cracks (in black) interspersed in a high-temperature (>1000 °C) silicate melt that quenches into glass after cooling (in mid-grey tone). When the melt fragments, the expansion of the gas originally pressurized in the bubbles drives the eruption. To have an explosive eruption, magma permeability must be low enough to prevent gas escape before fragmentation. Until recently, porous flow through interconnected bubbles was considered the only potential path for gas escape in low (<1000 Pa s) viscosity magmas like the basalt depicted here. However, the broken and fracture-cut crystals visible in the picture testify to the fracturing and healing of the melt[1]. These processes imply that part of the gas could have escaped via permeable flow through temporary cracks formed at fragmentation.

1. Taddeucci J. *et al.* (2021). *Nat. Geosci.*, 14(4):248.

Contact: Jacopo Taddeucci <jacopo.taddeucci@ingv.it>

Erupciones volcánicas explosivas expulsan partículas de magma fragmentadas que varían en tamaño, desde micrones hasta metros (en el recuadro, una columna eruptiva del volcán Etna, rica en partículas milimétricas de ceniza). Al escanear mediante micrografías electrónicas de barrido el interior de las partículas, se revela la naturaleza del magma en el momento de la erupción, con cristales sólidos (tonos gris oscuro y claro) y burbujas de gas y grietas (en negro) intercaladas en un silicato fundido a alta temperatura (>1000 °C) que se templó en forma vidriosa después de enfriarse (en gris intermedio). Cuando el material fundido se fragmenta, la expansión del gas originalmente presurizado en las burbujas impulsa la erupción. Para que ocurra una erupción explosiva, la permeabilidad del magma debe ser baja para evitar que el gas escape antes de la fragmentación. Hasta hace poco, se consideraba que el flujo poroso a través de las burbujas interconectadas era el único camino posible para que escapara el gas en magmas de viscosidad baja (<1000 Pa s) como el basalto que se muestra aquí. Sin embargo, los cristales rotos y con líneas de corte visibles en la imagen atestiguan el fracturamiento y reagrupamiento del material fundido[1]. Estos procesos implican que parte del gas podría haber escapado en forma de flujo permeable por las grietas temporales formadas durante la fragmentación.

E. F. Médici and A. D. Otero (eds.), *Album of Porous Media*,
https://doi.org/10.1007/978-3-031-23800-0_29

A VIEW OF THE INTERNAL PORE STRUCTURE OF A VOLCANIC ROCK: XALTIPAN IGNIMBRITE

Jonathan De la Rosa Maldonado and Sandra Vega
Laboratorio Universitario de Microtomografía de Rayos X (LUMIR), Centro de Geociencias, Universidad Nacional Autónoma de México

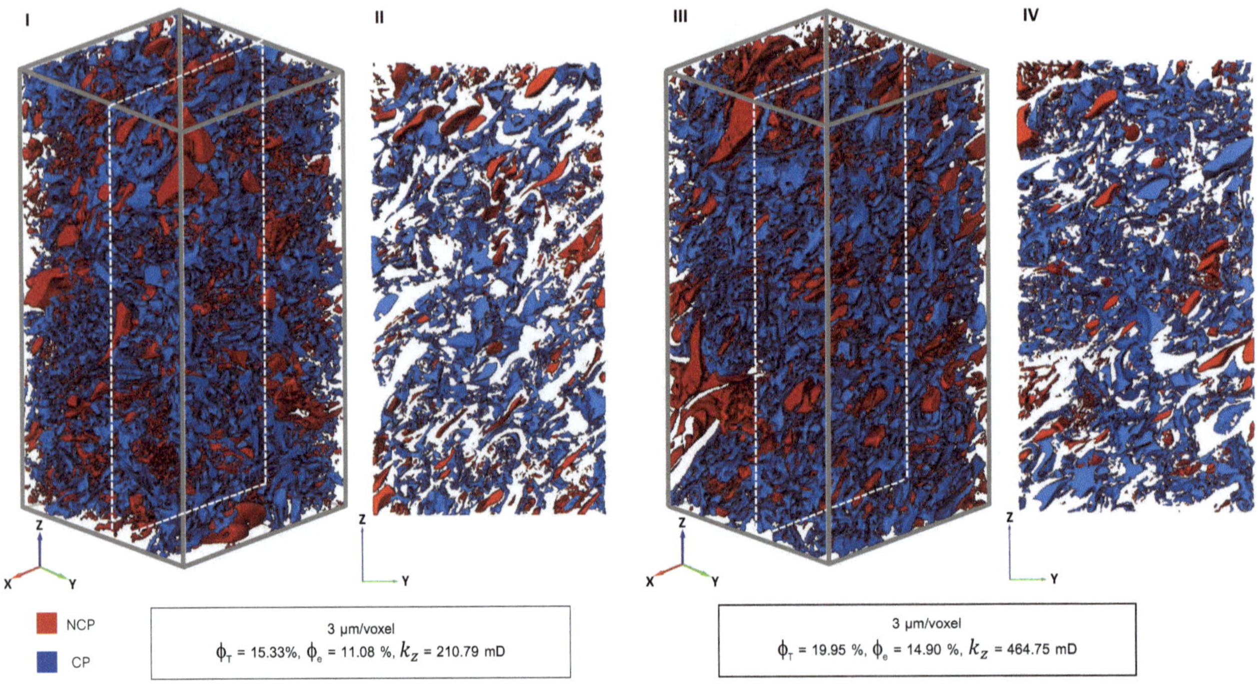

The present figure shows an example of the Xaltipan Ignimbrite from an outcrop located at 674359 (x) 2184836 (y) coordinates (UTM) in Los Humeros Caldera in Mexico[1, 2]. The four pictures shown here are microtomographic images taken at 3 μm/voxel of resolution; processed with anisotropic diffusion, unsharp masking filters and Otsu segmentation method. The solid part of this rock sample has been removed to appreciate more clearly its pore structure. The colors blue and red represent the connected (CP) and non-connected pores (NCP), respectively. The (I) and (III) images are 3D internal views of two close locations, showing the high textural heterogeneity of this rock sample. The (II) and (IV) images show 2D sections from (I) and (III) images, respectively. The 2D section positions are indicated by white dot lines on the 3D images. The total porosity (Φ_T), effective or connected porosity (Φ_e), and permeability in the Z direction (k_Z) calculated in each 3D image are displayed on the figure. It is to note that the major difference here is the distribution of the pore space (in both, connected and non-connected pores).

Esta figura muestra un ejemplo de ignimbrita Xaltipán proveniente de un afloramiento rocoso ubicado en las coordenadas (UTM) 674359 (x) 2184836 (y) en la caldera de Los Humeros, en México[1, 2]. Las cuatro visualizaciones son imágenes microtomográficas tomadas con una resolución de 3 μm/vóxel; procesadas con difusión anisotrópica, filtros de máscara de enfoque y el método de segmentación Otsu. La parte sólida de esta muestra de roca ha sido retirada para apreciar con más claridad su estructura porosa. Los colores azul y rojo representan los poros conectados (PC) y no conectados (PNC) respectivamente. Las imágenes (I) y (III) son vistas internas en 3D de dos ubicaciones cercanas, que muestran la gran heterogeneidad textural de esta muestra de roca. Las imágenes (II) y (IV) muestran secciones en 2D de las imágenes (I) y (III) respectivamente. Las posiciones de las secciones en 2D están indicadas con líneas de puntos blancos en las imágenes 3D. Esta figura muestra en cada imagen 3D los cálculos de la porosidad total (Φ_T), la porosidad efectiva o conectada (Φ_e) y la permeabilidad en la dirección Z (k_Z). Cabe indicar que la mayor diferencia se encuentra aquí en la distribución del espacio poroso (en ambos, poros conectados y no conectados).

1. Ferriz H. and Mahood G.A. (1984). *J. Geophys. Res.*, 89(B10):8511.
2. URL `https://doi.org/10.22201/igg.terradigitalis.2017.2.24.78`.

Contact: Jonathan De la Rosa Maldonado <jhon_delarosam93@outlook.es>

E. F. Médici and A. D. Otero (eds.), *Album of Porous Media*,
https://doi.org/10.1007/978-3-031-23800-0_30

IN SITU MICROTOMOGRAPHY OF THE TENSILE FRACTURE AND FRAGMENTATION OF PORPHYRY COPPER ORES

Nicolas Francois[1], *Yulai Zhang*[1], *Richard Henley*[1], *Ron Cruikshank*[1], *Ajay Limaye*[2], *Michael Turner*[1], *Levi Beeching*[1], *Andrew Kingston*[1], *Mohammad Saadatfar*[1], *and Mark Knackstedt*[1]

[1] ARC Training Centre for M3D Innovation, Research School of Physics, The Australian National University
[2] National Computational Infrastructure, The Australian National University

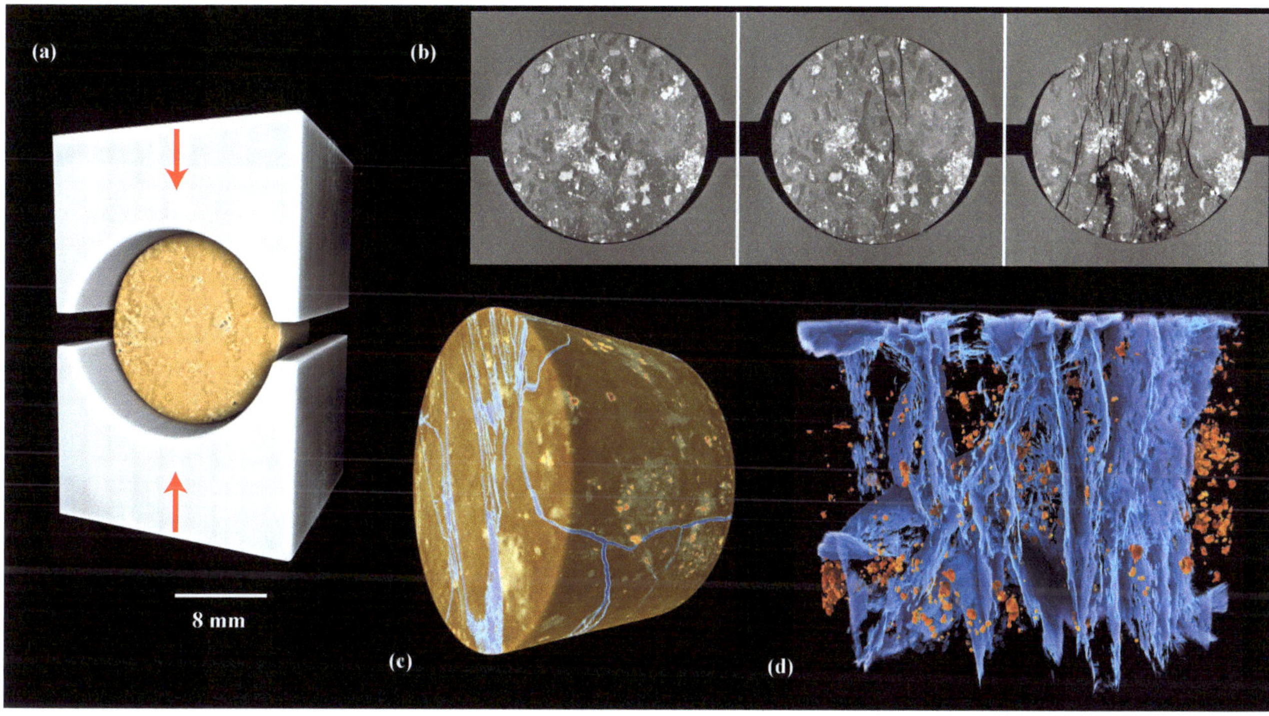

A better understanding of ore breakage and fragmentation is a key to meeting important industrial challenges as the mining industry is exploring and extracting more low-grade ore deposits[1]. This image illustrates how high-resolution X-ray μCT imaging of tensile failure and onset of fragmentation of a Cu-rich ore has the potential to enhance industrial comminution processes. By enabling *in situ* microstructural characterization, μCT will inform studies of the impact of cracks, porosity and ore texture heterogeneity on rock strength, fragment distribution and mineral liberation[2]. (a) 3D reconstruction of a Cu-rich ore sample (volcaniclastic breccia) placed in the jaws of a high-pressure ore-crusher apparatus. This instrument is fully integrated to an X-ray μCT platform at the Australian National University[3] (Red arrows indicate external loads, Scale bar = 8 mm). (b) Midplane tomographic sections showing the rock undergoing successive failure and fragmentation. (c) 3D visualization of the fragmented rock: fractures (blue) and Cu-rich minerals (orange) are highlighted. (d) 3D view of the fracture network and the Cu-rich grains.

1. Tavares L.M. (2021). *Minerals*, 11(5):445.
2. Francois N. *et al.* (2022). URL `https://arxiv.org/abs/2206.12763`.
3. Francois N. *et al.* (2022). *Review of Scientific Instruments*, 93(8):083704.

Comprender mejor la ruptura y fragmentación mineral es clave para superar importantes retos industriales a medida que la industria minera explora y extrae más desde depósitos minerales de baja calidad[1]. La imagen ilustra cómo las imágenes de μTC de alta resolución con rayos X de falla por tensión y el comienzo de la fragmentación de mineral de cobre tienen el potencial de mejorar los procesos industriales de conminución. Al permitir la caracterización de la microestructura *in situ*, las μTC informarán en los estudios sobre el impacto de las grietas, porosidad y heterogeneidad de la textura mineral en la resistencia de la roca, distribución de los fragmentos y liberación mineral[2]. (a) Reconstrucción 3D de una muestra de mineral rico en cobre (brecha volcaniclástica) colocada en la mandíbula de un aparato triturador de mineral de alta presión. Este instrumento se encuentra completamente integrado a una plataforma μTC de rayos X en la Australian National University[3] (las flechas rojas indican las cargas externas, escala = 8 mm). (b) Tomografías de las secciones de los planos medios muestran la roca sometida a sucesivos quiebres y fragmentaciones. (c) Visualización 3D de la roca fragmentada: se resaltan las fracturas (azul) y los minerales ricos en cobre (naranja). (d) Vista 3D de la red de fracturas y de los granos ricos en cobre.

Contact: Nicolas Francois <nicolas.francois@anu.edu.au>

E. F. Médici and A. D. Otero (eds.), *Album of Porous Media*,
https://doi.org/10.1007/978-3-031-23800-0_31

FRACTURED POROUS MEDIA: A CRETACEOUS CARBONATE OUTCROP SAMPLE

Sotirios Kokkalas[1], *Syed Nizamuddin*[2], *and Sandra Vega*[3]

[1] Department of Geology, University of Patras, Greece
[2] Khalifa University, United Arab Emirates
[3] Laboratorio Universitario de Microtomografía de Rayos X (LUMIR), Centro de Geociencias, Universidad Nacional Autónoma de México

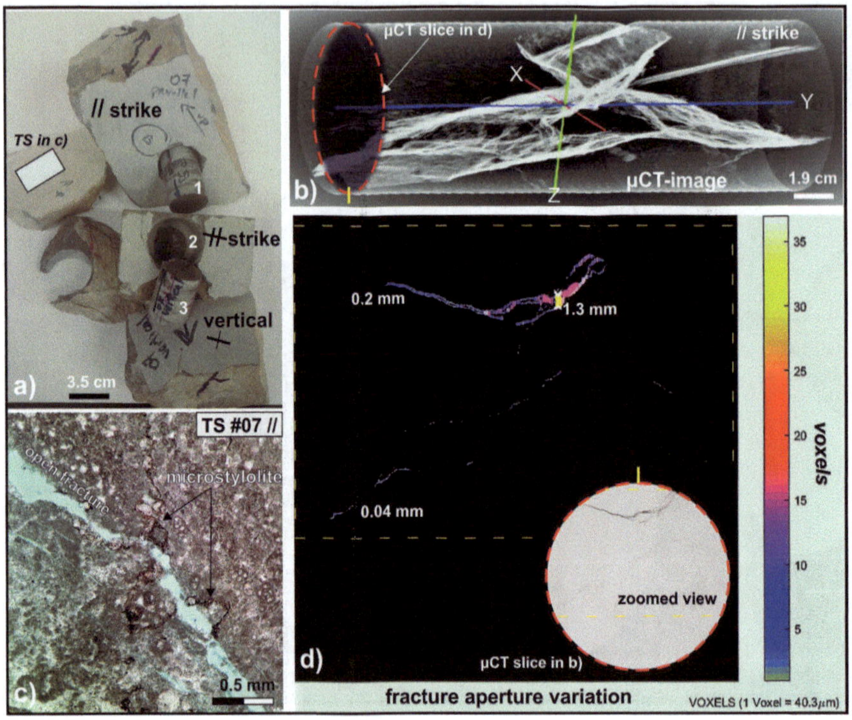

The present figure shows an example of a naturally micro-fractured mud-dominated packstone from Wasia Formation in Ras Al-Khaimah (UAE) and its fracture aperture variation. Images display: (a) a three-orthogonal plug-size cores drilled from the bulk rock sample collected in Ras Al-Khaimah outcrop. The cores are numbered and marked according to their orientation with the bed strike; parallel (//) to bedding along strike (1), parallel to bedding normal (≠) to strike (2) and vertical (+) to the bedding (3). The microscopy thin section (TS) image, shown in c), is marked on the rock sample. (b) a high-resolution 3D micro-CT image from one of the examined core plugs (plug 1 in fig. a), showing the detailed geometry of the microfractures and their highly overlapping to linkage setting. The micro-CT image section shown in (d) is marked on the full CT rendering (core plug dimension: 1.5" diameter × 3" length). (c) an optical microscopic image from a thin section from the same studied rock sample, showing a blue epoxy-filled open fracture cross-cutting a brown-black colored microstylolite, and (d) the aperture variation along strike of a fracture as is calculated from the 2D section of the micro-CT image, which is also shown in (b) for reference.

Contact: Sandra Vega <sandravega@geociencias.unam.mx>

La presente figura muestra un ejemplo de caliza lodogranular microfracturada (packstone) y rica en lodo de la formación Wasia en Ras Al-Khaimah (EAU), y su variación de apertura de fractura. Visualización de imágenes: (a) tres tapones (cilindros) de núcleo ortogonales entre sí provenientes de un afloramiento en Ras Al-Khaimah . Los núcleos están numerados y marcados según su orientación de arrumbamiento, en paralelo (//) a la estratificación y en dirección del rumbo (1), en paralelo a la estratificacion y perpendicular (≠) al rumbo (2) y vertical (+) a la estratificación (3). La imagen de microscopio óptico (TS) de la lamina delgada, mostrada en c), está marcada sobre la muestra de roca en (a). (b) imagen de microtomografía de rayos X de alta resolución de uno de los núcleos estudiados (cilindro 1 de la imagen a), mostrando en detalle la geometría de las microfracturas y sus uniones altamente superpuestas. La sección mostrada en (d) está marcada sobre la imagen de microtomografía (dimensión del núcleo: 1,5 pulgadas de diámetro × 3 pulgadas de largo). (c) imagen de microscopio óptico de una lámina delgada de la misma muestra de roca estudiada, en la que se observa una fractura abierta de corte transversal llena de resina con epoxi azul, un microestilolito marrón-negro y (d) la variación de la apertura a lo largo del rumbo de una fractura a medida que se calcula la sección de la microtomografía que también se muestra en (b) como referencia.

E. F. Médici and A. D. Otero (eds.), *Album of Porous Media*,
https://doi.org/10.1007/978-3-031-23800-0_32

FRACTURE NETWORK IN A SHALE CUBE HYDRAULICALLY FRACTURED IN THE LABORATORY

Mei Li, Earl Magsipoc, Aly Abdelaziz, Johnson Ha, Karl Peterson, and Giovanni Grasselli
University of Toronto

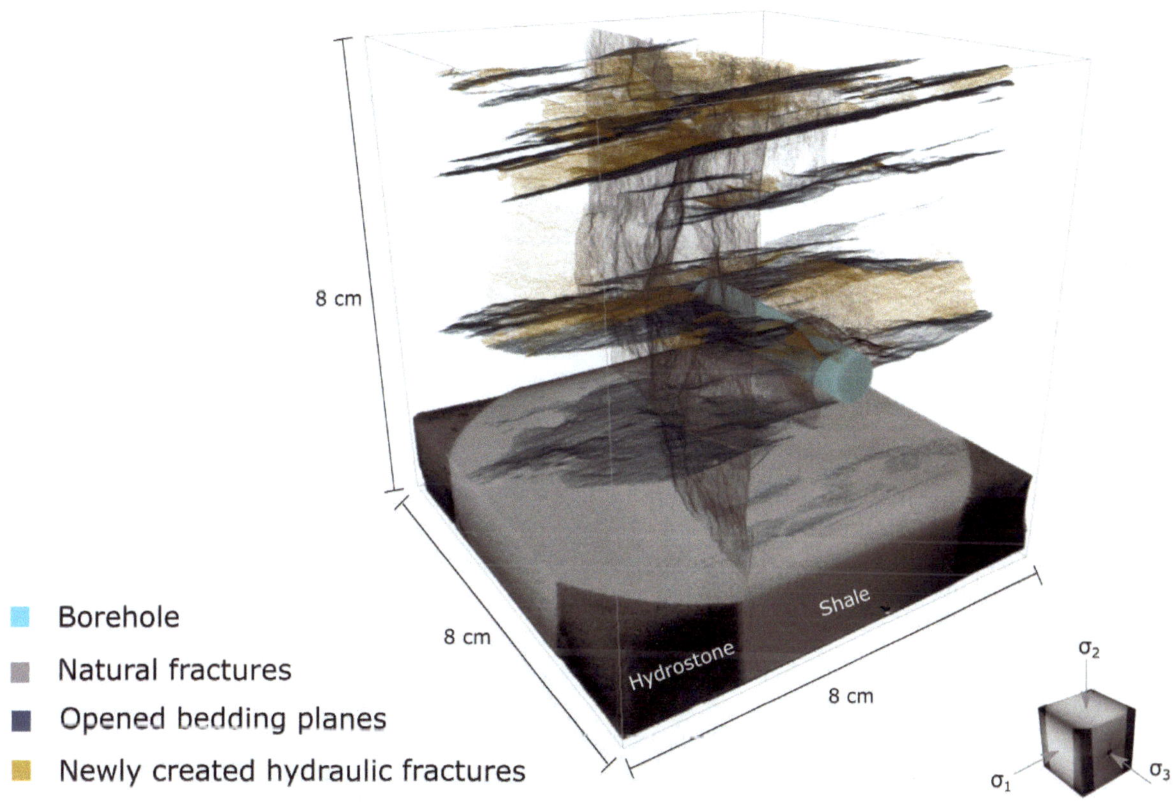

The figure shows the fracture network in a fractured middle Montney shale core sample visualized at a spatial resolution of $39 \times 39 \times 50\ \mu m^3$ utilizing the serial-section reconstruction method. The shale cube was made by casting gypsum cement around a full-diameter shale core and drilling a horizontal borehole (light green) to the middle of the sample. The fracture network formed by laboratory hydraulic fracturing test was complex due to the impact of the intrinsic anisotropy of the shale sample contributed by natural fractures and bedding planes. The hydraulic fracturing test not only generated new hydraulic fractures (yellow) dipping toward σ_3, but also opened subvertical natural fractures (dark gray) and activated bedding planes (blue) dipping 17° to horizontal. The opened natural fractures and bedding planes dominate in the volume of the fracture network, where the natural fractures cross-cut most of the bedding planes - encouraging fracture connectivity and accordingly fluid transmissivity. This fracture network pattern can be expected to be a reliable proxy for the fracturing process occurring near the wellbore in similar shale formations[1].

La figura presenta una red de fracturas en una muestra fracturada de lutita de Montney medio visualizada con una resolución espacial de $39 \times 39 \times 50\ \mu m^3$ utilizando el método de reconstrucción de secciones seriales. El cubo de lutita se construyó depositando completamente cemento de yeso alrededor de un núcleo de lutita y haciendo una perforación horizontal (verde claro) en el centro de la muestra. La red de fracturas producida por el ensayo de fractura hidráulica en laboratorio fue compleja debido al impacto de la anisotropía intrínseca aportadas por las fracturas naturales y los estratos de la muestra de lutita. La prueba de fractura hidráulica no solo generó fracturas hidráulicas (amarillo) inclinadas hacia σ_3, sino que también abrió fracturas naturales subverticales (gris oscuro) y activó estratos (azul) inclinados a 17° respecto de la horizontal. Las fracturas naturales abiertas y los estratos dominan el volumen de la red de fracturas, en el que las fracturas naturales cortan transversalmente la mayoría de los estratos - alentando la conexión entre fracturas y, en consecuencia, la transmisividad del fluido. Se puede esperar que este patrón de la red de fracturas sea una representación fiable del proceso de fractura que ocurre cerca de los pozos en formaciones similares de lutita[1].

1. Li M. *et al.* (2021). *Rock Mech Rock Eng*, 55(5):2937.

Contact: Mei Li <meili.li@mail.utoronto.ca>

E. F. Médici and A. D. Otero (eds.), *Album of Porous Media*,
https://doi.org/10.1007/978-3-031-23800-0_33

GEOLOGICAL CARBON STORAGE AT THE UNIVERSITY OF BERGEN, NORWAY

Kristoffer Eikehaug, Malin Haugen, Benyamine Benali, Zachary Paul Alcorn, Atle Rotevatn, Jan Martin Nordbotten, and Martin Fernø
University of Bergen

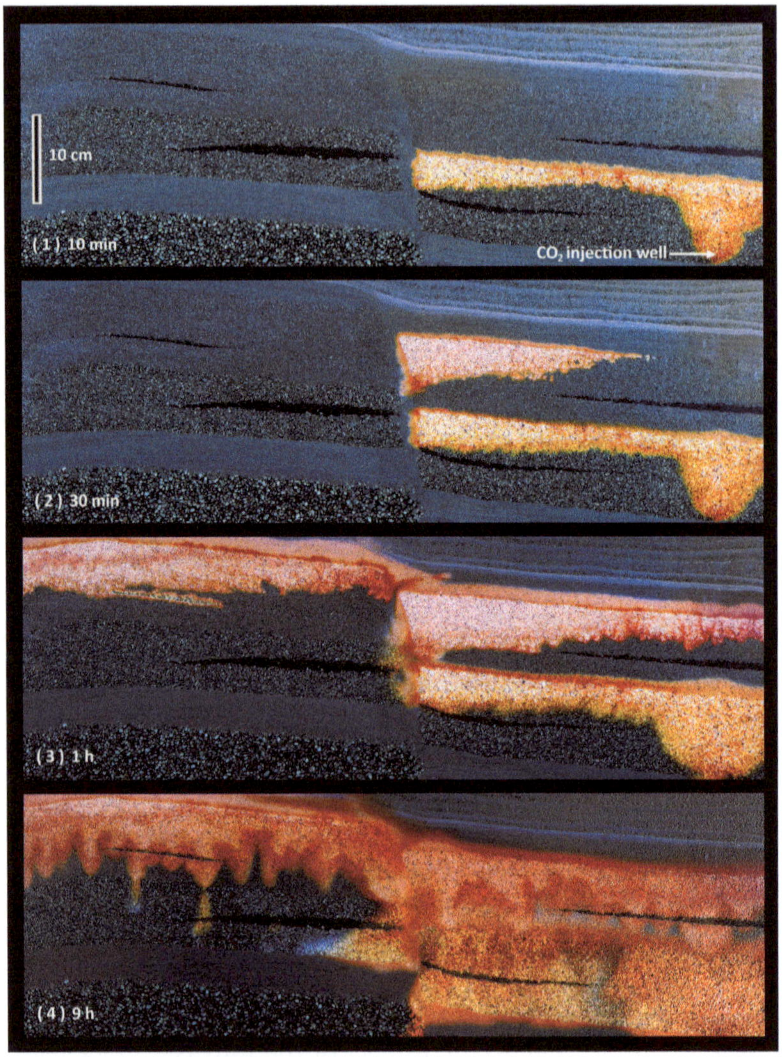

The time-lapsed image series show CO_2 injection in a layered sandstone system modelled using unconsolidated sand of various grain sizes. This laboratory experiment mimics an upwards fining reservoir offset by a central clay-smear fault. The pore space was initially saturated with water and pH-sensitive dyes. Gaseous CO_2 in the system appear bright, whereas other colors reflect the CO_2 concentration in the aqueous phase (green = zero; red = high). The injection rate was 10 ml/min of gaseous CO_2 for 1 hour.
(1) The CO_2 migrated upwards along the inclined interface and became trapped by the low-permeable fault.
(2) The CO_2 flowed along the fault when the capillary entry pressure of the layer above was exceeded.
(3) With increasing gas pressure CO_2 entered the fault, displacing partially CO_2-saturated water into overlying layers.
(4) Convective processes continue long after injection ceased.
The experiment demonstrates important trapping mechanisms for subsurface CO_2 storage, including dissolution, structural, and residual trapping. Repeated injections on the same geometry allow for improving digital simulation tools for multiphase flow in porous media.

La serie de imágenes secuenciales muestra la inyección de CO_2 en un sistema de capas de arenisca basado en arena no consolidada de varios tamaños de grano. Este experimento de laboratorio imita un reservorio granocreciente con un desfasaje en la parte central debido a una falla de arcilla remoldeada.
El espacio poroso fue saturado inicialmente con agua y tintas sensibles al pH. El CO_2 en estado gaseoso aparece brillante en el sistema, mientras que otros colores reflejan la concentración de CO_2 en la fase acuosa (verde = cero; rojo = alto). El caudal de inyección fue de 10 ml/min de CO_2 gaseoso por 1 hora.
(1) El CO_2 migró hacia arriba a lo largo de la interfaz inclinada y quedó atrapado por la falla de baja permeabilidad.
(2) El CO_2 fluyó a lo largo de la falla cuando se excedió la presión de entrada capilar de la capa superior.
(3) Con una presión del gas en aumento, el CO_2 entró en la falla, desplazando parcialmente el agua saturada de CO_2 hacia las capas superiores.
(4) Los procesos de convección prosiguen mucho después del cese de la inyección.
El experimento demuestra la importancia de los mecanismos de captura para el almacenamiento de CO_2 en el subsuelo, incluyendo disolución, y retención estructural y residual. Inyecciones repetidas en la misma geometría permiten mejorar las herramientas de simulación digital para flujo multifásico en medios porosos.

Contact: Kristoffer Eikehaug <kristoffer.eikehaug@uib.no>

E. F. Médici and A. D. Otero (eds.), *Album of Porous Media*,
https://doi.org/10.1007/978-3-031-23800-0_34

III. TWO PHASE FLOW

Pieces of pottery found in the Xianrendong Cave, China from approximately 20000 years ago. This is the earliest evidence of a culture that mastered the technique of baking clay to make usable (non-permeable) containers.

Pedazos de alfarería encontrados en la cueva de Xianrendong en China de hace aproximadamente 20000 años. Esta es la evidencia más temprana de una cultura que dominó la técnica de hornear arcilla para hacer recipientes utilizables (impermeables).

From Wu, X., et al. (2012). Early pottery at 20,000 years ago in Xianrendong Cave, China. Science 336.6089: 1696-1700. Reprinted with permission from AAAS.

CAPILLARY RISE OF WATER IN NANOPOROUS GLASS AS REVEALED BY X-RAY TOMOGRAPHY-BASED LATTICE-BOLTZMANN SIMULATIONS

Guido Dittrich[1], *Juliana Martins de Souza e Silva*[2,3], *Cristine Santos de Oliveira*[3], *Sahar Bakshian*[4], *Andriy V. Kityk*[5], *Martin Steinhart*[6], *Dirk Enke*[7], *Ralf Wehrspohn*[2], *Nima Shokri*[8], *and Patrick Huber*[1]

1 Hamburg University of Technology and Deutsches Elektronen-Synchrotron DESY
2 Institute of Physics, Martin Luther University of Halle-Wittenberg
3 Fraunhofer Institute for Microstructure of Materials and Systems IMWS, Halle, Germany
4 Bureau of Economic Geology, Jackson School of Geosciences, The University of Texas at Austin
5 Faculty of Electrical Engineering, Czestochowa University of Technology
6 Institute of Chemistry of New Materials, Osnabrück University
7 Institute of Chemical Reaction Technology, Leipzig University
8 Institute of Geo-Hydroinformatics, Hamburg University of Technology

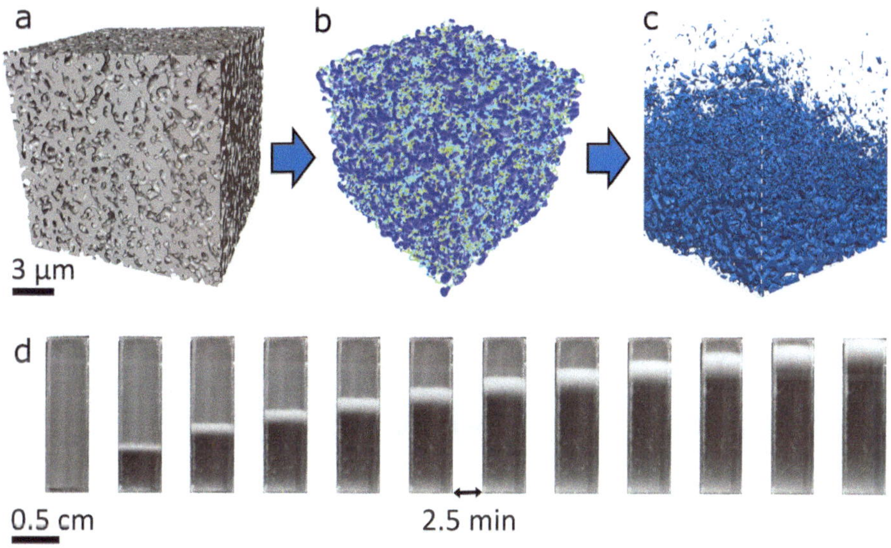

Imbibition in a nanoporous solid is governed by a complex interplay of the liquid's viscosity, the liquid-solid interaction and the geometry of the porous medium. Here a tomography of a controlled pore glass (Vycor, CPG) provides the base for Lattice-Boltzmann computer simulations (L-B) of this process from the single-pore up to the Darcy scale. (a) CPG with 400 nm-sized pores reconstructed from X-ray tomography. (b) Representation of the empty pore space. The color shades indicate the pore size distribution. (c) L-B capillary-rise simulation snapshot based on the tomography[1]. (d) Photographic snapshots during water imbibition into 4 nm-CPG[2, 3]. The imbibition is characterized by a Lucas-Washburn square-root-of-time front advancement along with a front broadening that follows in agreement with the L-B simulations also a square-root-of-time dynamics, see white light scattering front[3]. Note that the X-ray tomography is 100 times scaled down to represent the 4 nm-CPG in the simulation, as justified by the scale invariant spinodal demixing-based pore formation in CPGs.

1. Bakhshian S. *et al.* (2020). *Geophys. Res. Lett.*, 47(14).
2. Gruener S. *et al.* (2009). *Phys. Rev. E*, 79(6).
3. Gruener S. *et al.* (2012). *Proc. Natl. Acad. Sci. U.S.A.*, 109(26):10245.

La imbibición en un sólido nanoporoso es gobernada por una compleja interacción entre la viscosidad del líquido, la interacción líquido-sólido y la geometría del medio poroso. Aquí, la tomografía de un vidrio con tamaño de poro controlado (Vycor, CPG por sus siglas en inglés) proporciona la base para simulaciones computacionales de Lattice-Boltzmann (L-B) de este proceso desde un único poro hasta la escala Darcy. (a) CPG con poros del tamaño de 400 nm reconstruidos de una tomografía de rayos X. (b) Representación del espacio poroso hueco. Las sombras de colores indican la distribución de los tamaños de poro. (c) Visualización de la simulación de L-B del ascenso capilar según la tomografía[1]. (d) Imagen fotográfica durante la imbibición del agua en CPG de 4nm[2, 3]. La imbibición está caracterizada por un desplazamiento del frente tipo raíz cuadrada del tiempo de Lucas-Washburn junto con un ensanchamiento del frente que sigue de acuerdo con las simulaciones de L-B, las cuales también siguen la dinámica de la raíz cuadrada del tiempo, ver dispersión del frente de luz blanca[3]. Obsérvese que la tomografía de rayos X se ha reducido 100 veces para representar el CPG de 4 nm en la simulación, tal y como se justifica por la formación de poros en CPG basados en la separación espinodal invariante en escala.

Contact: Guido Dittrich <guido.dittrich@tuhh.de>

E. F. Médici and A. D. Otero (eds.), *Album of Porous Media*,
https://doi.org/10.1007/978-3-031-23800-0_35

CAPILLARY FINGERING PATTERN IN SLOW DRAINAGE FLOWS

Marcel Moura[1], *Knut Jørgen Måløy*[1], *Eirik Flekkøy*[1], *and Renaud Toussaint*[2]

[1] PoreLab, the Njord Centre, Department of Physics, University of Oslo, Norway

[2] CNRS, Université de Strasbourg, France and PoreLab, the Njord Centre, Department of Physics, University of Oslo, Norway

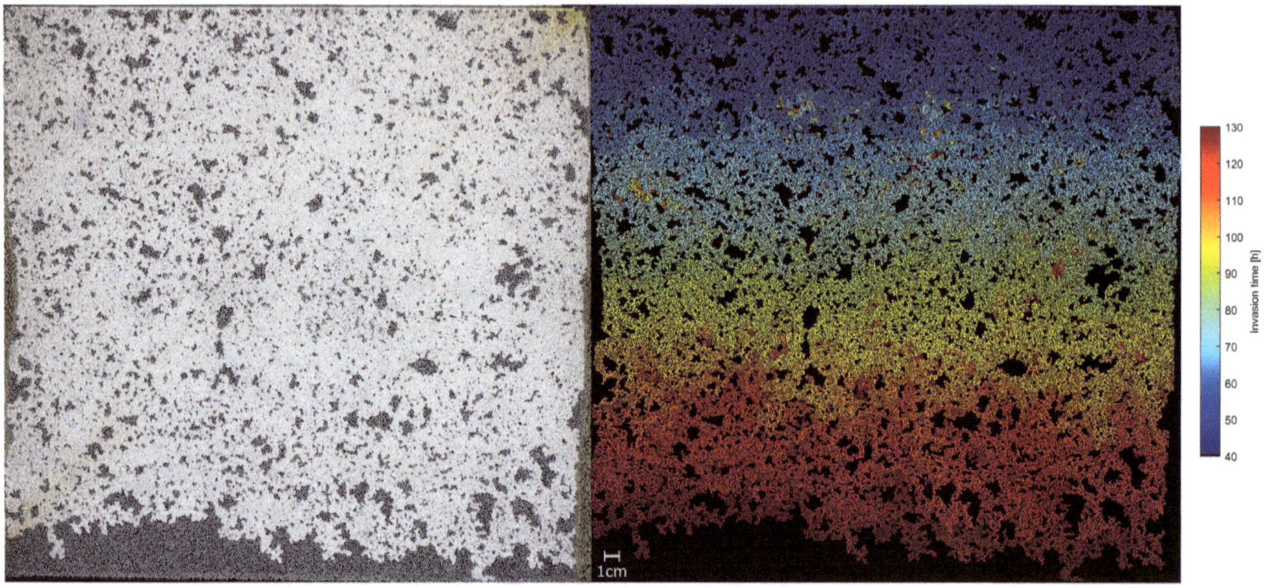

Soils that seem dry can contain vast amounts of water trapped under the surface. This image shows the result of an experiment that exemplifies this. A transparent porous material (a monolayer of glass beads) was initially filled with a liquid (a glycerol–water mixture colored blue), and then dried by removing the liquid from the bottom while keeping the top open to the atmospheric pressure[1, 2, 3]. If the experiment is done slowly enough, we are left with a capillary fingering pattern[4], like the one observed here. This invasion pattern is characterized by the presence of many trapped clusters of the defending wetting fluid, as we can see in the left side of the image (raw experimental image). In the right side of the image we see a spatiotemporal map of the invasion, in which colors are added to visualize the passage of time: blue denotes points close to the surface, that get dried first, while red shows the areas deep underground, which take longer to dry.

Los suelos aparentemente secos pueden contener grandes cantidades de agua atrapada bajo la superficie. La imagen muestra el resultado de un experimento que lo ejemplifica. Un material poroso transparente (una capa de esferas de vidrio) fue llenado inicialmente con un líquido (mezcla de glicerol y agua de color azul) y luego secado, retirando el líquido por la parte inferior, mientras que se deja la parte superior abierta a la presión atmosférica[1, 2, 3]. Si el experimento se realiza lo suficientemente lento, queda un patrón de digitaciones capilares[4], como el que aquí se observa. El patrón de invasión está caracterizado por la presencia de muchos cúmulos atrapados del fluido mojante desplazado, visible en la mitad izquierda de la imagen (imagen experimental sin modificar). En la mitad derecha de la imagen, vemos un mapa espaciotemporal de la invasión en el que se han añadido colores para visualizar el paso del tiempo: el azul denota puntos cercanos a la superficie que se secan primero, mientras que el rojo muestra las zonas subterráneas profundas que tardan más en secarse.

1. Moura M. *et al.* (2017). *Phys. Rev. Lett.*, 119(15).
2. Moura M. *et al.* (2019). *Phys. Rev. Fluids*, 4(9).
3. Moura M. *et al.* (2020). *Front. Phys.*, 7.
4. Lenormand R. and Zarcone C. (1989). *Transp. Porous Med.*, 4(6).

Contact: Marcel Moura <marcelmoura@yahoo.com.br>

E. F. Médici and A. D. Otero (eds.), *Album of Porous Media*,
https://doi.org/10.1007/978-3-031-23800-0_36

DISPERSION PATTERNS WITH TRACER FLOW IN FRACTURES

Le Xu[1], *Knut Jørgen Måløy*[1], *Renaud Toussaint*[2], *Eirik Grude Flekkøy*[1], *and Benjy Marks*[4]

[1] PoreLab, the Njord Centre, Department of Physics, University of Oslo, Norway
[2] CNRS, Université de Strasbourg, France and PoreLab, the Njord Centre, Department of Physics, University of Oslo, Norway
[3] The University of Sydney

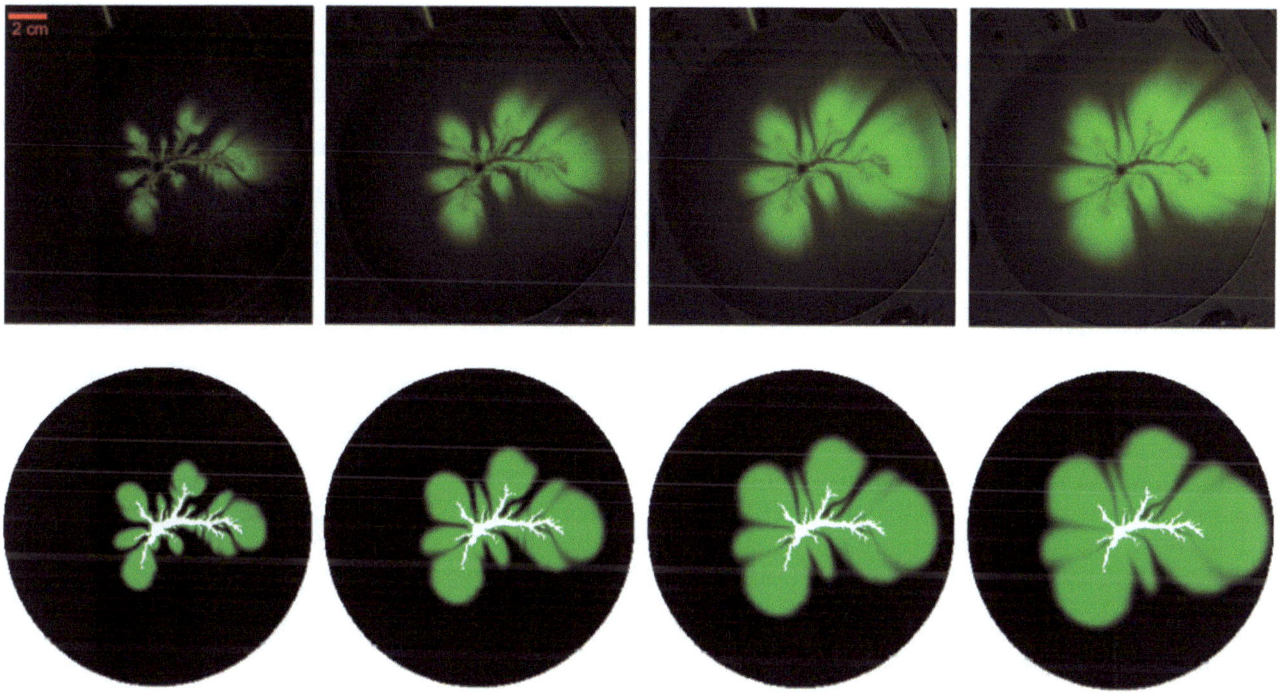

When a reactive fluid is injected into an open fracture, it modifies the fracture surface locally and creates a ramified structure around the injection point. Because of its increased permeability, this structure will have a significant impact on the dispersion of the injected fluid. In this study, we injected a fluorescent tracer fluid into a transparent artificial fracture with such a ramified structure. Because of the model's transparency, the detailed dispersion of the tracer concentration can be followed. The experiments were compared to two-dimensional (2D) computer simulations of convective motion and molecular diffusion. The attached images depict this work and compare experimental results with simulations.
The top images are experimental images subtracted by the initial state image, while the bottom images are simulations. Before fluid flow begins, the boundary conditions are extracted from the experimental image. The green intensity represents the normalized tracer concentration in the experiments and the simulations. The vertical pairs of images show comparison after an injection that lasted 10/20/30/40 minutes, providing a qualitative description of the similarity between the experiments and simulations.

Contact: Knut Jørgen Måløy <k.j.maloy@fys.uio.no>

Cuando se inyecta un fluido reactivo en una fractura abierta, este modifica localmente la superficie de la fractura y crea una estructura ramificada alrededor del punto de inyección. Debido al aumento de permeabilidad, esta estructura tendrá un impacto significativo en la dispersión del fluido inyectado. En este estudio, inyectamos un fluido trazador fluorescente en una fractura artificial transparente con la estructura ramificada mencionada. Por la transparencia del modelo, se puede seguir la dispersión detallada de la concentración del trazador. Los experimentos se compararon con simulaciones computacionales en 2D del movimiento convectivo y la difusión molecular. Las imágenes adjuntas representan este trabajo y comparan resultados experimentales con simulaciones.
Las imágenes superiores son imágenes experimentales a las que les fue sustraída la imagen del estado inicial, mientras que las imágenes inferiores son simulaciones. Antes de que empiece a fluir el fluido, se extraen las condiciones de contorno de la imagen experimental. La intensidad verde representa la concentración normalizada del trazador en los experimentos y en las simulaciones. Los pares de imágenes verticales muestran la comparación después de una inyección que duró 10/20/30/40 minutos, lo que proporciona una descripción cualitativa de la similitud entre los experimentos y las simulaciones.

E. F. Médici and A. D. Otero (eds.), *Album of Porous Media*,
https://doi.org/10.1007/978-3-031-23800-0_37

GPU-ENHANCED PORE-NETWORK SIMULATION OF INSTABILITY DURING FLOW AND TRANSPORT IN POROUS MEDIA

Senyou An and Vahid Niasar
University of Manchester

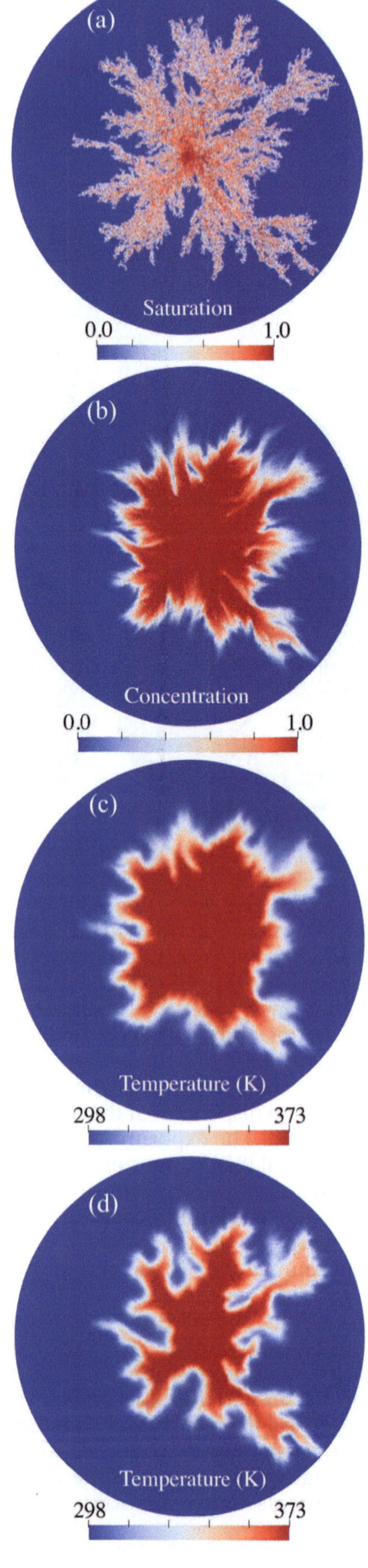

Snapshots of instabilities in two-phase flow and transport simulated using in-house developed pore-network modelling enhanced by the graphics processing unit. The network in all four cases had a radius of 50 mm and thickness of 0.2 mm, comprising about one million pore bodies and four million pore throats, with an average pore radius of 0.0257 mm and variance of 2.12×10^{-4} mm.
(a) Saturation distribution: viscous fingering during two-phase flow, the viscosity of invading and receding phases were 1 mPa s and 50 mPa s, respectively. The interfacial tension was 50 mN/m. The flow rate at the inlet boundary (from the centre) was 1 mm^3/s. (b) Concentration distribution: instability in solute transport in a shear-thinning fluid (Xanthan gum) visualised at the breakthrough time for the injection rate of 0.1 mm^3/s. (c) Temperature distribution: instability in heat transfer in water as a Newtonian fluid for the injection rate of 0.1 mm^3/s. (d) Temperature distribution: instability in heat transfer in crude oil as a non-Newtonian fluid with temperature and local shear-dependent rheology for the injection rate of 0.1 mm^3/s.

Imágenes de inestabilidades de flujo de dos fases y transporte, simulados utilizando un modelo de red de poros de desarrollo propio acelerado por la unidad de procesamiento gráfico. En los cuatro casos, la red tenía un radio de 50 mm y un grosor de 0,2 mm, y comprendía alrededor de un millón de poros y cuatro millones de gargantas, con un promedio de radio de poro de 0,0257 mm y una varianza de 2.12×10^{-4} mm.
(a) Distribución de la saturación: digitación viscosa durante el flujo de dos fases, la viscosidad de las fases inyectada y desplazada fue respectivamente de 1 mPa s y 50 mPa s. La tensión superficial fue de 50 mN/m. El caudal en el contorno de entrada (en el centro) fue de 1 mm^3/s. (b) Distribución concentraciones: inestabilidad del transporte de solutos en un fluido sujeto a pseudoplasticidad (goma xantana) visualizada en el momento que el agua logra cruzar el dominio para un caudal de inyección de 0,1 mm^3/s. (c) Distribución de temperatura: inestabilidad en la transferencia de calor en el agua, fluido newtoniano, para un caudal de inyección de 0,1 mm^3/s. (d) Distribución de la temperatura: inestabilidad en la transferencia de calor en el petróleo, fluido no newtoniano, con temperatura y reología localmente dependiente de las tensiones de corte para un caudal de inyección de 0,1 mm^3/s.

Contact: Vahid Niasar <vahid.niasar@manchester.ac.uk>

E. F. Médici and A. D. Otero (eds.), *Album of Porous Media*,
https://doi.org/10.1007/978-3-031-23800-0_38

VISCOUS FINGERING FRACTAL IN A POROUS MEDIUM

Marcel Moura and Knut Jørgen Måløy
PoreLab, the Njord Centre, Department of Physics, University of Oslo, Norway

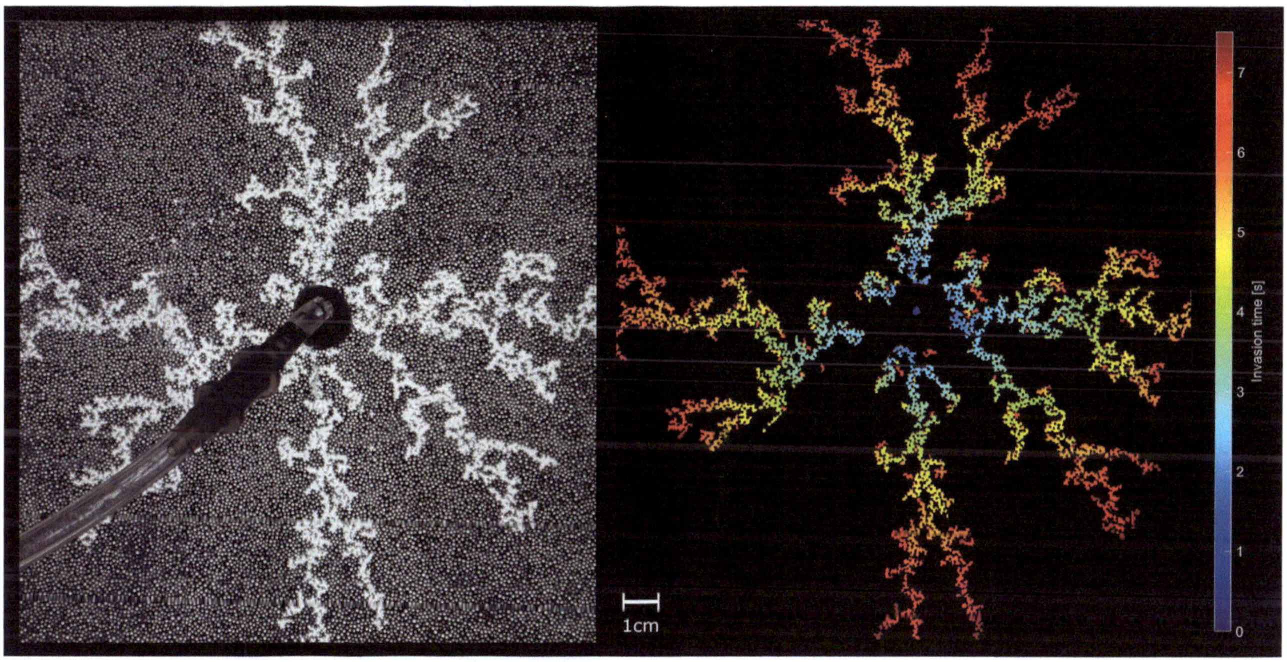

A striking viscous fingering fractal pattern arises when a less-viscous fluid displaces a more-viscous fluid from a quasi-2D porous network[1]. On the left we show an experimental image of the process and on the right a digitally colored reconstruction, where the colors map the invasion time (blue for the start of the experiment and red for the end). The experiment is performed by the fast injection of air into a glycerol-water mixture that initially saturates a quasi-2D porous medium. The medium is composed by a monolayer of glass beads (diameter 1mm) placed in the gap of a modified Hele-Shaw cell. The air injection rate was 35 ml/h.

1. Måløy K.J. *et al.* (1985). *Phys. Rev. Lett.*, 55(24):2088.

Un llamativo patrón fractal de digitaciones viscosas aparece cuando un fluido menos viscoso desplaza un fluido más viscoso en una red porosa cuasibidimensional[1]. A la izquierda mostramos una imagen experimental del proceso y a la derecha una reconstrucción coloreada digitalmente, en la que los colores indican el tiempo de invasión (azul para el inicio del experimento y rojo para el final). El experimento se realiza con la inyección rápida de aire en una mezcla de glicerol y agua que inicialmente satura el medio poroso cuasibidimensional. El medio está compuesto por una sola capa de esferas de vidrio (de 1 mm de diámetro) colocadas en el espacio de una celda de Hele-Shaw modificada. El caudal de la inyección de aire fue de 35 ml/h.

Contact: Marcel Moura <marcelmoura@yahoo.com.br>

E. F. Médici and A. D. Otero (eds.), *Album of Porous Media*,
https://doi.org/10.1007/978-3-031-23800-0_39

STRONG IMBIBITION IN PATTERNED MICROFLUIDICS

Benzhong Zhao[1], *Chris MacMinn*[2], *and Ruben Juanes*[3]

[1] McMaster University
[2] University of Oxford
[3] Massachussets Institute of Technology

Displacement of a viscous defending fluid (μ_{def} = 340 mPa s by a less viscous invading fluid (μ_{inv} = 0.99 mPa s) in a radial micromodel with a diameter of 10 cm. It is patterned with cylindrical posts of height 100 μm and median diameter 1.1 mm. The invading fluid is strongly wetting to the micromodel (θ = 7°), and it advances by sequentially coating the perimeters of neighboring posts via corner flow at the low capillary number corresponding to this experiment (Ca=0.0058). Post coating via corner flow is followed by filling of the pore bodies. Characteristics of the post coating process resemble invasion percolation: (1) There is no directionality—the chain of coated posts can wrap around itself. (2) Post coating occurs in bursts—clusters of closely spaced posts are rapidly coated, which is followed by quiescent periods. We record the displacement process with a scientific camera (Orca Flash 4.0, Hamamatsu) mounted directly above the center of the flow cell, while an LED light panel (Knema LLC) illuminates the cell from below. The invading fluid is seeded with a light-absorbing dye (Brilliant Blue G, Alfa Aesar), which allows us to probe the invading fluid saturation (illustrated by the colormap) across the micromodel gap via a calibration curve [1, 2, 3, 4].

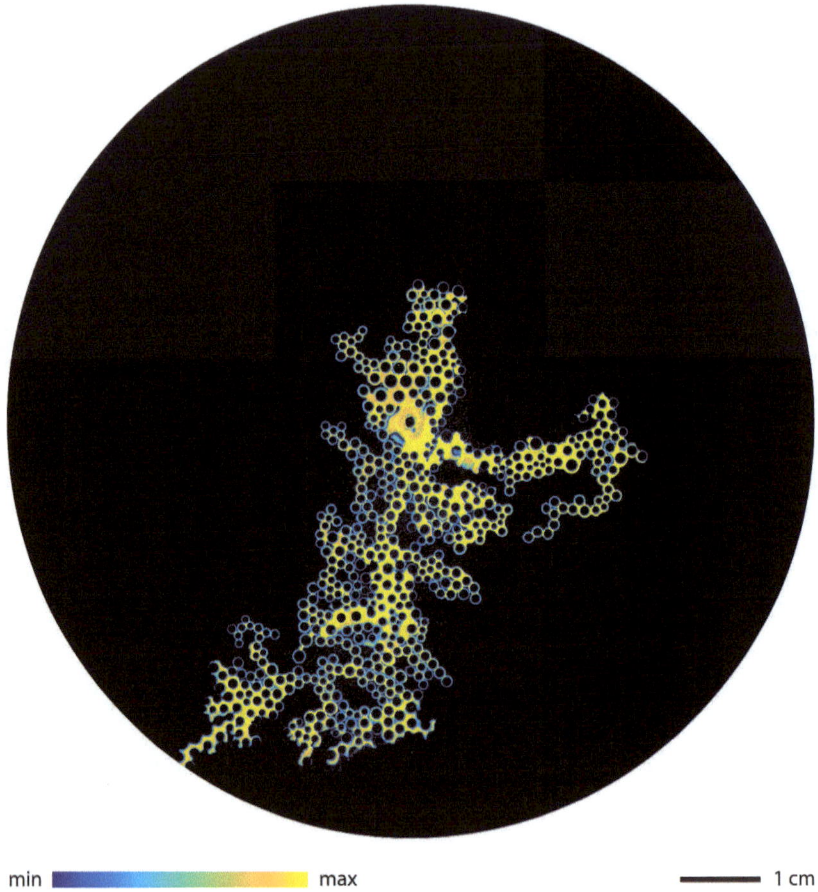

Movimiento de un fluido viscoso desplazado (μ_{def} = 340 mPa s) por la inyección de un fluido menos viscoso (μ_{inv} = 0,99 mPa s) en un micromodelo radial con un diámetro de 10 cm. El micromodelo tiene un patrón de postes cilíndricos de 100 μm de altura y un diámetro medio de 1,1 mm. El fluido inyectado tiene una alta mojabilidad en el micromodelo (θ = 7°), y avanza recubriendo secuencialmente los perímetros de los postes vecinos debido al flujo a lo largo de las esquinas en este experimento realizado a una tasa de inyección correspondiente a un número capilar bajo (Ca = 0,0058). Al recubrimiento por flujo de esquinas, le sigue el llenado de los espacios porosos. Las características del proceso posterior al recubrimiento se parecen a un proceso de invasión por percolación: (1) No hay direccionalidad: la cadena de postes recubiertos puede rodearse a sí misma. (2) La etapa posterior al recubrimiento ocurre en ráfagas: cúmulos de postes cercanos se recubren rápidamente, seguido de periodos de inactividad. Grabamos el proceso de desplazamiento con una cámara científica (Orca Flash 4.0, Hamamatsu) montada directamente sobre el centro de la celda de flujo, mientras un panel con luces LED (Knema LLC) ilumina la celda desde abajo. El fluido inyectado está teñido con un pigmento miscible que absorbe la luz (Brilliant Blue G, Alfa Aesar), permitiendo de ese modo medir la saturación del fluido inyectado (ilustrado por la escala de colores) a lo largo del espacio del micromodelo, por medio de una curva de calibración [1, 2, 3, 4].

1. Zhao B. *et al.* (2016). *Proc. Natl. Acad. Sci. U.S.A.*, 113(37):10251.
2. Zhao B. *et al.* (2019). *Proc. Natl. Acad. Sci. U.S.A.*, 116(28):13799.
3. Primkulov B.K. *et al.* (2021). *J. Fluid Mech.*, 923.
4. Primkulov B.K. *et al.* (2022). *Commun Phys*, 5(1).

Contact: Ruben Juanes <juanes@mit.edu>

E. F. Médici and A. D. Otero (eds.), *Album of Porous Media*,
https://doi.org/10.1007/978-3-031-23800-0_40

RISING INVADER

Joachim Falck Brodin, Knut Jørgen Måløy, Marcel Moura, and Per Arne Rikvold
PoreLab, the Njord Centre, Department of Physics, University of Oslo, Norway

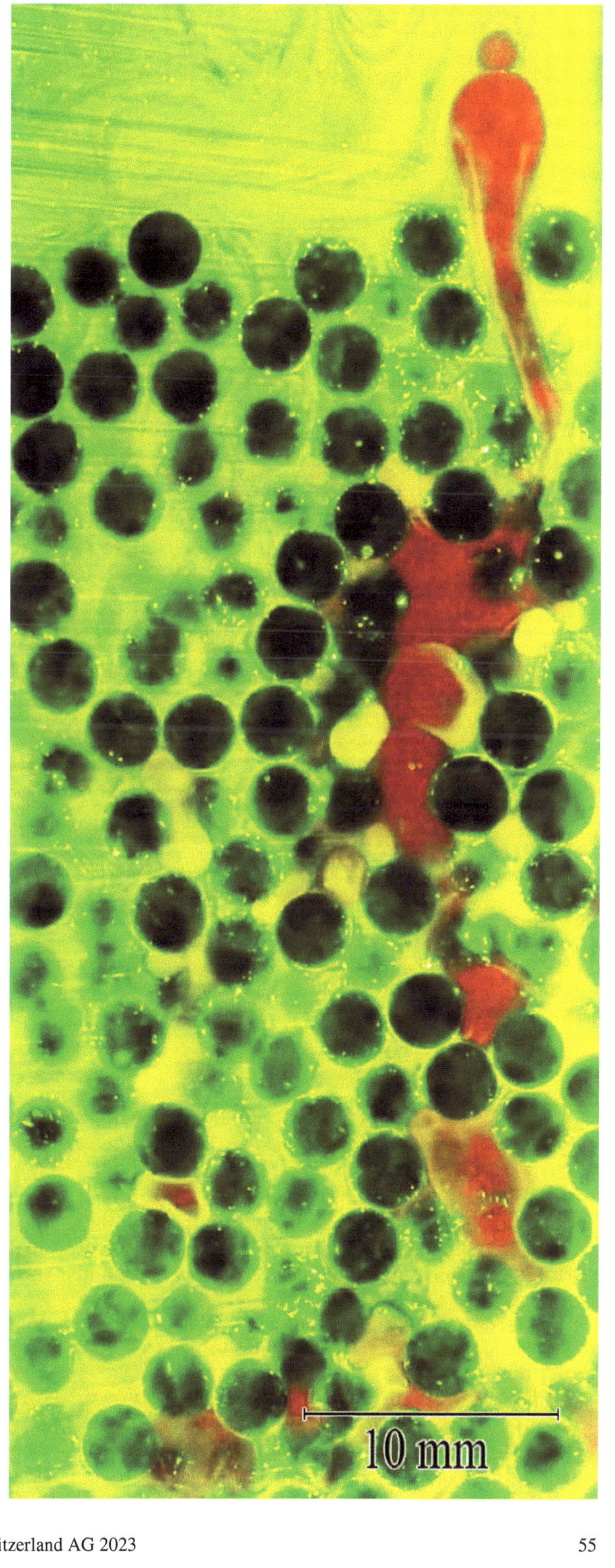

Experimental method: development of an optical 3D scanner based on index matching. First trial experiment of the working 3D-scanner. We see inside a cell filled with 3 mm glass beads. The cell is illuminated by a laser sheet parallel to the imaging plane. The green defending fluid, glycerol, is illuminated by green fluorescence (Fluorescein 549), in red, coming in from below, we see the invading fluid, canola oil, dyed with red (Pyrromethene 650).

Método experimental: desarrollo de un escáner óptico de 3D basado en el ajuste del índice de refracción. Primera prueba del experimento con el escáner de 3D funcional. Vemos el interior de una celda llena de esferas de vidrio de 3 mm. La celda está iluminada por una lámina láser paralela al plano de la imagen. El fluido verde desplazado, glicerol, está iluminado de color verde fluorescente (fluoresceína 549); en color rojo, vemos el fluido inyectado desde abajo, aceite de colza, teñido de rojo (pirrometeno 650).

Contact: Joachim Falck Brodin <j.f.brodin@fys.uio.no>

E. F. Médici and A. D. Otero (eds.), *Album of Porous Media*,
https://doi.org/10.1007/978-3-031-23800-0_41

SIMULATION OF TWO-PHASE FLOW IN BEREA SANDSTONE

Christian Hinz
Math2Market GmbH

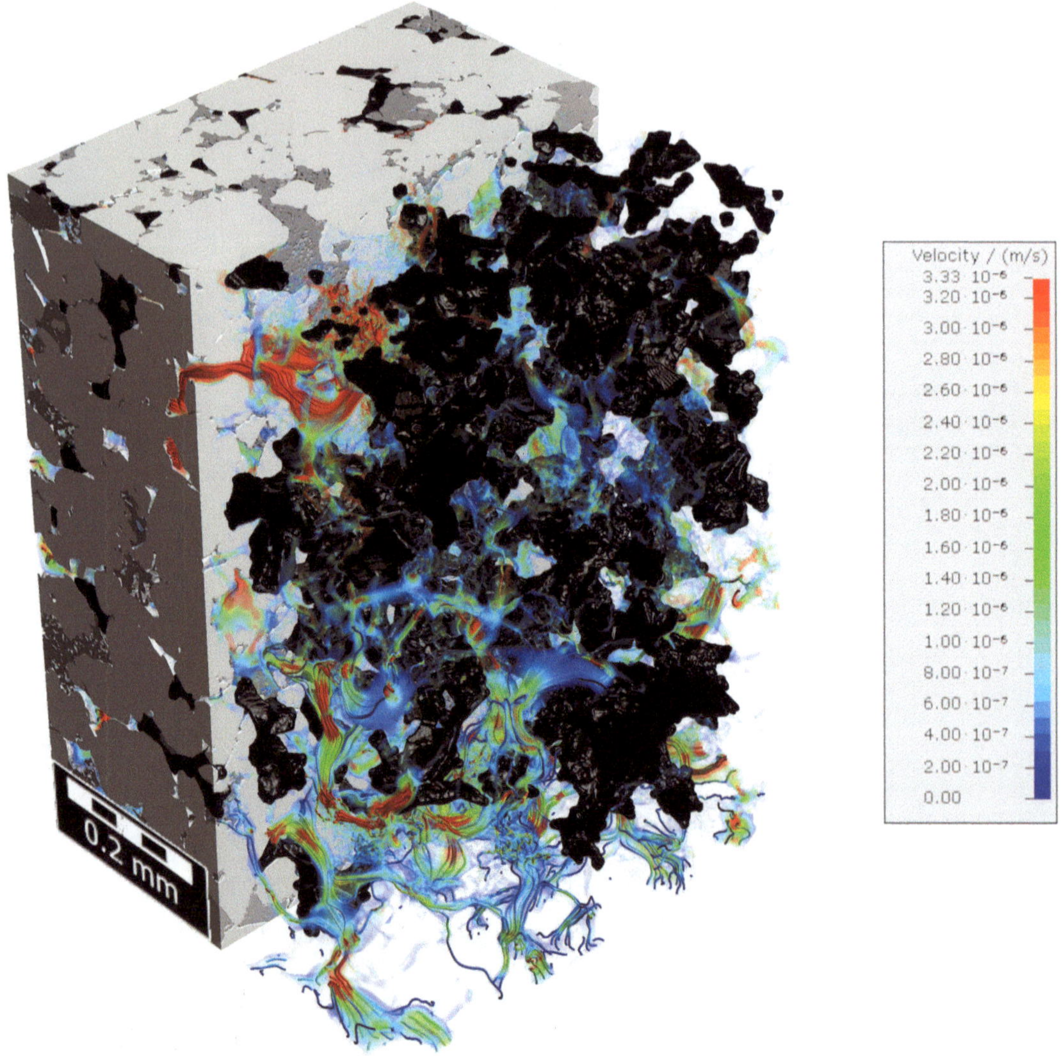

Two-phase flow is simulated in a 3D microstructure of Berea sandstone with the GeoDict software. The image shows the Berea sandstone[1] at dimensions of 720×720×1024 voxels cut in half to visualize an intermediate step of a relative permeability computation of a primary drainage in a water-wet system performed with the GeoDict software package. The flow direction is from left to right of the image. The black color represents the invading oil phase, and blue to red colors depict the increasing velocity in the water phase at 60 % water saturation. The importance of this type of simulation is the possibility to upscale it to determine how much oil can be produced from the rock.

El flujo de dos fases de una estructura 3D de arenisca de Berea es simulado con el software GeoDict. La imagen muestra la arenisca de Berea[1] con dimensiones de 720×720×1024 vóxeles cortada por la mitad para visualizar un paso intermedio en el cálculo de la permeabilidad relativa del drenado principal en un sistema húmedo con agua realizado con el paquete de software GeoDict. La dirección del flujo es de izquierda a derecha de la imagen. El color negro representa la fase de petróleo inyectado y del azul al rojo se ilustra el aumento de la velocidad en la fase acuosa con una saturación del 60 % de agua. La importancia de este tipo de simulación reside en la posibilidad de escalarlo para determinar cuánto petróleo puede producirse de la roca.

1. Andrä H. *et al.* (2013). *Computers & Geosciences*, 50:33.

Contact: Barbara Planas <barbara.planas@math2market.de>

E. F. Médici and A. D. Otero (eds.), *Album of Porous Media*,
https://doi.org/10.1007/978-3-031-23800-0_42

WETTABILITY EFFECTS ON TWO-PHASE FLOW IN HETEROGENEOUS POROUS MEDIA

Sahar Bakhshian[1], *Harris Rabbani*[2], *and Nima Shokri*[3]

[1] Bureau of Economic Geology, Jackson School of Geosciences, The University of Texas at Austin, USA
[2] Texas A&M University at Qatar
[3] Institute of Geo-Hydroinformatics, Hamburg University of Technology, Germany

Understanding of multiphase flow in natural porous media is of essential importance in many applications including geological storage of CO_2, enhanced oil recovery (EOR), groundwater remediation, and nuclear waste storage. The complex interplay between surface wettability and structural heterogeneity has profound impacts on the nature of fluid flow in porous media. This figure represents a pore-scale description of flow regime transitions under different wetting conditions in a 3D natural sandstone extracted by X-ray microtomography. Two-phase flow simulations were carried out using the lattice Boltzmann (LB) method[1, 2]. The equilibrium contact angle θ is reported with respect to the defending fluid. The invasion pattern is colored based on the distance from the inlet boundary. As the wetting condition changes from drainage to imbibition, we observe a transition from burst to corner flow leading to a compact displacement in the heterogeneous porous medium.

1. Bakhshian S. *et al.* (2020). *Geophys. Res. Lett.*, 47(14).
2. Bakhshian S. *et al.* (2021). *Transport in Porous Media*, 140(1):85.

Comprender el flujo multifásico en los medios porosos naturales es esencial para muchas aplicaciones, incluidas la de almacenamiento geológico del CO_2, la recuperación mejorada del petróleo (EOR, por sus siglas en inglés), la remediación de agua subterránea y el almacenamiento de desechos nucleares. La compleja interacción entre la mojabilidad de la superficie y la heterogeneidad estructural tiene impactos profundos en la naturaleza del flujo en medios porosos. Esta figura representa la descripción a escala del poro de las transiciones del régimen de flujo bajo distintas condiciones de mojabilidad en arenisca natural 3D obtenida mediante microtomografía de rayos X. Las simulaciones de flujo de dos fases se llevaron a cabo utilizando el método Lattice Boltzmann (LB)[1, 2]. El ángulo de contacto en equilibrio θ es representado con respecto al fluido desplazado. El patrón de invasión está coloreado según la distancia al contorno de entrada. A medida que la condición de mojabilidad cambia de drenado a imbibición, observamos una transición de un flujo de irrupción violenta a un flujo de esquina conduciendo a un desplazamiento compacto en el medio poroso heterogéneo.

Contact: Sahar Bakhshian <sahar.bakhshian@beg.utexas.edu>

E. F. Médici and A. D. Otero (eds.), *Album of Porous Media*,
https://doi.org/10.1007/978-3-031-23800-0_43

OIL DISPLACEMENT BY INJECTION OF COMPLEX DISPERSIONS

Raphael Ribeiro and Marcio Carvalho
Pontifícia Universidade Católica do Rio de Janeiro (PUC-Rio), Brasil

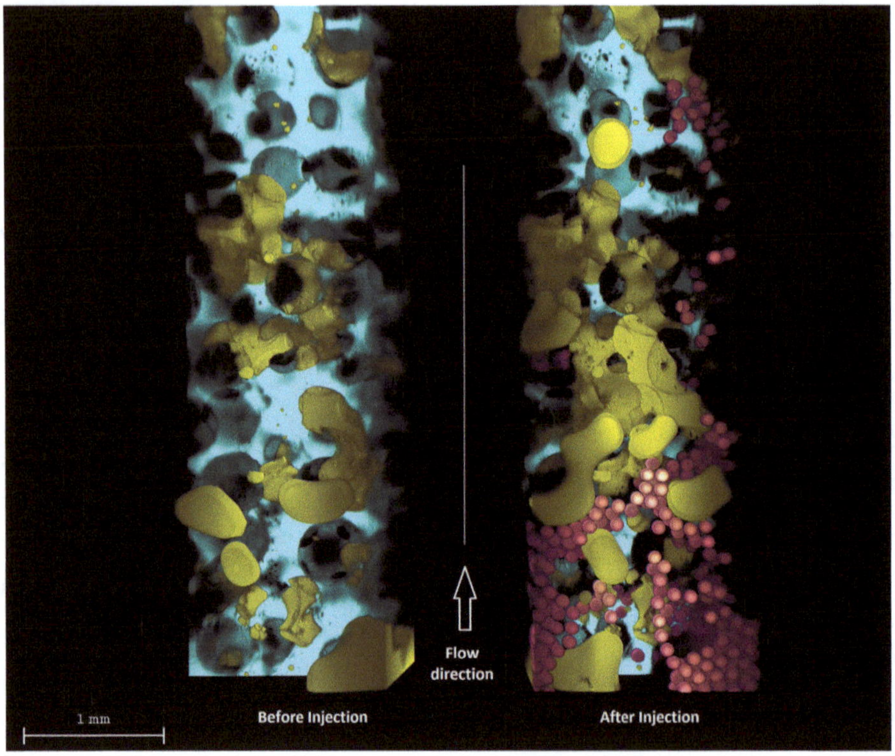

Complex dispersions, such as emulsions and microcapsule suspensions, can be used to partially block pores already swept by the water phase during oil recovery to divert the injected phase towards oil-filled pores[1, 2]. The fundamental understanding of the pore scale distribution of the pore blocking agent and the oil displacement process is essential to the optimization of the process and design of the complex dispersions. We use Confocal Laser Scanning Microscopy[3] to visualize the phase distribution and flow of water, oil and microcapsules through a consolidated glass bead porous media. To allow the simultaneous visualization of the three phases, different fluorescent dyes were used in each phase and each 2D image was scanned in three different channels, each with a different wavelength detection. The resonant scanner attached in the microscope allowed a high-velocity acquisition, a frame of a focal plane (2D image) with 512 lines only needs about 20 milliseconds to acquire[4]. The image shows the water (blue), oil ganglia (yellow) and microcapsules (pink) distribution before and after the dispersion injection. The empty volume corresponds to the glass beads.

Las dispersiones complejas, como las emulsiones y las suspensiones de microcápsulas, pueden utilizarse para bloquear parcialmente poros que ya han sido barridos por la fase acuosa durante la recuperación de petróleo, y así desviar la fase inyectada hacia los poros repletos de petróleo[1, 2]. La comprensión de la distribución del agente bloqueador y el proceso de desplazamiento del petróleo a escala de poro es esencial para la optimización del proceso y diseño de dispersiones complejas. Utilizamos un microscopio confocal[3] para visualizar la distribución de fases y flujo de agua, petróleo y microcápsulas a través de un medio poroso consolidado de esferas de vidrio. Para permitir una visualización simultánea de las tres fases, se emplearon distintas tintas fluorescentes para cada fase, y cada imagen 2D fue escaneada en tres canales distintos, cada uno con una detección de longitud de onda distinta. El escáner de resonancia unido al microscopio permitió una adquisición de alta velocidad: la imagen de un plano focal (imagen 2D) con 512 líneas solo necesita unos 20 milisegundos para ser obtenida[4]. La imagen muestra la distribución de agua (azul), los ganglios de petróleo (amarillo) y las microcápsulas (rosa) antes y después de la inyección de la dispersión. El volumen vacío corresponde a las esferas de vidrio.

1. Guillen V.R. *et al.* (2012). *International J. Multiphase Flow*, 43:62.
2. Guillen V.R. *et al.* (2012). *Transp Porous Med*, 94(1):197.
3. Datta S.S. *et al.* (2013). *Phys. Rev. Lett.*, 111(6).
4. Chalhub O. S. Ribeiro R. (2020). In *Annual Technical Conference and Exhibition*. SPE.

Contact: Marcio Carvalho <msc@puc-rio.br>

E. F. Médici and A. D. Otero (eds.), *Album of Porous Media*,
https://doi.org/10.1007/978-3-031-23800-0_44

INVASION PATTERN

Joachim Falck Brodin, Per Arne Rikvold, Marcel Moura, and Knut Jørgen Måløy

PoreLab, the Njord Centre, Department of Physics, University of Oslo, Norway

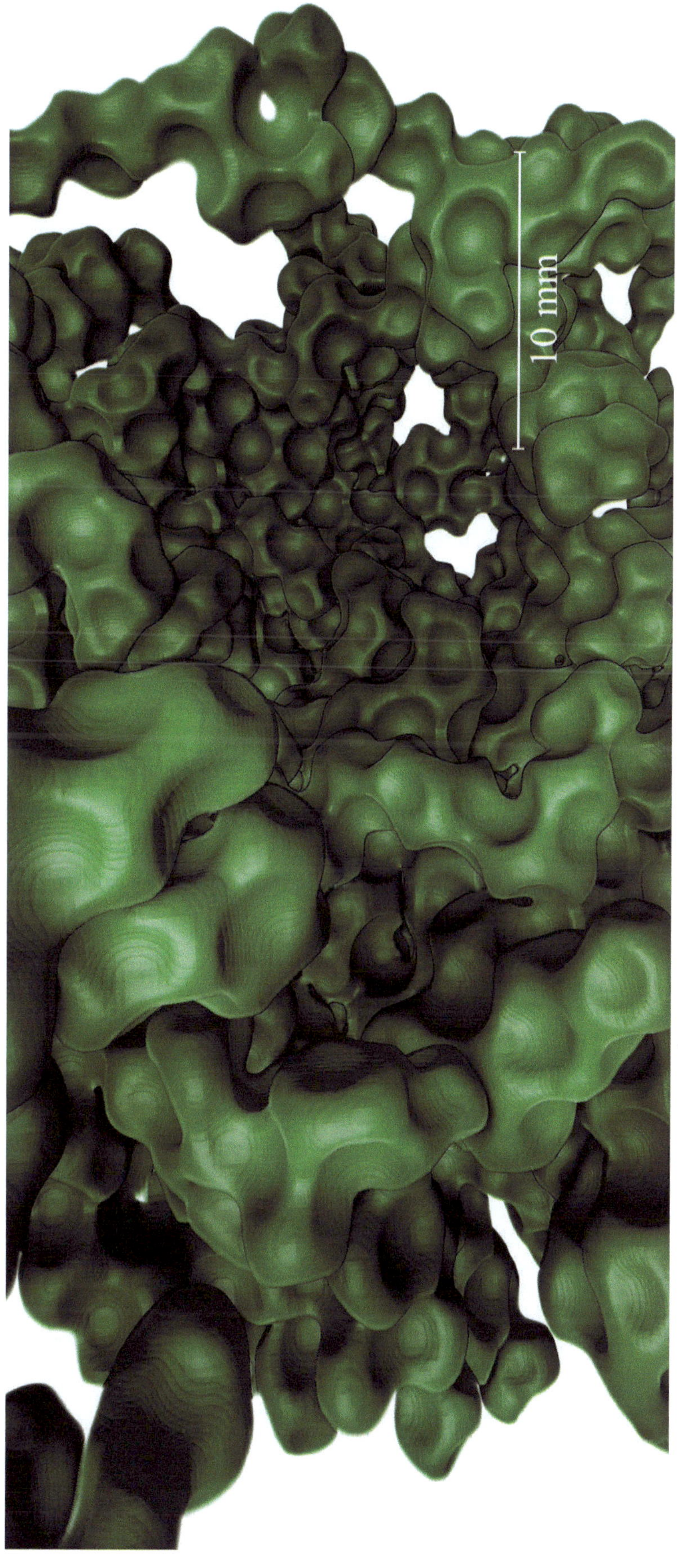

Experimental result – optical 3D-scanner, based on index matching. From an experiment with two immiscible fluids in a porous medium made of glass beads, glycerol and canola oil, on the interplay between capillary, viscous and gravitational forces. Here we see a 3D rendering of the segmented invading fluid phase, glycerol. The porous medium and the defending fluid have been removed, leaving the invading structure suspended alone. The beads used were uniform, with diameter 3 mm, the experiment was conducted in a cubic cell of 8 cm on each side, the structure we see is about 4 cm across.

Resultado experimental: escáner óptico 3D basado en el ajuste del índice de refracción. Experimento con dos fluidos inmiscibles en un medio poroso hecho de esferas de vidrio, glicerol y aceite de colza, interacción entre fuerzas capilares, viscosas y gravitacionales. Aquí vemos una representación en 3D de la fase fluida invasora segmentada del glicerol. Se han extraído el medio poroso y el líquido desplazado, por lo que queda solamente la estructura inyectada suspendida. Las esferas utilizadas eran uniformes, con un diámetro de 3 mm. El experimento se realizó en una celda cúbica de 8 cm de lado y la estructura visible tiene unos 4 cm de ancho.

Contact: Joachim Falck Brodin <j.f.brodin@fys.uio.no>

E. F. Médici and A. D. Otero (eds.), *Album of Porous Media*,
https://doi.org/10.1007/978-3-031-23800-0_45

FORMATION OF A PENDULAR LIQUID RING

Navneet Kumar[1], *Jaywant H. Arakeri*[2], *and M. S. Bobji*[2]

[1] Indian Institute of Technology Jammu
[2] Indian Institute of Science Bangalore

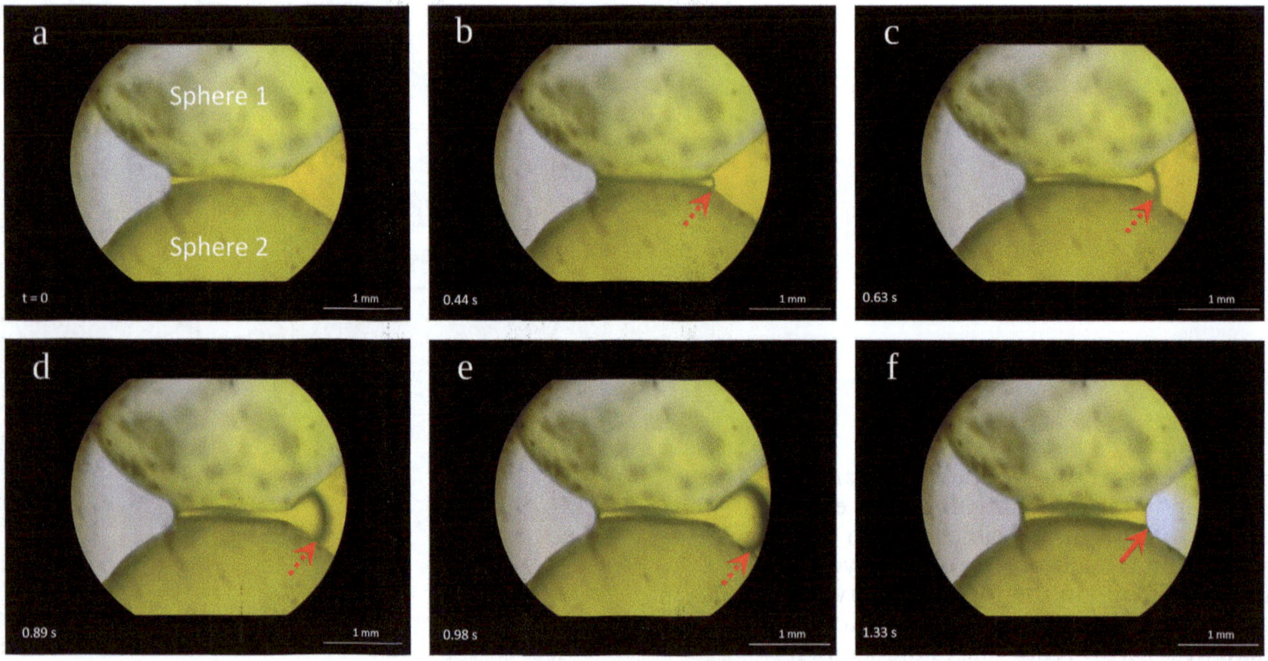

A series of snapshots, captured under a microscope showing the formation of a pendular ring (trapped liquid at the contact of two solid particles) from an initially flooded condition, during evaporation of water, between two 3 mm diameter glass balls in contact. The balls were rigidly held in contact on a horizontal glass slide. Optical contrast was enhanced by adding fluorescein dye to water. Evaporation leads to the formation of a meniscus on the left side (a) while an interface, near the spheres' bottom, is seen moving (b-e). Water at the spheres' top and bottom is connected through a capillary film which is stretched with time. At its limit (e), the film snapped, like a rubber band, thereby forming another meniscus (f). The two menisci thus formed a 'pendular' ring at the contact (separated particles never forms a pendular ring). This mechanism is relevant to the petroleum industries and is crucial in understanding drainage, imbibition, and evaporation from a porous medium[1, 2].

Serie de imágenes captadas con un microscopio que muestran la formación de un anillo pendular (líquido atrapado en contacto con dos partículas sólidas) desde un estado inicial saturado, durante la evaporación de agua, entre dos esferas de vidrio de 3 mm de diámetro en contacto. Las esferas se sostuvieron rígidamente en contacto en un portaobjetos de cristal horizontal. El contraste óptico se aumentó añadiendo tinta de fluoresceína al agua. La evaporación condujo a la formación de meniscos en la parte izquierda (a) mientras se ve una interfaz moviéndose cerca de la parte inferior de las esferas (b-e). El agua en las partes superior e inferior de las esferas está conectada mediante una película capilar que se estira con el tiempo. En el límite (e), la película se rompe, como una goma elástica, formando otro menisco (f). Por lo tanto, los dos meniscos forman un anillo «pendular» en la zona de contacto (las partículas separadas nunca forman un anillo pendular). Este mecanismo es relevante para la industria del petróleo y es crucial para entender el drenaje, imbibición y evaporación desde un medio poroso[1, 2].

1. Lenormand R. *et al.* (1983). *J. Fluid Mech.*, 135(-1):337.
2. Kudrolli A. (2008). *Nature Mater*, 7(3):174.

Contact: Navneet Kumar <navneet.kumar@iitjammu.ac.in>

E. F. Médici and A. D. Otero (eds.), *Album of Porous Media*,
https://doi.org/10.1007/978-3-031-23800-0_46

VISUALIZATION OF THE ROLE OF WETTABILITY, PORE SHAPE AND CORNER EFFECTS IN POROUS MEDIA

Richard Dawe[1], Carlos Grattoni[2], and Kamal Idris[3]

[1] Imperial College / University of the West Indies
[2] Imperial College / University of Leeds
[3] Imperial College / Universiti Teknologi, Petronas, Malaysia

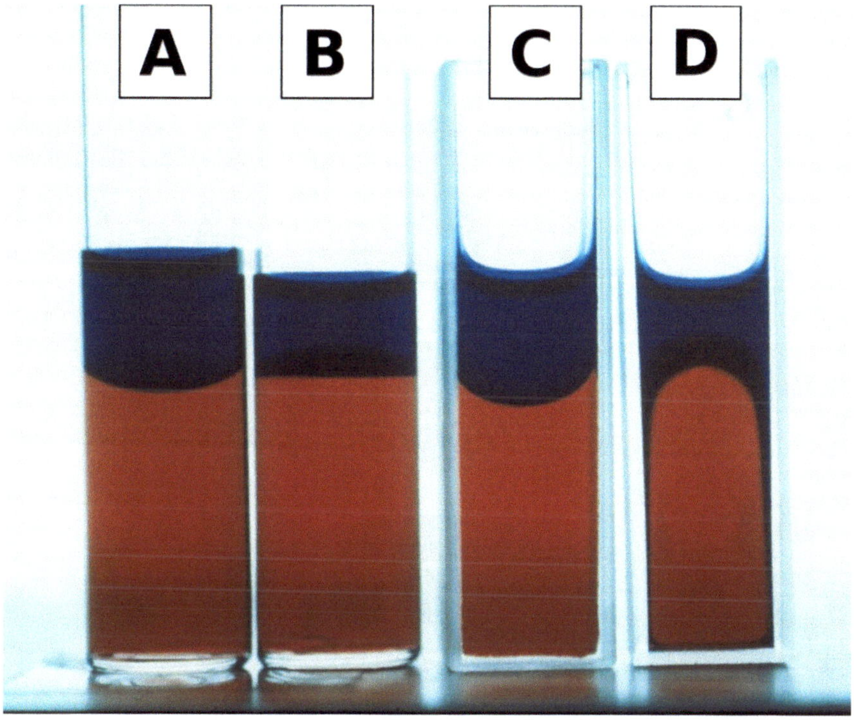

The effects of pore shape and wettability on menisci in porous media can be clearly demonstrated using circular and square tubes containing oil (blue, top layer) and water (red, bottom layer). Two water wet tubes (A, C) and two oil-wet tubes (B, D) are needed. We used diameter 10 mm and length 100 mm tubes. The shape of the interface curvatures with the solid defines the capillary forces between water-oil and oil-air (transparent) and are different in each case, plus there is film flow in the corners of the square tubes. These differences in meniscus behaviour are remarkable and controlled by the differences in pore shape (circular and square tubes) and wettability (the balance between the cohesion and adhesion forces of the fluids and a solid). In the circular tubes (A, B) the interfaces have low curvature which reverses with wettability, but no fluid movement. In contrast, in the square tubes (C, D) the capillary forces at the corners causes the fluid to rise or fall, which is notable and very significant at the oil-water interface in the oil-wet tube. We show,

1. The wettability affects the meniscus shape.
2. The corners affect capillary pressures.
3. The differences in the geometry of the tubes, square and circular, demonstrate effects of pore shape.

Los efectos de la forma del poro y mojabilidad sobre las características de los meniscos en medios porosos pueden demostrarse claramente utilizando tubos circulares y cuadrados que contienen aceite (capa superior azul) y agua (capa inferior roja). Se necesitan dos tubos con mojabilidad preferencial al agua (A, C) y dos tubos con mojabilidad preferencial al aceite (B, D). Utilizamos tubos de 10 mm de diámetro y 100 mm de largo. La forma de la curvatura de la interfase con el sólido define las fuerzas capilares entre las fases agua y aceite y entre aceite y aire (transparente) y son diferentes en cada caso, además de haber una película de fluido en las esquinas de los tubos cuadrados. Estas diferencias en el comportamiento de los meniscos son destacables y controladas por las diferencias en la forma de los poros (tubos circulares y cuadrados) y la mojabilidad (el equilibrio entre las fuerzas de cohesión y de adhesión de los líquidos y el sólido). En los tubos circulares (A, B), las interfases tienen una curvatura reducida que se revierte con la mojabilidad, pero sin movimiento del fluido. Por el contrario, en los tubos cuadrados (C, D) las fuerzas capilares actuando en las esquinas causan que el fluido suba o baje, lo que es notable y muy significativo en la interfase entre el aceite y el agua en el tubo con mojabilidad preferencial al aceite. Demostramos que:

1. La mojabilidad afecta la forma del menisco.
2. Las esquinas afectan la presión capilar.
3. Las diferencias en la geometría de los tubos, cuadrado y circular, demuestran los efectos de la forma de los poros.

The original version of this chapter was revised: The city and country name in the affiliation of Professor Richard Dawe has been updated in the chapter XML. The correction to this chapter can be found at https://doi.org/10.1007/978-3-031-23800-0_115

Contact: Richard Dawe <radawe@yahoo.com>

E. F. Médici and A. D. Otero (eds.), *Album of Porous Media*,
https://doi.org/10.1007/978-3-031-23800-0_47

THE INTERPLAY OF WATER SPREADING, EVAPORATION AND IMBIBITION AT NANOPOROUS SILICON SURFACES

Laura Gallardo[1,2,3], *Juan Sanchez*[1,2,3], *Hugo Bellezza*[4], *Olivier Vincent*[4], *and Patrick Huber*[1,2,3]

[1] Institute for Materials Physics and X-Ray Physics, Hamburg University of Technology, Germany
[2] Deutsches Elektronen-Synchrotron DESY
[3] Center for Hybrid Nanostructures CHyN, Hamburg University, Germany
[4] CNRS, Univ. Lyon, Univ. Claude Bernard Lyon 1, Institut Lumière Matière, France

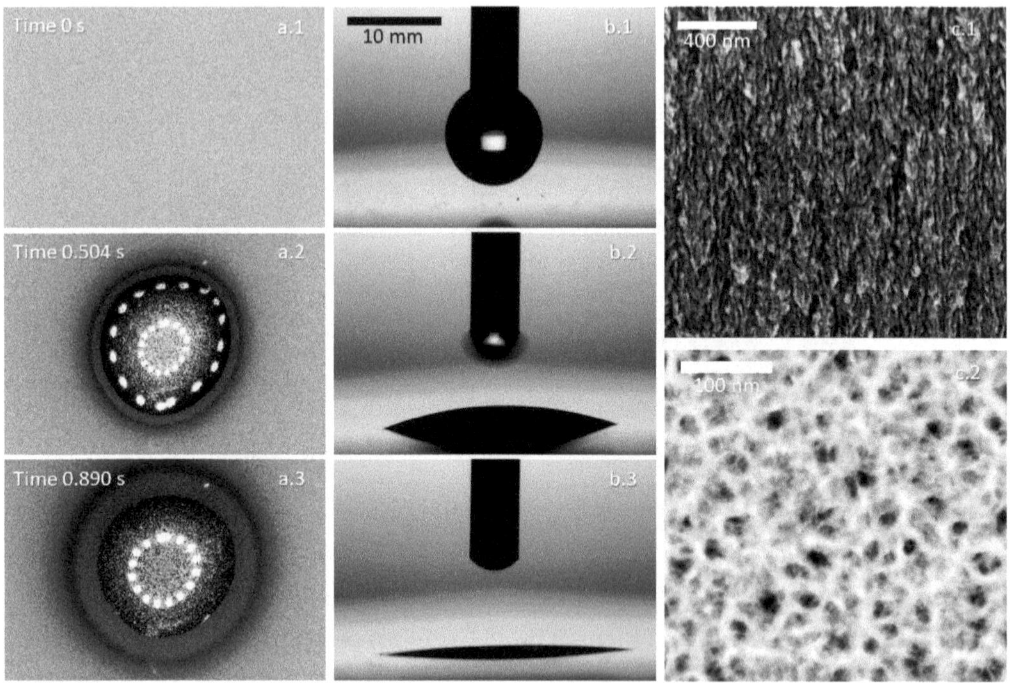

When a droplet is deposited on a porous substrate, three simultaneous processes take place: spreading at the surface, evaporation, and imbibition into the porous medium[1, 2, 3]. Two examples of this interplay are shown in a top and a side visualization ((a) and (b) at room temperature and 20 % RH). The processes are described with three matching frames: in (a.1), (b.1) a water droplet is deposited. In (a.2)-(a.3), and (b.2)-(b.3) the droplet volume decreases due to the water imbibition into the porous substrate and evaporation. Meanwhile the contact line increases as the wetted area of the droplet increases. The porous substrates are obtained through anodization of a silicon wafer in hydrofluoric acid[4], followed by an oxidation process of the surface to stabilize the layer. The sample used in (a) was obtained with a lightly doped silicon wafer with <111>(vector is orthogonal to the crystallographic planes) crystal orientation, resulting in smaller and more branched pores compared to the sample used in (b) (<100>crystal orientation, stronger doping). Panel (c) shows scanning electron micrographs of the sample in (b).

Cuando se deposita una gota en un sustrato poroso, ocurren simultáneamente tres procesos: extensión en la superficie, evaporación e imbibición en el medio poroso[1, 2, 3]. Se muestran dos ejemplos de esta interacción vistas desde arriba y de perfil ((a) y (b) a temperatura ambiente y 20 % HR). Los procesos están descritos con tres imágenes coincidentes: en (a.1), (b.1) una gota de agua es depositada. En (a.2)-(a.3) y (b.2)-(b.3) el volumen de la gota disminuye debido a la imbibición del agua en el sustrato poroso y a la evaporación. Mientras tanto, la línea de contacto crece a medida que el área mojada de la gota aumenta. Los sustratos porosos se obtienen mediante la anodización de una oblea de silicio en ácido fluorhídrico[4], seguido de un proceso de oxidación de la superficie para estabilizar la capa. La muestra utilizada en (a) se obtuvo con una oblea de silicio ligeramente dopada con una orientación del cristal <111>(el vector es ortogonal a los planos cristalográficos), por lo que se obtuvieron poros más pequeños y más ramificados en comparación con la muestra utilizada en (b) (orientación del cristal <100>, mayor dopaje). El panel (c) muestra micrografías electrónicas de barrido de la muestra (b).

1. Xiao J. *et al.* (2012). *Langmuir*, 28(9):4208.
2. Gruener S. and Huber P. (2008). *Phys. Rev. Lett.*, 100(6).
3. Vincent O. *et al.* (2017). *Langmuir*, 33(7):1655.
4. Sailor M.J. (2011). *Porous Silicon in Practice: Preparation, Characterization and Applications.* Wiley-VCH, Weinheim.

Contact: Laura Gallardo <laura.gallardo@tuhh.de>

E. F. Médici and A. D. Otero (eds.), *Album of Porous Media*,
https://doi.org/10.1007/978-3-031-23800-0_48

IV. PHASE CHANGE AND DIFFUSION

The first known depiction of a device used to remove impurities from a fluid dates back to the ancient Egypt. Wall engraving from the tomb of Amenophis II in Thebes circa 1450 B.C[1]. Public Domain/Ancient World Digital Library.

La primera representación conocida de un dispositivo utilizado para la eliminación de impurezas de un fluido se remonta al antiguo Egipto. Grabado del muro de la tumba de Amenofis II en Tebas hacia 1450 a.C.[1]. Dominio Publico/Ancient World Digital Library

1. Wilkinson J.G. (1878). *The manners and customs of the ancient Egyptians*, volume 2. London, John Murray.

EVAPORATION OF WATER IN A GLASS BEAD-PACK

Armin Afrough and Karen Feilberg
Danish Offshore Technology Centre, Technical University of Denmark

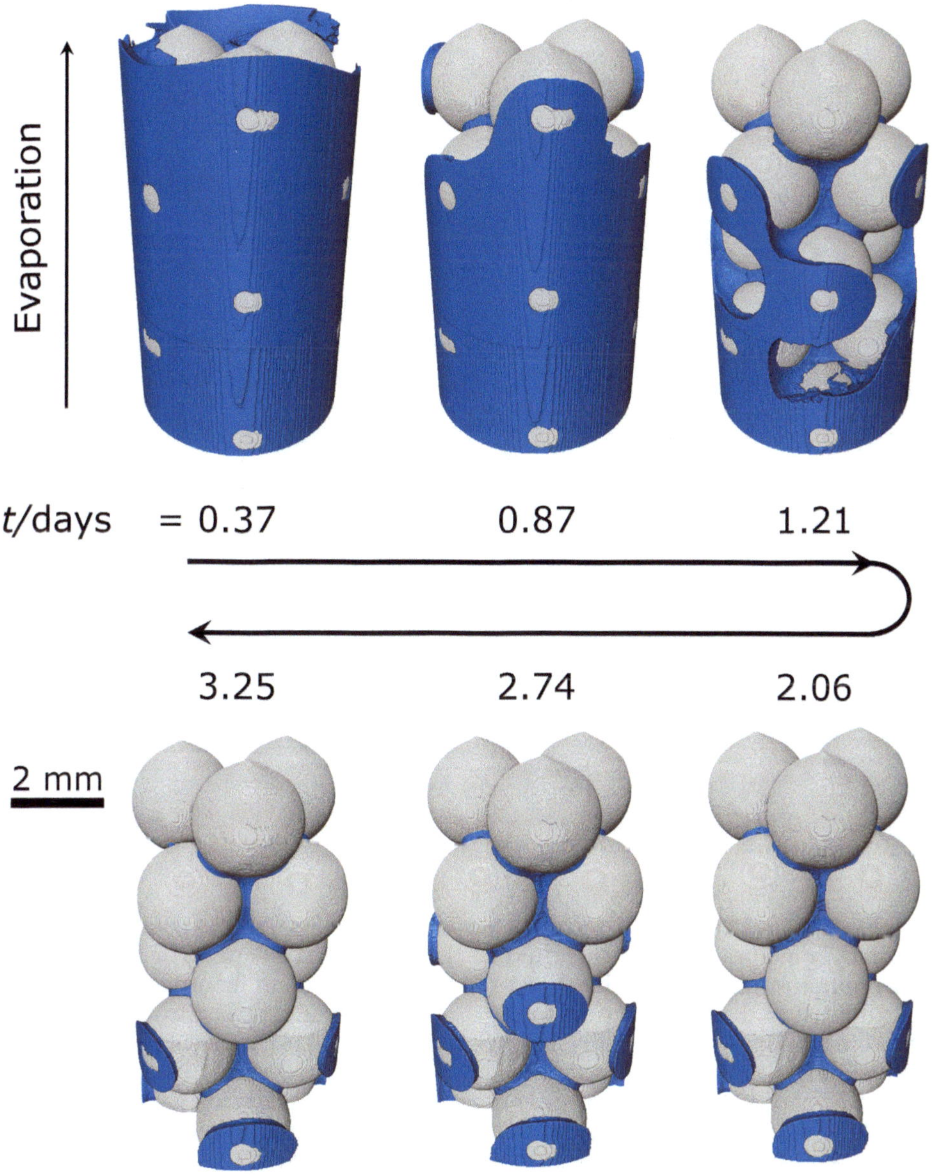

Water evaporation in porous media exhibits the complex evolution of air/liquid interfaces. Six segmented images demonstrate deionized water (in blue) evaporating from a glass bead (gray, d_{bead} = 2 mm) pack at different time points[1]. Watershed segmentation separated water, air, glass beads, and the vial ($d_{vial,internal}$ = 4.6 mm, $h_{vial,shown}$ = 8.5 mm). The container and air are transparent. The images were acquired by a custom-built X-ray tomography system with a reconstructed voxel size of 11.1 μm in a chamber of 21 °C and 45 % relative humidity.

La evaporación del agua en medios porosos exhibe una evolución compleja de las interfaces de aire y líquido. Seis imágenes segmentadas muestran el agua desionizada (en azul) evaporándose de un empaquetamiento de esferas de vidrio (en gris, d_{bead} = 2 mm) en distintos momentos[1]. La transformación divisoria separó el agua, el aire, las esferas de vidrio y el vial ($d_{vial,interno}$ = 4,6 mm, $h_{vial,mostrado}$ = 8,5 mm). El recipiente y el aire son transparentes. Las imágenes fueron tomadas con un sistema de tomografía de rayos X hecho a medida con un tamaño de vóxel reconstruido de 11,2 μm en una cámara a 21 °C y 45 % de humedad relativa.

The original version of this chapter was revised: The author's name has been updated in the chapter XML. The correction to this chapter can be found at https://doi.org/10.1007/978-3-031-23800-0_115

1. URL `https://doi.org/10.11583/DTU.21371892`.

Contact: Armin Afrough <armin.afrough@gmail.com>

E. F. Médici and A. D. Otero (eds.), *Album of Porous Media*,
https://doi.org/10.1007/978-3-031-23800-0_49

A FLOATING SALT CRUST OVER EVAPORATING SAND

Uri Nachshon
ARO - Volcani center

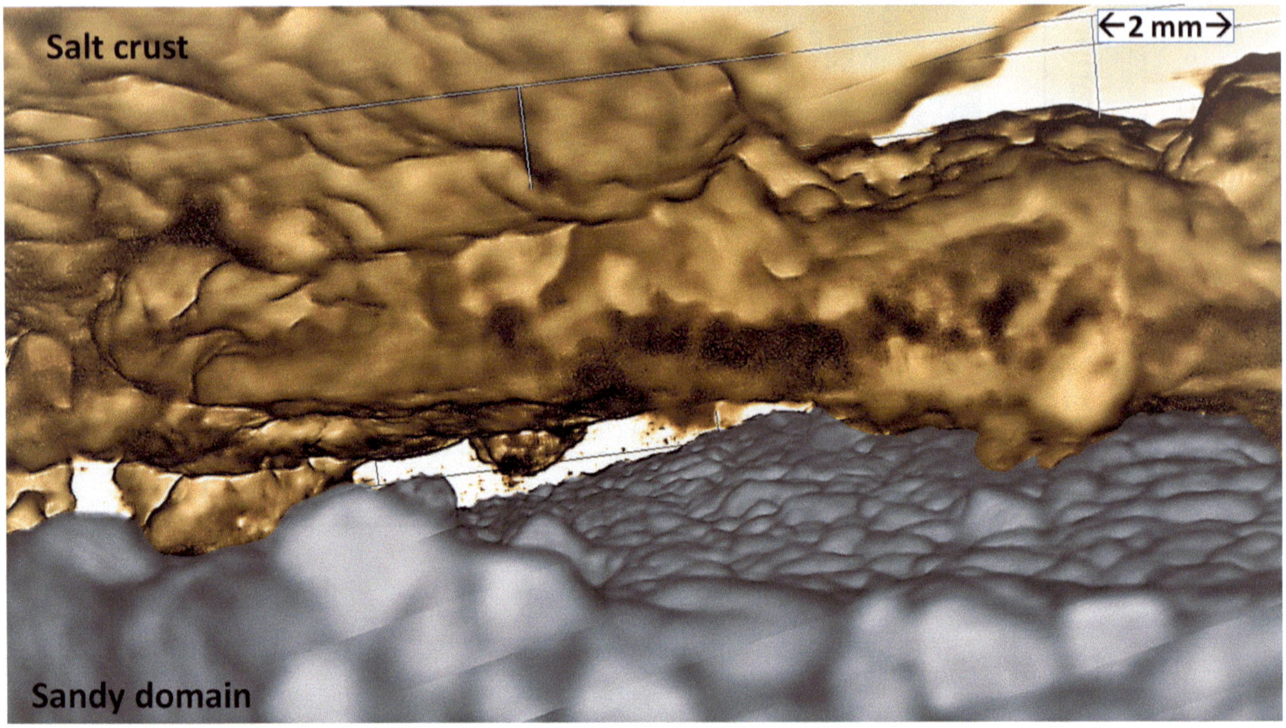

Salt crust formation over evaporating porous media is a well known phenomenon. However, the structural nature of the salt crust and its connection to the underlying media were explored only recently, by the use of micro CT imagining. Studies have shown that a porous salt crust is growing out of a limited number of nucleation centers and that most of the crust is "floating" above the media. This feature has great influence on important processes, such as water evaporation and gas exchange between the soil and atmosphere.
In the image a precipitated NaCl salt crust (top brownish layer) is observed over a sandy domain (grey domain), with typical grain diameter of 100 μm. The sandy domain was saturated with NaCl solution, and the salt crust was formed due to evaporation of the saline solution. It is seen that a 1-2 mm gap is formed between the salt crust and the sand, where a limited number of salt pillars connect between the two domains[1].

La formación de cortezas de sal sobre medios porosos que se evaporan es un fenómeno muy conocido. Sin embargo, la naturaleza de la estructura de la corteza de sal y su conexión con el medio subyacente solo se llegaron a comprender completamente hace muy poco, gracias a la visualización mediante micro TC. Los estudios han mostrado que una corteza de sal porosa crece sobre un número limitado de centros de nucleación y que la mayoría de corteza «flota» sobre el medio. Esta característica tiene una gran influencia en procesos importantes como la evaporación del agua y el intercambio de gas entre el suelo y la atmósfera.
En la imagen, una precipitación de corteza de sal NaCl (capa superior marrón) se observa sobre un medio arenoso (zona gris), con un diámetro de grano de alrededor de 100 μm. El medio arenoso fue saturado con una solución de NaCl, y la corteza de sal se formó debido a la evaporación de la solución salina. Se observa la formación de una brecha de 1-2 mm entre la corteza de sal y la arena, donde un número limitado de pilares de sal conectan los dos medios[1].

The original version of this chapter was revised: The author's name has been updated in the chapter XML. The correction to this chapter can be found at https://doi.org/10.1007/978-3-031-23800-0_115

1. Nachshon U. *et al.* (2018). *Geophys. Res. Lett.*

Contact: Uri Nachshon <urina@agri.gov.il>

E. F. Médici and A. D. Otero (eds.), *Album of Porous Media*,
https://doi.org/10.1007/978-3-031-23800-0_50

INFLUENCE OF THERMAL GRADIENTS ON INVASION PATTERNS DURING DRYING OF GRANULAR POROUS MEDIUM

Divyansh Pandey[1], *Kaustubh Desai*[1], *Debashis Panda*[1], *Abdolreza Kharaghani*[2], *and Vikranth Kumar Surasani*[1]

[1] Birla Institute of Technology and Science, Pilani, Hyderabad Campus
[2] Otto-von-Guericke University Magdeburg, Germany

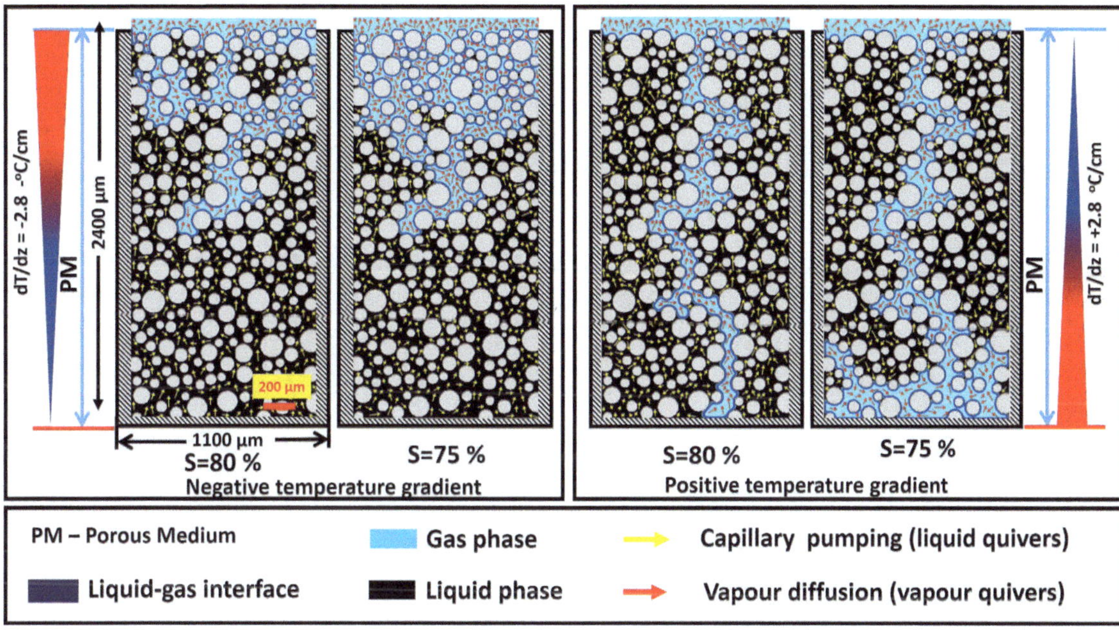

Drying in porous media exhibits complex multiphase phenomena involving liquid-gas interactions[1]. Invasion occurs at the liquid-gas interface having least capillary pressure. For a given porous medium, temperature dependency on surface tension plays a significant role on invasion patterns (IP). To display temperature effects, Shan Chen Lattice Boltzmann Method (SC LBM)[2, 3, 4] is applied to simulate drying in a granular porous medium (1100 μm×2800 μm) subjected to constant positive and negative temperature gradients. Invasion patterns at saturations 80 % and 75 % and the capillary flow within the liquid and the vapor diffusion in the gas are presented with quivers. Capillary fingering is dominant in both cases. However, the negative gradient exhibits a stabilizing drying front or IP. In positive gradient, higher temperatures towards the depth of porous medium lead to lower capillary and higher liquid pressures at the receding front. The pressure gradients on the interface lead to capillary pumping within the liquid clusters towards the top drying front. Positive thermal gradient leads to capillary effects in destabilizing IP leading to the formation of "two" drying fronts at the bottom and top of the porous medium.

El secado de medios porosos presenta fenómenos multifásicos complejos que involucran interacciones entre el líquido y el gas[1]. La invasión ocurre en la interfaz entre el líquido y el gas que tiene la menor presión capilar. Para un determinado medio poroso, la dependencia a la temperatura de la tensión superficial desempeña un papel significativo en los patrones de invasión (PI). Para mostrar los efectos de la temperatura, aplicamos el método Shan Chen Lattice Boltzmann (SC LBM)[2, 3, 4] para simular el secado en un medio poroso granular (1100 μm×2800 μm) sujeto a gradientes de temperatura constante positiva y negativa. Los patrones de invasión con saturaciones del 80 % y del 75 % y el flujo capilar del líquido y la difusión de vapor en el gas se presentan con flechas. Las digitaciones capilares son dominantes en ambos casos. Sin embargo, el gradiente negativo muestra una estabilización del frente de secado o PI. En el gradiente positivo, las temperaturas elevadas en lo profundo del medio poroso conllevan una menor capilaridad y una mayor presión del líquido en el frente recesivo. Los gradientes de presión en la interfaz conducen a un bombeo capilar dentro de los cúmulos de líquido hacia la parte superior del frente de secado. Un gradiente térmico positivo conduce a efectos capilares al desestabilizar los PI en la formación de «dos» frentes de secado, en la parte superior e inferior del medio poroso.

The original version of this chapter was revised: The author's name has been updated in the chapter XML. The correction to this chapter can be found at https://doi.org/10.1007/978-3-031-23800-0_115

1. Metzger T. *et al.* (2008). In *Micro-Macro-interaction*, (187–204). Springer Berlin Heidelberg.
2. Zachariah G.T. *et al.* (2019). *Chemical Eng. Sci.*, 196:310.
3. Panda D. *et al.* (2020). *Chemical Eng. Sci.*, 220:115634.
4. Panda D. *et al.* (2020). *Physics of Fluids*, 32(12):122116.

Contact: Vikranth Kumar Surasani <surasani@hyderabad.bits-pilani.ac.in>

E. F. Médici and A. D. Otero (eds.), *Album of Porous Media*,
https://doi.org/10.1007/978-3-031-23800-0_51

PREFERENTIAL AIR INVASION IN AN EVAPORATING MONOLAYER POROUS MEDIUM

Navneet Kumar[1] *and Jaywant H. Arakeri*[2]

[1] Indian Institute of Technology Jammu
[2] Indian Institute of Science Bangalore

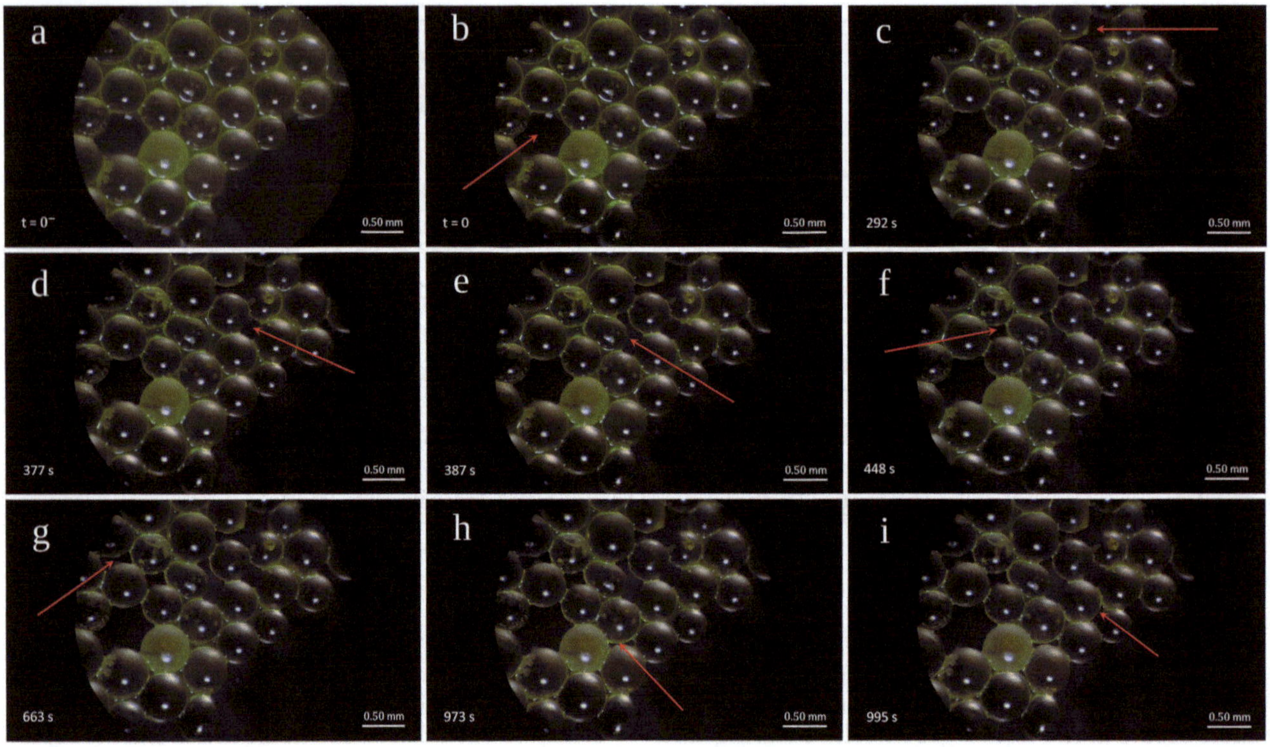

A series of snapshots showing the preferential air-invasion in the simplest form of a wet porous medium consisting of a monolayer of glass beads (0.40-0.50 mm diameter). The liquid used here was water which was turned green using the fluorescein dye particles to enhance visibility. The porous medium rested on a horizontal glass slide and the process of evaporation, in the ambient air, was filmed by a camera mounted on a microscope. Initially, the porous medium is in capillary state (a). With evaporation, air chooses the largest wet pore and invades it completely while the other pores remained wet leading to funicular state (consisting of a mixture of unsaturated and saturated pores). The process continued (b-i) until the last pore was eventually invaded leading to the pendular state (independent trapped water between two particles). Red arrows in each panel show the air-invaded largest wet pores at that instant. Such a simple experimental demonstration of the preferential air-invasion is crucial in modelling and understanding evaporation from a porous medium such as foams, soils, etc[1, 2, 3].

Serie de imágenes que muestran la invasión preferencial de aire en el más simple medio poroso mojado formado por una única capa de esferas de vidrio (0,40-0,50 mm de diámetro). El líquido empleado aquí fue agua teñida de verde mediante partículas de fluoresceína para mejorar la visibilidad. El medio poroso reposaba en un portaobjetos de cristal y el proceso de evaporación, al aire, fue filmado con una cámara montada en un microscopio. Inicialmente, el medio poroso se encontraba en un estado capilar (a). Con la evaporación, el aire elige los poros mojados más grandes y los invade completamente mientras que el resto de poros permanecen mojados, lo que causa un estado funicular (que consiste en una mezcla de poros saturados y no saturados). El proceso prosiguió (b-i) hasta que el último poro eventualmente fue invadido, lo que llevó a un estado pendular (agua desconectada atrapada entre dos partículas). Las flechas rojas de cada panel muestran los poros mojados más grandes invadidos por el agua en cada instante. Esta demostración experimental tan simple de la invasión preferencial del aire es crucial para el modelado y la comprensión de la evaporación de medios porosos como espumas, suelo, etc[1, 2, 3].

The original version of this chapter was revised: The author's name has been updated in the chapter XML. The correction to this chapter can be found at https://doi.org/10.1007/978-3-031-23800-0_115

1. Kumar N. and Arakeri J.H. (2018). *Transp Porous Med*, 125(2):311.
2. Kumar N. and Arakeri J.H. (2018). *Water Resour. Res.*, 54(10):7670.
3. Kumar N. and Arakeri J.H. (2020). *International J. Thermal Sciences*, 148:106151.

Contact: Navneet Kumar <navneet.kumar@iitjammu.ac.in>

E. F. Médici and A. D. Otero (eds.), *Album of Porous Media*,
https://doi.org/10.1007/978-3-031-23800-0_52

VISUALIZATION OF APPLICATION OF ULTRASONIC DRYING FOR MOIST POROUS PAPER

Zahra Noori O'Connor, Jamal Yagoobi, and Burt Tilley
Worcester Polytechnic Institute

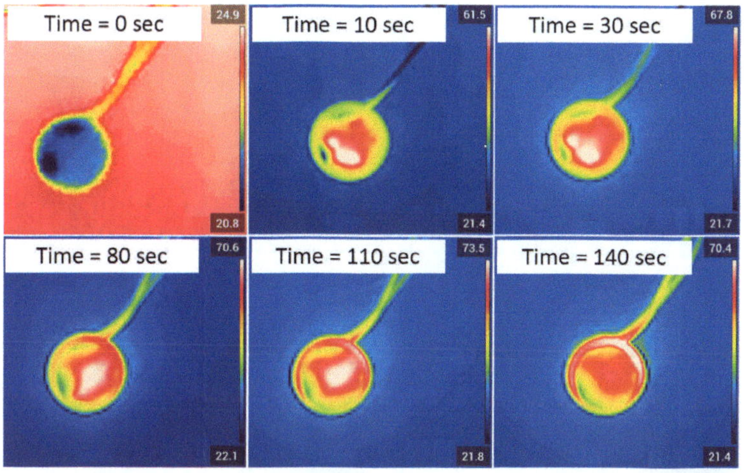

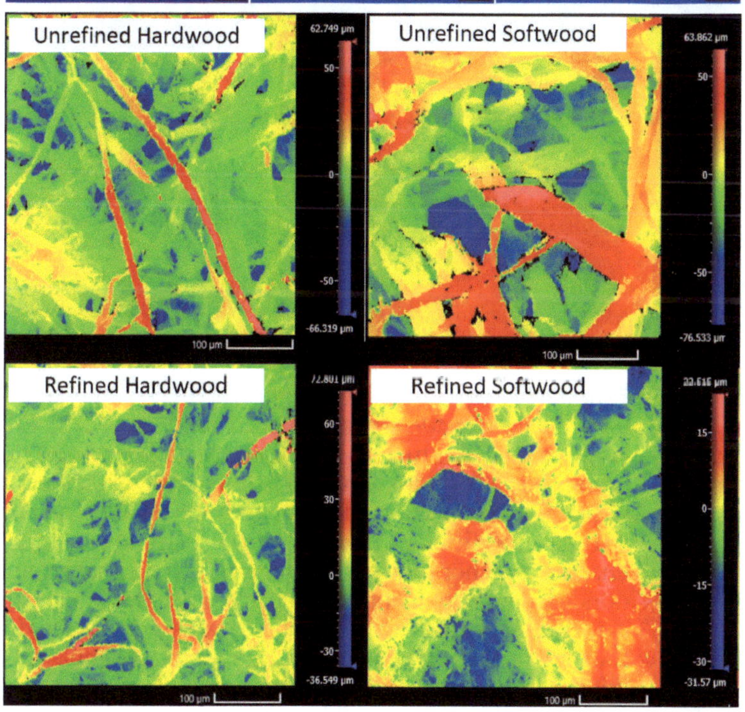

La tecnología de ultrasonido es una tecnología innovadora y verde en cuanto al secado del papel se refiere. En comparación con las tecnologías de secado actuales, tiene beneficios como un menor tiempo de secado, una mayor tasa de secado y una mayor eficiencia energética. El papel húmedo se coloca sobre la superficie de un transductor que vibra, y debido a las vibraciones mecánicas, el agua es extraída de la superficie del medio poroso. Al principio, con un contenido de humedad inicial alto, el mecanismo dominante para el secado por ultrasonido es la atomización. La imagen superior muestra la evolución de las temperaturas de la superficie de un papel de madera dura (con 1,9 cm de diámetro) durante un experimento de secado por ultrasonidos, captado mediante una cámara IR. Debido al energizado del transductor, la temperatura del transductor, así como la del papel que está en contacto con él, aumenta con el tiempo, por lo que en las etapas finales de secado el recalentamiento también es importante. La imagen inferior representa la microestructura de distintas muestras secadas con ultrasonido captadas con un perfilómetro óptico 3D ZYGO (altura de la rugosidad). Los experimentos han mostrado que la estructura del papel poroso tienen un impacto significativo en el comportamiento del secado por ultrasonido. En la estructura de papel, las fibras se apilan aleatoriamente y estas imágenes revelan que los distintos papeles tienen diferentes tamaños de poros y porosidades.

Ultrasound technology is an innovative and green technology for paper drying. Comparing to the current drying technologies, it has benefits such as lower drying time, higher drying rate, and higher energy efficiency. The moist paper is placed on the surface of a vibrating transducer and due to mechanical vibrations, water is removed from the surface of the porous media. Initially, at high initial moisture content, the dominant mechanism for ultrasonic drying is atomization. The top image shows the temperature evolution of the surface of a refined hardwood paper (with diameter of 1.9 cm) during ultrasonic drying experiment, captured using IR Camera. Due to energizing the transducer, the temperature of the transducer as well as the paper in contact increase by time and hence, at the final stages of drying, heating is also important. The bottom image represents the microstructure of different ultrasonically dried samples captured by ZYGO 3D optical profilometer (roughness height). The experiments have shown that the structure of the porous paper has a significant impact on ultrasonic drying behavior. In a paper structure, fibers are randomly packed, and these images reveal that different papers have different pore sizes and porosities.

The original version of this chapter was revised: The author's name has been updated in the chapter XML. The correction to this chapter can be found at https://doi.org/10.1007/978-3-031-23800-0_115

Contact: Zahra Noori O'Connor <znoori@wpi.edu>

E. F. Médici and A. D. Otero (eds.), *Album of Porous Media*,
https://doi.org/10.1007/978-3-031-23800-0_53

CRACKS IN WET SAND

Bjørnar Sandnes, James Campbell, and Deren Ozturk
Swansea University

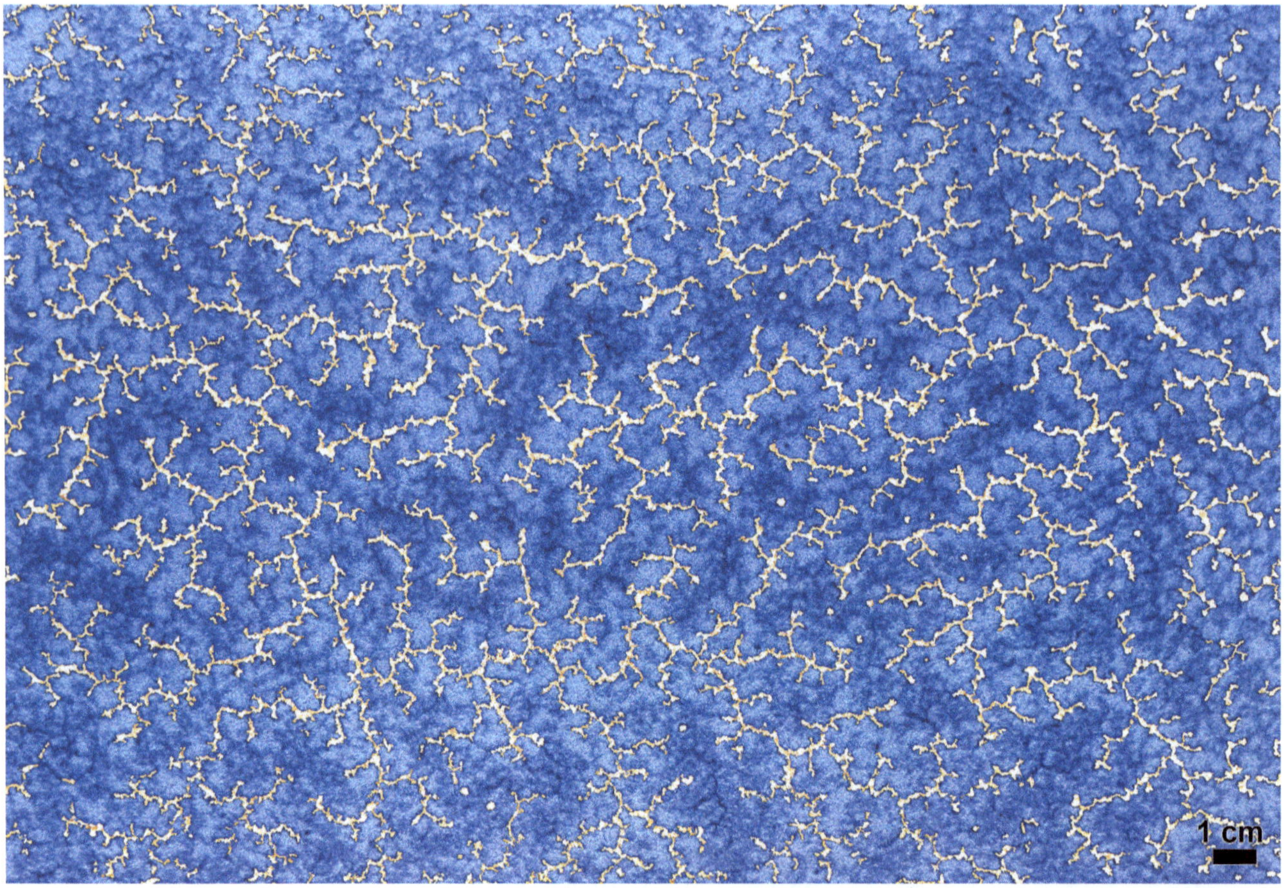

Spectacular fracture patterns appear when wet sand in a Hele-Shaw cell dries out[1]. At the tip of the fracture, the invading meniscus pushes the grains aside. When the fracture expands, the material compacts, and friction eventually balances the surface tension. The fractures take the path of least resistance and branch out into a maze where the branches are kept apart by the compacted material between them. In nature, gas-driven fracturing controls methane venting in sediments and magma degassing, and in industry, a potential application is the generation of permeability for accelerated treatment of contaminated soil. The image shows air fractures in yellow (contrasted by image subtraction), penetrating a packing of glass beads (~100 μm) in water. The blue color originates from the LED light source which has been oversaturated for visual effect. The cell was slowly dried out (over several days) by evaporation through the open edge of the cell. The fractures grew in a stick-slip fashion: long periods of quiescence followed by sudden fracturing events. After the cell was completely dried out, this picture was taken with a camera placed underneath the cell.

Espectaculares patrones de fracturas aparecen en una celda de Hele-Shaw de arena húmeda cuando se seca[1]. En la punta de la fractura, el menisco inyectado empuja los granos hacia los lados. Cuando la fractura se expande, el material se compacta y la fricción termina equilibrando la tensión superficial. La fractura toma el camino en el que encuentra menor resistencia, y se ramifica en un laberinto donde las ramas se mantienen separadas debido al material compactado entre ellas. En la naturaleza, la propagación de fracturas debido a la presencia de gas controla la ventilación del metano en sedimentos y la desgasificación del magma, y en la industria, una posible aplicación es la generación de permeabilidad para el tratamiento acelerado de tierra contaminada. La imagen muestra fracturas inducidas por aire en amarillo (contrastadas por sustracción de imágenes) que penetran el empaquetamiento de esferas de vidrio (~100 μm) en agua. El color azul se origina por la fuente de luz LED que ha sido sobresaturada para obtener un efecto visual. La celda fue secada gradualmente (durante varios días) mediante evaporación a través de los bordes abiertos de la misma. Las fracturas se propagan en modo oscilación de relajación: largos periodos estáticos seguidos de súbitos eventos de fractura. Después de que la celda se secara completamente, se tomó esta imagen con una cámara colocada debajo de la misma.

The original version of this chapter was revised: The author's name has been updated in the chapter XML. The correction to this chapter can be found at https://doi.org/10.1007/978-3-031-23800-0_115

1. Campbell J.M. *et al.* (2017). *Phys. Rev. Applied*, 8(6).

Contact: Bjornar Sandnes <b.sandnes@swansea.ac.uk>

E. F. Médici and A. D. Otero (eds.), *Album of Porous Media*,
https://doi.org/10.1007/978-3-031-23800-0_54

GRANULAR LABYRINTH

Bjørnar Sandnes[1], *Knut Jørgen Måløy*[2], *and Eirik Grude Flekkøy*[2]

[1] Swansea University

[2] PoreLab, the Njord Centre, Department of Physics, University of Oslo, Norway

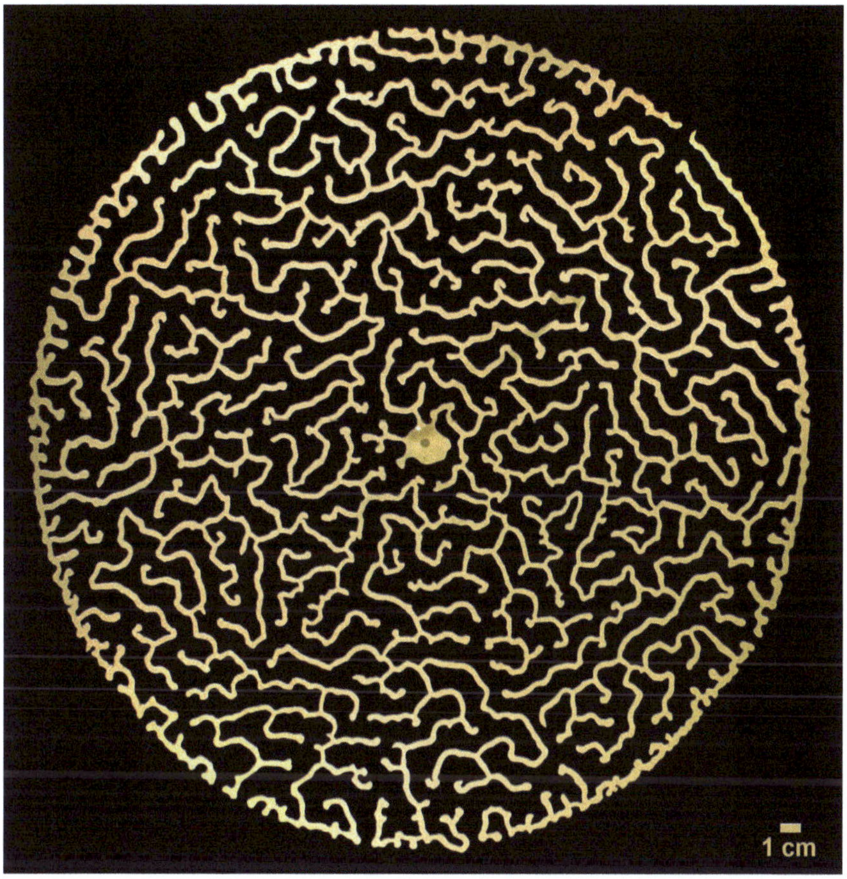

The drying of a granular-liquid mixture in a Hele-Shaw cell leaves behind a labyrinth residue[1]. In the picture, the golden walls are glass beads, bulldozed aside by invading fingers of air (black) carving out paths in the granular layer. The fingers never snap off, and the resulting structure is a branching binary tree. A characteristic finger width arises from the tug-of-war between two dominating forces: Surface tension acts to minimise surface by reducing finger-tip curvature, however a widening of the fingers is resisted by the friction in the bulldozed granular front. The "frictional finger" width - seen here as the separation distance between opposing labyrinth walls - minimises the total work required to create new surface and push against friction. In this experiment a Hele-Shaw cell filled with 20 % volume fraction glass beads (~80 μm diameter) in glycerol-water was slowly drained using a syringe pump set to 0.01 ml/min withdrawal rate. The image was taken with a camera placed underneath the cell, against a black background, and with light coming in from the sides (darkfield illumination).

El secado una mezcla de líquido y granos en una celda de Hele-Shaw deja un residuo con forma de laberinto[1]. En la imagen, las paredes doradas son esferas de vidrio empujadas a los lados por digitaciones de aire (color negro) que labran caminos en la capa granular. Las digitaciones nunca se interrumpen, por lo que la estructura termina siendo un árbol con ramas binarias. El ancho característico de las digitaciones resulta del balance entre dos fuerzas dominantes y opuestas: la tensión superficial intenta minimizar la superficie reduciendo la curvatura de la punta de la digitación, mientras que el ensanchamiento de las digitaciones es resistido por la fricción que resulta de empujar el frente granular. El ancho de la «digitación por fricción» que se observa aquí como la distancia que separa las dos paredes opuestas del laberinto, minimiza el trabajo total necesario para crear una superficie nueva y empujar contra la fricción. En este experimento, una celda de Hele-Shaw con un 20 % del volumen ocupado por las esferas de vidrio (~80 μm de diámetro) en agua y glicerol fue drenada lentamente utilizando una bomba de infusión de jeringa a una razón de 0,01 ml/min. La imagen se tomó colocando una cámara debajo la celda, contra un fondo negro y con luz proyectada desde los lados (iluminación de campo oscuro).

The original version of this chapter was revised: The author's name has been updated in the chapter XML. The correction to this chapter can be found at https://doi.org/10.1007/978-3-031-23800-0_115

1. Sandnes B. *et al.* (2007). *Phys. Rev. Lett.*, 99(3).

Contact: Bjornar Sandnes <b.sandnes@swansea.ac.uk>

E. F. Médici and A. D. Otero (eds.), *Album of Porous Media*,
https://doi.org/10.1007/978-3-031-23800-0_55

FRICTIONAL FINGER LABYRINTH

Kristian Stølevik Olsen[1], *Eirik Grude Flekkøy*[1], *Knut Jørgen Måløy*[1], *and James Campbell*[2]

[1] PoreLab, the Njord Centre, Department of Physics, University of Oslo, Norway
[2] College of Engineering, Swansea University, United Kingdom

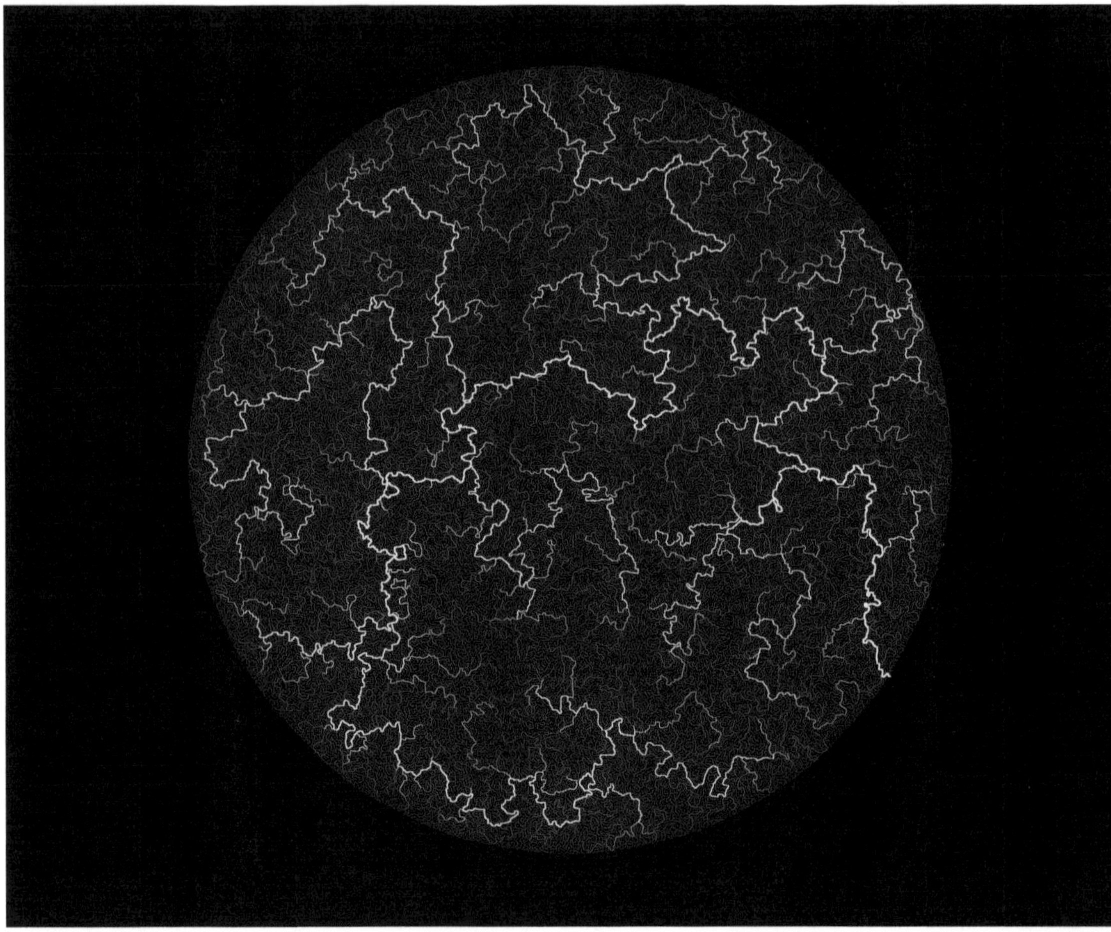

Nature is full of patterns that exhibit complex and intricate geometric features. When a liquid phase is drained from a liquid-grain mixture, invading paths of air result in a branching labyrinth structure[1]. This pattern, dubbed frictional finger labyrinths, shares features with several bifurcating patterns found in Nature, and has only recently been studied extensively[2, 3]. The image shows a labyrinth that has been generated numerically. In the numerical framework developed, the interface between the liquid-grain mixture and air is discretised before a sequence of small deformations are made at the point of least resistance. Paths in the image are ordered and colored according to a Horton-Strahler scheme. Our studies show that the fractal dimension of shortest paths connecting any two points is close to 1.22, a value characteristic of so-called 2D minimum spanning trees. In terms of physical dimensions the simulated pattern would be approximately 5 meters in diameter.

La naturaleza está llena de patrones que exhiben características geométricas complejas e intricadas. Cuando se drena una fase líquida de una mezcla de líquido y granos, los caminos del aire inyectado crean una estructura laberíntica ramificada[1]. Este patrón, apodado laberintos de digitaciones por fricción, comparte características con varios patrones de fenómenos naturales que se bifurcan, y solo recientemente ha sido estudiado extensamente[2, 3]. La imagen muestra un laberinto generado numéricamente. En el sistema numérico desarrollado, la interfaz entre la mezcla de líquido, granos y aire se discretiza antes de la creación de una secuencia de deformaciones pequeñas en el punto de menor resistencia. Los caminos de la imagen están ordenados y coloreados según el esquema de Horton-Strahler. Nuestros estudios muestran que la dimensión fractal de los caminos más cortos que conectan dos puntos cualquiera es cercana a 1,22, un valor característico de los llamados árboles en 2D de mínima expansión. En cuanto a las dimensiones físicas, el patrón simulado tendría unos 5 metros de diámetro.

The original version of this chapter was revised: The author's name has been updated in the chapter XML. The correction to this chapter can be found at https://doi.org/10.1007/978-3-031-23800-0_115

1. Sandnes B. *et al.* (2007). *Phys. Rev. Lett.*, 99(3).
2. Olsen K.S. *et al.* (2019). *New J. Phys.*, 21(6):063020.
3. Olsen K.S. and Campbell J.M. (2020). *Front. Phys.*, 8.

Contact: Kristian Stølevik Olsen <k.stolevik.olsen@gmail.com>

E. F. Médici and A. D. Otero (eds.), *Album of Porous Media*,
https://doi.org/10.1007/978-3-031-23800-0_56

ASSEMBLY OF PARTICULATE PATTERNED ARRAYS USING LIQUID TEMPLATES

Diego Barba Maggi[1], *Román Martino*[2], *Marcelo Piva*[2], *and Alejandro Boschan*[2]

[1] Escuela Superior Politécnica de Chimborazo
[2] Grupo de Medios Porosos, Facultad de Ingeniería, Universidad de Buenos Aires

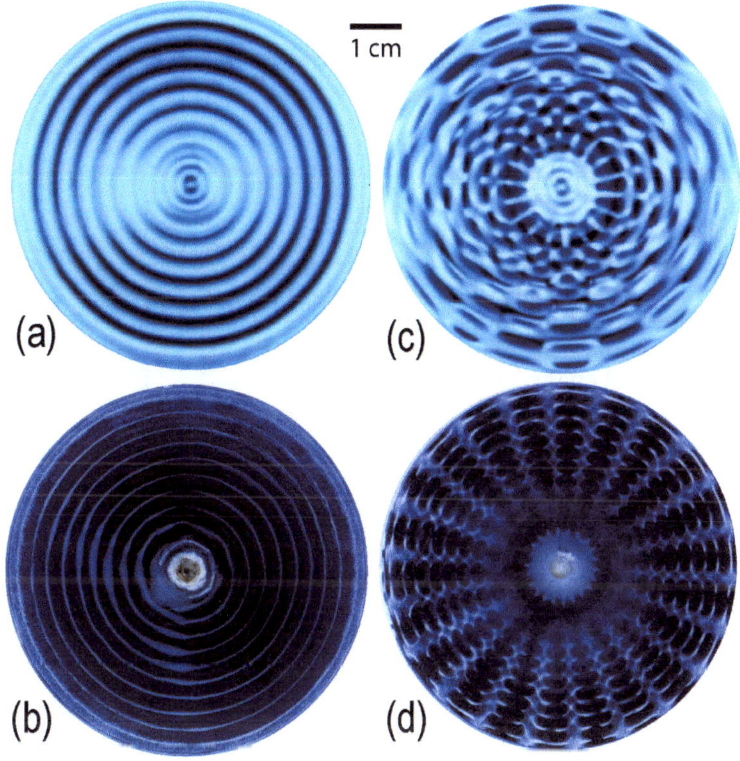

Assembly of small particles in patterned arrays can be used to obtain organic and inorganic advanced materials with a wide range of textures and porosities. Waves in a liquid can be used as a template to mobilize dispersed particles to target positions in the array. In the experiment shown in the image, the waves are generated by vertically vibrating a cylindrical vessel containing a 1.88 mm thick layer of a water-glycerol mixture (ρ_l = 1.03 g/cm^3). Heavy polystyrene particles (20 μm in radius, ρ_p = 1.05 g/cm^3), initially dispersed in the liquid bulk, interact with the waves while settling by gravity to end-up forming the patterned array at the bottom of the vessel. Left column (vibration frequency = 65 Hz, acceleration = 1.03 g): (a) Capillary waves in the liquid free surface (bright and dark rings obtained using a beam-splitter), and (b) the associated ring-shaped particle patterns at the bottom of the vessel. Right column (vibration frequency = 65 Hz, acceleration = 1.2 g): (c) Superposition of capillary and Faraday waves, and (d) the associated star-shaped particle patterns.

Contact: Diego Barba Maggi <dbarba@espoch.edu.ec>

El ensamblaje de partículas pequeñas en patrones ordenados puede utilizarse para obtener materiales avanzados, orgánicos e inorgánicos, con un amplio rango de texturas y porosidades. Las ondas en un líquido pueden utilizarse como plantillas para movilizar partículas dispersas hacia lugares predeterminados dentro del patrón. En el experimento que se muestra en la imagen, las ondas son generadas mediante la vibración vertical de un recipiente cilíndrico que contiene una capa de una mezcla de glicerol y agua (ρ_l = 1,03 g/cm^3), de 1,88 mm de espesor. Las partículas pesadas de poliestireno (20 μm de radio, ρ_p = 1,05 g/cm^3), dispersas inicialmente en el volumen del líquido, interactúan con las ondas, mientras sedimentan por gravedad, para terminar formando un patrón en el fondo del recipiente. Columna izquierda (frecuencia de vibración = 65 Hz, aceleración = 1,03 g): (a) Ondas capilares en la superficie libre del líquido (anillos brillantes y oscuros obtenidos con un divisor de haces), y (b) Los correspondientes patrones de partículas, formando anillos concéntricos, en el fondo del recipiente. Columna derecha (frecuencia de vibración = 65 Hz, aceleración = 1.2 g): (c) Superposición de ondas capilares y Faraday, y (d) Los correspondientes patrones de partículas, en forma de estrella.

E. F. Médici and A. D. Otero (eds.), *Album of Porous Media*,
https://doi.org/10.1007/978-3-031-23800-0_57

CHARACTERIZATION OF UNSATURATED SOLUTE TRANSPORT WITH X-RAY SYNCHROTRON MICROTOMOGRAPHY

Sharul Nizam Hasan[1], *Vahid Niasar*[1], *Nikolaos Karadimitriou*[2], *Jose R. da Assuncao Godinho*[3], *Nghia T. Vo*[4], *Senyou An*[1], *Arash Rabbani*[1], *Dongwon Lee*[2], *and Holger Steeb*[2]

[1] University of Manchester
[2] Institute of Mechanics (CE), Stuttgart University
[3] Helmholtz-Zentrum Dresden-Rossendorf
[4] Diamond Light source

Unsaturated solute transport was directly characterized with the use of Synchrotron-based X-ray Computed Tomography (s-XRCT). The spatial and temporal resolutions were 3.25 μm and 6 seconds, respectively. Detailed information on the fate of solute transport in a glass-bead packing under various unsaturated conditions was obtained for the first time in the literature. The solute concentration, and its rate of change, was visualized at any given location within the field of view, creating a data set bigger than 50 TB. A processed snapshot of the distribution of phases at a given time is shown in the Figure where the porous structure (glass beads) is depicted in light gray, the concentration of potassium iodine (KI) in the wetting phase (water) in shades of green, and the non-wetting phase (Fluorinert) in dark gray. In this pore-scale resolved investigation we clearly showed the impact of the fluid's topology on the evolution of non-Fickian transport, and that this should be included in the Darcy-scale solute transport models. We also showed that the assumption of fully mixing on the pore scale is conditionally valid, and in this case even after the introduction of many pore volumes the concentration within a single pore is not uniform. Additionally, we showed that dispersivity is changing with saturation, being two-fold larger at intermediate saturations compared to this of the fully saturated porous medium[1].

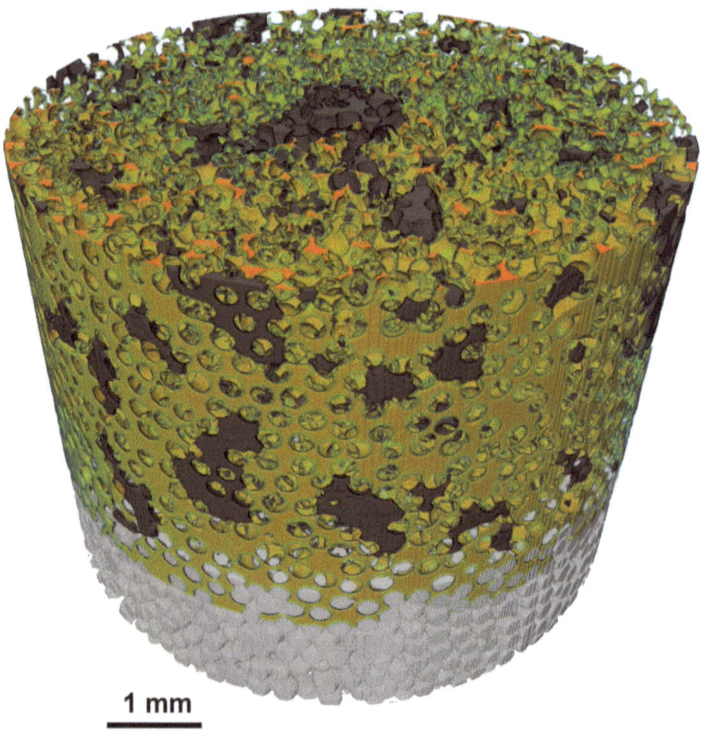

Un transporte de solutos no saturado fue caracterizado directamente con el uso de tomografía computarizada de rayos X (s-XRCT) de sincrotrón. Las resoluciones espacial y temporal fueron de 3,25 μm y 6 segundos, respectivamente. Se obtuvo por primera vez en la literatura información detallada sobre el destino del transporte de solutos en un empaquetamiento de esferas de vidrio bajo varias condiciones no saturadas. La concentración de soluto y su razón de cambio fueron visualizadas en cualquier ubicación dentro del campo de visión, creando un conjunto de datos de más de 50 TB. En la figura se muestra una captura instantánea procesada de la distribución de fases en un momento determinado, en el que la estructura porosa (esferas de vidrio) se observa en gris claro, la concentración de yoduro de potasio (KI) en la fase mojante (agua) en tonos verdes, y la fase no mojante (Fluorinert) en gris oscuro. En esta investigación resuelta a escala del poro, mostramos claramente el impacto de la topología de un fluido en la evolución de un transporte no fickiano, y que este debería estar incluido en los modelos de transporte de solutos de la escala de Darcy. También mostramos que la hipótesis de mezclado completo a escala del poro es condicionalmente válida, y en este caso incluso después de la introducción de muchos volúmenes de poros, la concentración de un único poro no es uniforme. Asimismo, mostramos que la dispersividad cambia con la saturación, siendo dos veces mayor en saturaciones intermedias en comparación con la de un medio poroso completamente saturado[1].

1. Hasan S. *et al.* (2020). *Proc. Natl. Acad. Sci. U.S.A.*, 117(38):23443.

Contact: Nikolaos Karadimitriou <nkaradimitriou@gmail.com>

E. F. Médici and A. D. Otero (eds.), *Album of Porous Media*,
https://doi.org/10.1007/978-3-031-23800-0_58

FLUID STRETCHING AND FOLDING IN 3D POROUS BEAD-PACKS

Joris Heyman[1], *Tanguy Le Borgne*[2], *Yves Méheust*[3], *and Daniel Lester*[4]

[1] CNRS, France
[2] University of Rennes, France
[3] Géosciences Rennes, Université de Rennes, CNRS, France
[4] Royal Melbourne Institute of Technology, Australia

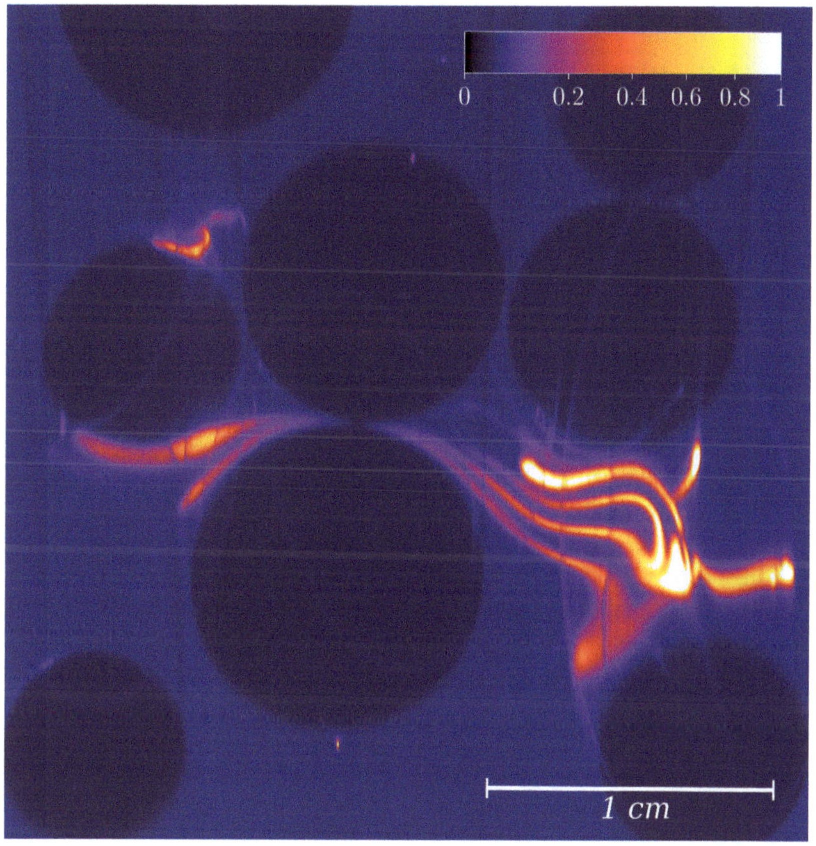

The image shows the deformation of a fluorescent dye (color scale indicates concentration of fluorescein) while advected by a viscous fluid (glycerine and water) through a 3D random bead-pack (blue disks). The mean flow direction is perpendicular to the image plane. It is obtained by the technique of refractive index matching and laser induced fluorescence. Stretching and folding of the dye in the plane transverse to the mean flow direction create striated lamellar structures, whose lengths increase exponentially with advection distance; the hallmark of chaotic advection. Due to exponential fluid compression, lamella thickness is fixed at a characteristic scale, the so-called Batchelor scale which is smaller than the pore scale for all Péclet number larger than 5. This first experimental quantification of pore-scale chaotic dynamics provides new insights for modelling solute transport, mixing and reactivity in porous media[1].

1. Heyman J. *et al.* (2020). *PNAS*, 117(24):13359.

La imagen muestra la deformación de una tinta fluorescente (la escala de colores indica concentración de fluoresceína) por la advección de un fluido viscoso (glicerina y agua) a través de un empaquetamiento aleatorio 3D de esferas (discos azules). La dirección del flujo medio es perpendicular al plano de la imagen. Esta imagen se obtuvo mediante la técnica de coincidencia del índice de refracción y fluorescencia inducida por láser. El estiramiento y plegamiento de la tinta en el plano transversal a la dirección del flujo medio crean estructuras laminares estriadas, cuyo largo aumenta exponencialmente con la distancia de advección, distintivo de la advección caótica. Debido a la compresión exponencial del fluido, el grosor de las láminas está fijado por una escala característica, llamada escala de Batchelor que es menor que la escala de poro para todos los números de Péclet superiores a 5. La primera cuantificación experimental de las dinámicas caóticas en la escala de poro proporciona nuevos conocimientos para modelar el transporte de solutos, mezclado y reactividad en medios porosos[1].

Contact: Joris Heyman <joris.heyman@univ-rennes1.fr>

E. F. Médici and A. D. Otero (eds.), *Album of Porous Media*,
https://doi.org/10.1007/978-3-031-23800-0_59

SPATIO-TEMPORAL PUSH-PULL IMAGING REVEALS THE FRACTAL NATURE OF CHAOTIC SOLUTE LANDSCAPES IN POROUS FLOWS.

Joris Heyman[1], *Daniel Lester*[2], *and Tanguy Le Borgne*[3]

[1] CNRS, France
[2] Royal Melbourne Institute of Technology, Australia
[3] University of Rennes, France

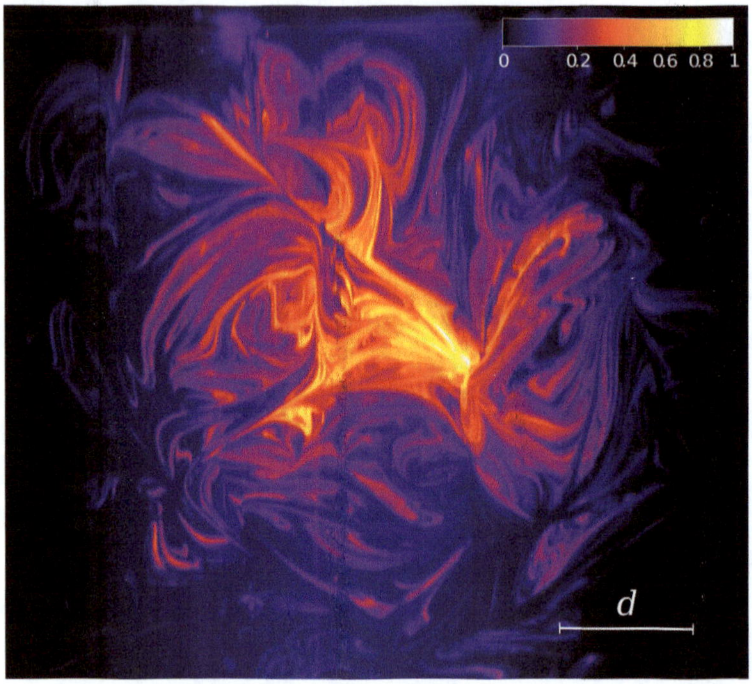

The image is obtained by pushing a fluorescent dye into a porous sample (gravel packing, mean diameter d=7 mm), then pulling it back and imaging the spatio-temporal return of dye concentrations with a laser sheet, transverse to the mean flow direction. The scalar signature (color scale indicates dye concentrations) of the echo provides key data about the mixing dynamics occurring inside the porous media. Although Stokes flows are fully reversible, the molecular diffusion of the dye (fluorescein) induces perturbations that are exponentially amplified by chaotic advection at pore scale. In the pull phase, dye molecules thus return to the injection plane taking a completely different path than in the push phase. They follow the unstable chaotic manifold of the porous flow, revealing its fractal structure. This image is taken in the pull phase after 20 advection time (the mean time to travel 20 pore diameters). At high Péclet numbers, the decay of scalar fluctuation with time can be related to the magnitude of the Lyapunov exponent of the chaotic porous flow. In principle, this generalized push-pull methodology enables a precise measure of mixing dynamics in any opaque porous materials[1].

1. Heyman J. *et al.* (2021). *Phys. Rev. Lett.*, 126(3).

Contact: Joris Heyman <joris.heyman@univ-rennes1.fr>

La imagen se obtuvo al empujar una tinta fluorescente en una muestra porosa (relleno de gravilla, diámetro promedio d=7 mm), luego extrayéndola y tomando imágenes espaciotemporales del retorno de las concentraciones de la tinta con una lámina láser, transversal a la dirección del flujo medio. La designación escalar (la escala de colores indica la concentración de tinta) del eco proporciona información clave sobre las dinámicas de mezclado que ocurren dentro de los medios porosos. A pesar de que los flujos de Stokes son completamente reversibles, la difusión molecular de la tinta (fluoresceína) induce perturbaciones que se amplifican exponencialmente por la advección caótica a escala del poro. Por lo tanto, en la fase de tracción las moléculas de la tinta vuelven al plano de inyección tomando un camino completamente distinto al de la fase de empuje. Éstas siguen la distribución caótica e inestable del flujo poroso, revelando su estructura fractal. Esta imagen fue tomada en la fase de tracción, después de 20 veces el tiempo de advección (tiempo promedio para viajar 20 diámetros de poro). Con números de Péclet altos, el decaimiento de la fluctuación escalar con el tiempo puede relacionarse con la magnitud del exponente de Lyapunov para flujo poroso caótico. En principio, esta metodología generalizada de empujar-traccionar permite una medición precisa de las dinámicas de mezclado en cualquier material poroso opaco[1].

E. F. Médici and A. D. Otero (eds.), *Album of Porous Media*,
https://doi.org/10.1007/978-3-031-23800-0_60

MIXING IN A 3D POROUS MEDIA

Mathieu Souzy[1], *Henri Lhuissier*[2], *Yves Méheust*[1], *Tanguy Le Borgne*[1], *Bloen Metzger*[2]
[1] Aix Marseille Université, CNRS, IUSTI, UMR 7343, France
[2] Geosciences Rennes, UMR 6118, Université de Rennes, CNRS, France

A layer of fluorescent dye (Rhodamine 6G) is injected in a 3D transparent porous media and visualized using a laser sheet[1]. The porous media is made of a random stack of monodisperse spherical 2 mm diameter PMMA beads. Using iso-optical index matching techniques, the refractive optic index of the interstitial fluid (Triton X-100) and the PMMA beads is adjusted to the 4th digits, allowing to visualize directly in the bulk of the 3D porous media. The flow goes from up to down. These experimental measurements highlight how the blob of dye is heterogenously elongated while flowing through the pore network, leading to a complex mixing process.

Una capa de tinta fluorescente (Rodamina 6G) es inyectada en un medio poroso transparente 3D y visualizado con una lámina láser[1]. El medio poroso está compuesto por un apilamiento aleatorio de esferas de PMMA monodispersas, de 2 mm de diámetro. Usando la técnica óptica de ajuste del índice de refracción, se ajusta el índice de refracción óptico del fluido intersticial (Tritón X-100) y de las esferas de PMMA hasta el 4° dígito, lo que permite visualizar directamente en el volumen del medio poroso en 3D. El flujo va desde abajo hacia arriba. Estas mediciones experimentales resaltan cómo la mancha de tinta se elonga heterogéneamente al mismo tiempo que fluye entre la red de poros, lo que conlleva un proceso de mezclado complejo.

1. Souzy M. *et al.* (2020). *J. Fluid Mech.*, 891.

Contact: Mathieu Souzy <mathieu.souzy@inrae.fr>

E. F. Médici and A. D. Otero (eds.), *Album of Porous Media*,
https://doi.org/10.1007/978-3-031-23800-0_61

DISPERSION OF A BLOB OF TRACERS IN A 3D POROUS MEDIA

Mathieu Souzy[1], *Henri Lhuissier*[2], *Yves Méheust*[1], *Tanguy Le Borgne*[1], *Bloen Metzger*[2]

[1] Aix Marseille Université, CNRS, IUSTI, UMR 7343, France
[2] Geosciences Rennes, UMR 6118, Université de Rennes, CNRS, France

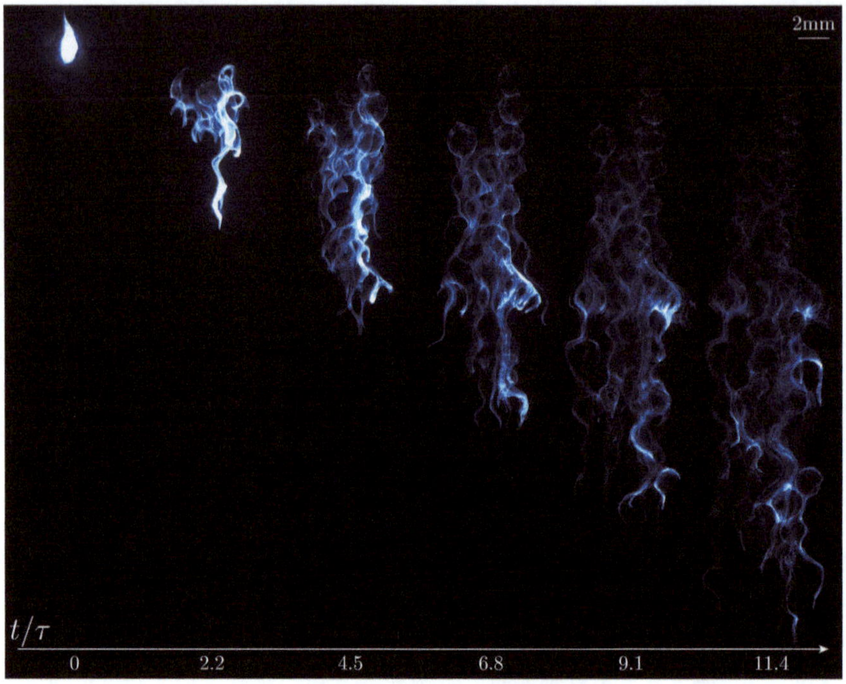

Successive experimental images of a blob of small non-diffusive tracers flowing through an index-matched 3D porous media made of a random bead-pack of monodisperse d = 2 mm diameter PMMA spheres. Time is normalized by the typical time $\tau = \langle U \rangle / d$ it takes for the interstitial fluid to travel one d. Note that the tracers here are solely submitted to dispersion, as the tracers are not submitted to molecular diffusion owing to their large size. The evolution of the blob is captured within one plane illuminated by a laser sheet (the field of view is 20d long in the streamwise direction and the fluorescent tracers have a diameter $d_{tracer} = d/620$). These measurements were used to investigate the dispersion process, by focusing on the evolution of the typical size of the enveloppe in which the tracers are contained. These measurements are obtained in a 3D porous media where the distribution of velocity is flat at low velocity. While such a distribution should lead to a persistent anomalous dispersion process for advected non-diffusive point particles, it was shown that the dispersion of non-diffusive tracers nonetheless becomes Fickian beyond a time set by the smallest effective velocity of the tracers. This low-velocity cutoff is always present owing to the finite size of the tracers or to their finite diffusivity[1].

1. Souzy M. *et al.* (2020). *J. Fluid Mech.*, 891.

Contact: Mathieu Souzy <mathieu.souzy@inrae.fr>

Sucesión de imágenes experimentales de una mancha pequeña de trazadores no difusivos fluyendo a través de un medio poroso 3D con coincidencia del índice de refracción compuesto por un relleno aleatorio de esferas de PMMA monodispersas con un diámetro d = 2 mm. El tiempo está normalizado por el tiempo característico $\tau = \langle U \rangle / d$ que tarda el fluido intersticial en desplazarse una distancia d. Obsérvese que aquí los trazadores están únicamente sometidos a dispersión, puesto que no están sometidos a difusión molecular debido a su gran tamaño. La evolución de la mancha está captada dentro de un plano iluminado por una lámina láser (el campo de visión es 20d de largo en el sentido de la dirección de la corriente y los trazadores fluorescentes tienen un diámetro $d_{tracer} = d/620$). Estas mediciones se utilizaron para investigar el proceso de dispersión, al centrarse en la evolución del tamaño típico del conjunto que contiene los trazadores. Estas mediciones se obtienen en un medio poroso 3D en el que la distribución de la velocidad es constante a baja velocidad. Si bien tal distribución debería conducir a un proceso de dispersión persistente anómalo en partículas puntuales advectivas no difusivas, se mostró que a pesar de ello, la dispersión de los trazadores no difusivos se convierte en fickiano pasado el tiempo establecido por la menor velocidad efectiva de los trazadores. Esta baja velocidad límite siempre se encuentra presente debido al tamaño finito de los trazadores o a su difusividad finita[1].

E. F. Médici and A. D. Otero (eds.), *Album of Porous Media*,
https://doi.org/10.1007/978-3-031-23800-0_62

MIXING OF A BLOB OF DYE IN A 3D POROUS MEDIA

Mathieu Souzy[1], Henri Lhuissier[2], Yves Méheust[1], Tanguy Le Borgne[1], Bloen Metzger[2]

[1] Aix Marseille Université, CNRS, IUSTI, UMR 7343, France
[2] Geosciences Rennes, UMR 6118, Université de Rennes, CNRS, France

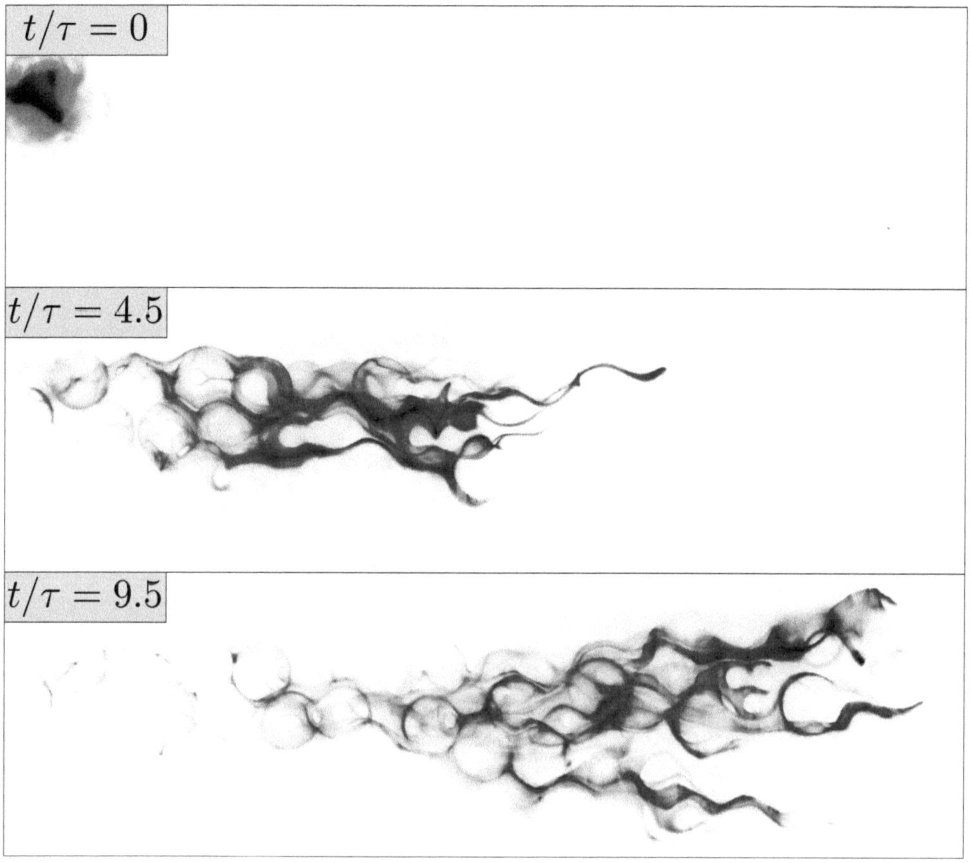

Successive experimental images of a blob of fluorescent dye (Rhodamine 6G) flowing through an index-matched 3D porous media made of a random bead-pack of monodisperse d=2 mm diameter PMMA spheres. Time is normalized by the typical time τ it takes for the fluid to travel one d in the streamwise direction. The evolution of the blob is captured within one plane illuminated by a laser sheet. These measurements were used to characterize the mixing process, which results from the coupling between both the dispersion process, and the molecular diffusion process. Mixing was investigated by characterizing how the blob of dye tends to form elongated lamellae structures which get stretched while being advected[1, 2]. By being stretched by the advection process, the dye concentration gradient in the direction transverse to the lamellae increases, which enhances the molecular diffusion process, therefore resulting in efficient homogenization of the dye concentration levels, thus in efficient mixing within the strongly stretched regions of the blob.

Sucesión de imágenes experimentales de una mancha de tinta fluorescente (Rodamina 6G) fluyendo a través de un medio poroso 3D con coincidencia del índice de refracción compuesto por un relleno aleatorio de esferas de PMMA monodispersas con un diámetro d=2 mm. El tiempo está normalizado por el tiempo representativo τ que le toma al fluido desplazarse una distancia d en la dirección de la corriente. La evolución de la mancha está captada dentro de un plano iluminado por una lámina láser. Estas mediciones se utilizaron para caracterizar el proceso de mezclado, que resulta del acoplamiento entre el proceso de dispersión y el de difusión molecular. El proceso de mezclado se investigó caracterizando cómo la mancha de tinta tiende a formar estructuras laminadas elongadas que se estiran con la advección[1, 2]. Al estirarse durante el proceso de advección, el gradiente de concentración de la tinta aumenta en la dirección transversal a las láminas, lo que aumenta el proceso de difusión molecular, por lo tanto resulta en una homogeneización eficiente de los niveles de concentración de la tinta, y por consiguiente, un mezclado eficiente dentro de las regiones fuertemente elongadas de la mancha.

1. Souzy M. *et al.* (2020). *J. Fluid Mech.*, 891.
2. Villermaux E. (2019). *Annu. Rev. Fluid Mech.*, 51(1):245.

Contact: Mathieu Souzy <mathieu.souzy@inrae.fr>

E. F. Médici and A. D. Otero (eds.), *Album of Porous Media*,
https://doi.org/10.1007/978-3-031-23800-0_63

FINGERPRINTS OF CHAOS IN A TWO-PHASE POROUS MEDIA FLOW

Gaute Linga[1], Joachim Mathiesen[2], Francois Renard[1], and Tanguy Le Borgne[3]

[1] University of Oslo
[2] University of Copenhagen
[3] University of Rennes

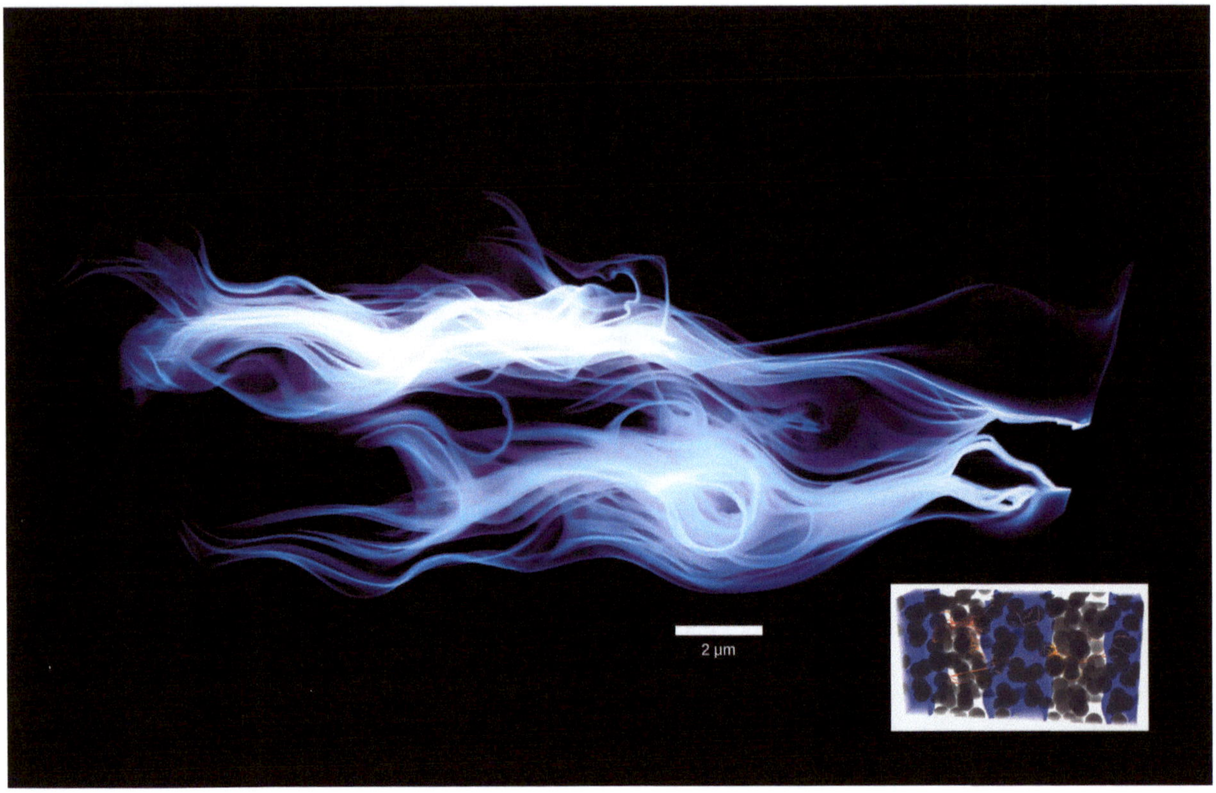

Solute mixing in porous media is a complex and often beautiful process. Here, a tiny strip of dye is injected numerically (at the right) into a simulated two-phase (leftward) flow through an array of solid spheres. The strip passively follows the velocity field and is stretched exponentially as it evolves in time; a signature of chaotic mixing.

At every time step, the one-dimensional line leaves a light imprint on the dark canvas, slowly drawing out a 2D sheet embedded in the 3D medium. The sheet reveals the flow paths of the medium in a striking way; if you look closely you will see the spheres!

The Lattice Boltzmann method[1, 2] was used to simulate flow driven by a uniform body force in a periodic domain. With a bead size of 2 μm, the system can be scaled[1] to a microcell filled with a silicone oil/water mixture (density ratio 1.25, kinematic viscosity ratio 0.57) with a mean streamwise velocity $\sim$0.2 m/s. Inset: The full model partially hiding the dye strip (solid beads are gray, oil is blue, water is transparent). The linear size of the dye structure is $\sim$25 μm.

1. Mathiesen J. *et al.* (2021). URL `https://arxiv.org/abs/2109.10985`.
2. URL `https://bitbucket.org/ulbm/felbm/`.

Contact: Gaute Linga <gaute.linga@gmail.com>

El mezcaldo de solutos en medios porosos es un proceso complejo y a menudo hermoso. Aquí, se inyecta numéricamente (derecha) una banda pequeña de tinta en un flujo de dos fases simulado (hacia la izquierda) a través de un arreglo de esferas sólidas. La banda sigue pasivamente el campo de velocidades y se estira exponencialmente a medida que evoluciona con el tiempo; una característica del mezclado caótico.

Con cada paso de tiempo, la línea unidimensional deja una marca clara en el lienzo oscuro, dibujando lentamente una lámina 2D embebida en un medio de 3D. La lámina revela los caminos del flujo en el medio de una manera impactante; si se mira de cerca, ¡se ven las esferas!

Se utilizó el método lattice Boltzmann[1, 2] para simular el flujo guiado por una fuerza volumétrica uniforme en un dominio periódico. Con un tamaño de esfera de 2 μm, el sistema puede escalarse[1] a una microcelda llena de una mezcla de aceite de silicona y agua (razón de densidad de 1,25, razón de viscosidad cinemática de 0,57) con una velocidad promedio en la dirección de la corriente de $\sim$0,2 m/s. Recuadro: Modelo completo, escondiendo parcialmente la banda de tinta (las esferas sólidas son grises, el aceite es azul y el agua es transparente). El tamaño lineal de la estructura de la tinta es de $\sim$25 μm.

E. F. Médici and A. D. Otero (eds.), *Album of Porous Media*,
https://doi.org/10.1007/978-3-031-23800-0_64

V. TRANSPORT IN MICROFLUIDIC DEVICES

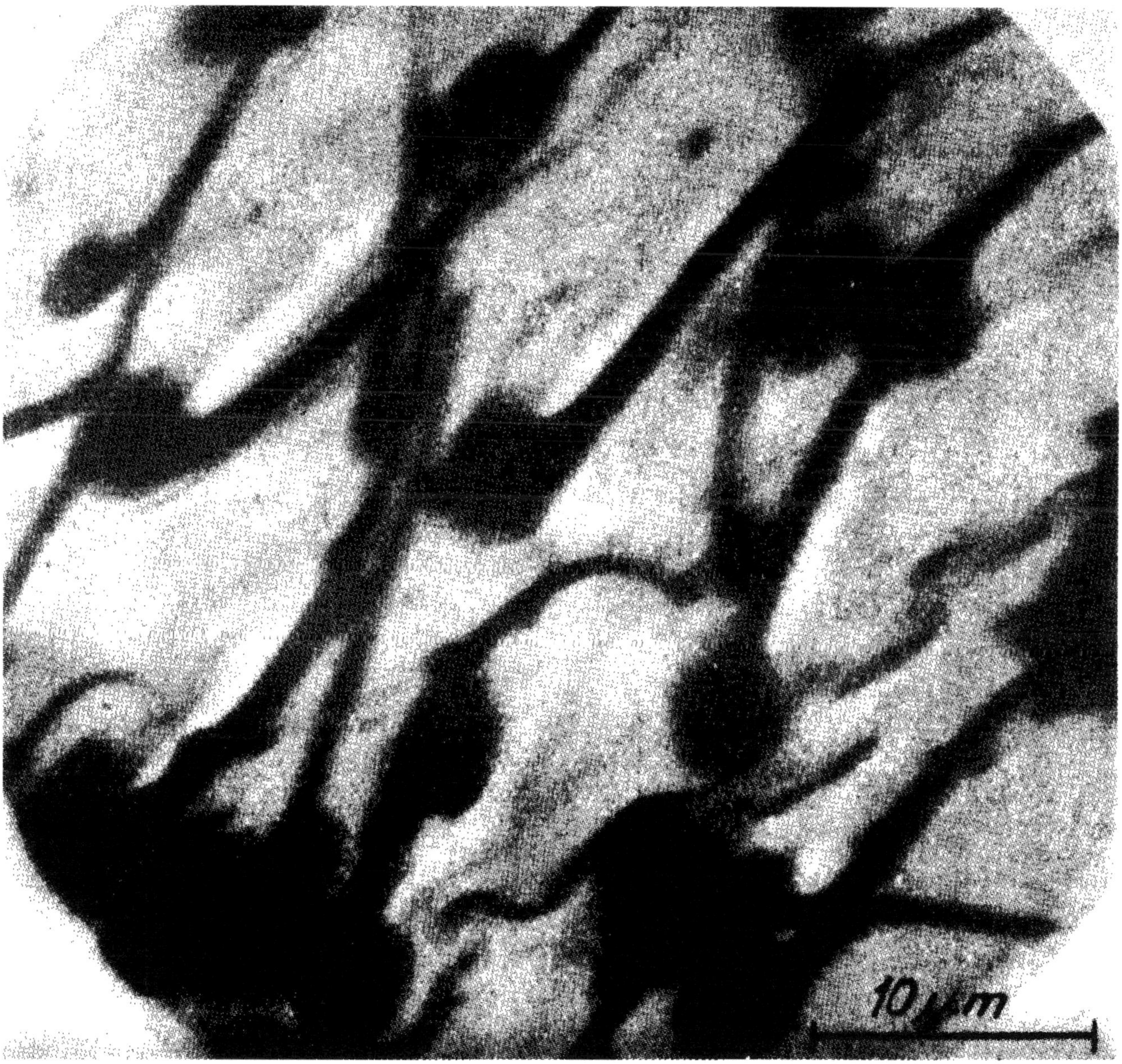

First scanning electron microscope (SEM) image: wing surface of the house fly. This image was taken in 1936 by the SEM inventor, and 1986 Nobel Prize in Physics, Ernst Ruska.

Primera imagen de un microscopio electrónico de barrido (MEB): superficie del ala de la mosca doméstica. Esta imagen fue tomada en 1936 por el inventor del MEB y premio Nobel de Física de 1986, Ernst Ruska.

TRANSITION TO DRYOUT IN DIABATIC MANIFOLD-MICROCHANNELS

David Deisenroth
University of Maryland, USA

Manifold-microchannels have shown excellent utility for high heat flux two-phase cooling of electronics[1]. Channels are embedded in the semiconductor chip by etching and then manifolding the channels with a series of alternating inlets and outlets. In this way, an array of short channels is formed with >10,000 channels/cm^2, allowing for low pumping power. This array of high aspect ratio microchannels may be thought of as non-stochastic porous media.
The flow was simultaneously imaged in the visible and infrared spectra. In this test with pentafluoropropane (R245fa), the mass flux was about 1,000 kg/m^{2}s with an average wall heat flux of 25 W/cm^2, and outlet quality of 20 %. The wall heat transfer coefficients (HTCs) were calculated with a 3D finite element steady-state conduction model using the thermal images and contact measurements as boundary conditions[2]. It was found that even with the highly wetting refrigerant (θ_C < 10°) at moderate heat flux, the combined effects of the inlet restriction and film boiling led to wall dryout at relatively low quality, resulting in an abrupt decline in average wall HTCs.

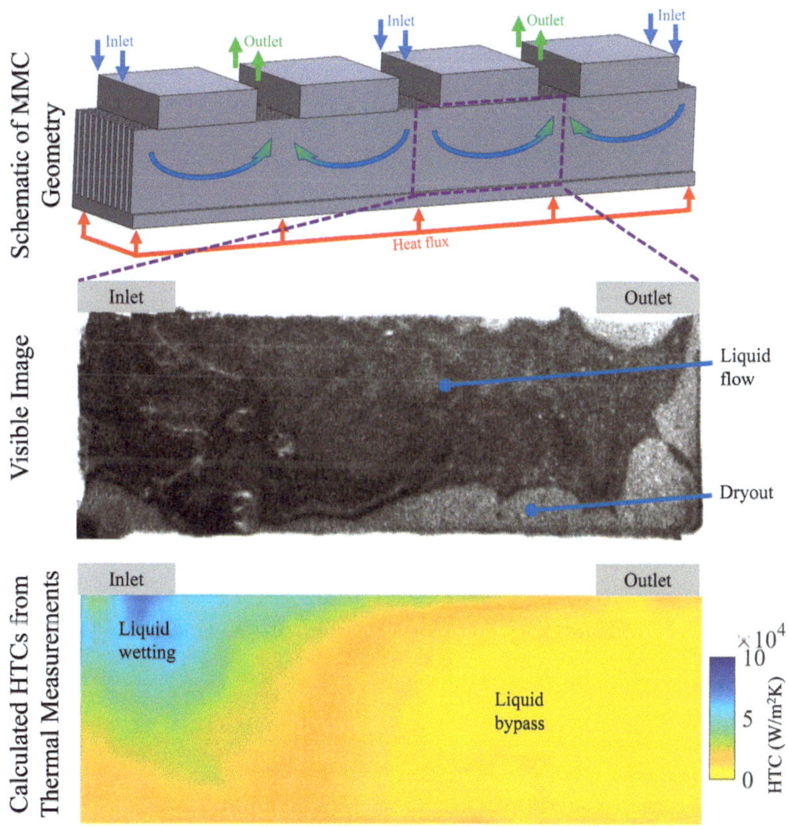

Los colectores de microcanales han mostrado ser de excelente utilidad en la refrigeración de electrónica mediante altos flujos de calor de dos fases[1]. Los canales se incorporan al circuito integrado del semiconductor grabándolos y luego se conectan los canales mediante una serie de entradas y salidas alternantes. De esta manera, se forma un arreglo de canales cortos con >10000 canales/cm^2, lo que requiere una potencia baja de bombeo. Este arreglo de microcanales de alta relación de aspecto puede considerarse como un medio poroso no estocástico.
Se tomaron simultáneamente imágenes del flujo en el espectro visible e infrarrojo. En esta prueba con pentafluoropropano (R245fa), el flujo de masa es de unos 1000 kg/m^{2}s con un promedio de flujo de calor en la pared de 25 W/cm^2 y una calidad de salida del 20 %. Los coeficientes de transferencia de calor de la pared (HTC, por sus siglas en inglés) se calcularon con un modelo de elementos finitos de conducción de calor estacionario en 3D utilizando las imágenes termales y las medidas de contacto como condiciones de contorno[2]. Se descubrió que incluso con el refrigerante altamente mojante (θ_C <10°) y un flujo de calor moderado, los efectos combinados de la restricción de entrada y la ebullición de la película condujo al secado de las paredes con una calidad relativamente baja, lo cual resultó en una disminución abrupta del HTC promedio de la pared.

1. Deisenroth D.C. (2018). URL `https://doi.org/10.13016/sebw-0zgv`.
2. Deisenroth D.C. *et al.* (2021). *IEEE Trans. Compon., Packag. Manufact. Tech.*, 11(10):1592.

Contact: David Deisenroth <dcdeisen@mtu.edu>

E. F. Médici and A. D. Otero (eds.), *Album of Porous Media*,
https://doi.org/10.1007/978-3-031-23800-0_65

CT IMAGING OF 3D PRINTED STEEL PARTICLE BEDS

Sondre Gjengedal[1], *Ole Tore Buset*[2], *and Vegard Brøtan*[3]

[1] The Norwegian Geotechnical Institute, Trondheim, Norway
[2] The Norwegian University of Science and Technology, Trondheim, Norway
[3] SINTEF Manufacturing AS, Trondheim, Norway

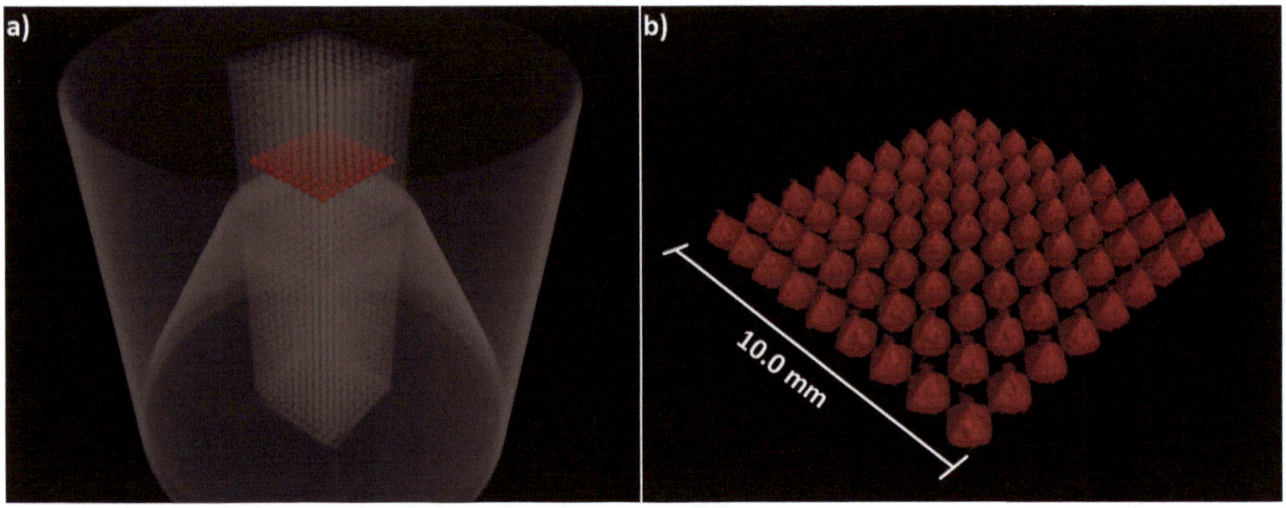

This CT image show a 3D printed porous structure containing 3500 individual steel octahedron particles (sample B3 of [1]). The shape and size dimensions of the individual octahedrons (1.0 mm diameter) are captured in detail by the CT scanner (Nikon XT H 225 ST) and the voxel data is analysed (VGSTUDIO MAX 3.1) to isolate each of the octahedron particles from the air-filled void cavities (black). Combining traditional fluid flow experiments with 3D printing technology and CT imaging allow us to study the influence of the geometry of a porous media in a controlled and accurate manner, which helps to improve the data acquisition and data interpretation of conventional fluid flow experiments. *a)* In this particular case, the entire steel sample is ghosted (grey) and a section of the particle bed (red) is isolated to visualize and analyse the internal structure of the cubical octahedron packing. *b)* Individual particles and pores are evaluated and their geometric properties, such as the porosity, the specific surface area, and the pore tortuosity, are estimated with high level of precision compared to bulk measurements on whole core samples.

1. Gjengedal S. *et al.* (2020). *Transp Porous Med*, 134(1):1.

Contact: Sondre Gjengedal <sondre.gjengedal@ngi.no>

Esta imagen de TC muestra una estructura porosa impresa en 3D que contiene 3500 partículas individuales octaedrales de acero (muestra B3 de [1]). La forma y las dimensiones de cada octaedro individual (1,0 mm de diámetro) son captadas en detalle con un escáner TC (Nikon XT H 225 ST) y los datos de los vóxeles son analizados (VGSTUDIO MAX 3.1) para aislar cada una de las partículas de los octaedros de las cavidades huecas, llenas de aire (negro). La combinación de experimentos tradicionales de flujo de fluidos con tecnología de impresión 3D e imágenes de TC nos permite estudiar la influencia de la geometría de un medio poroso de una manera controlada y precisa, lo que ayuda a mejorar la adquisición y la interpretación de datos experimentales de fluidos convencionales. *a)* En este caso en particular, toda la muestra de acero es hecha transparente (gris) y la sección de la capa de partículas (rojo) está aislada para visualizar y analizar la estructura interna del empaquetamiento cúbico de octaedros. *b)* Cada partícula y poro son analizados y sus propiedades geométricas, tales como porosidad, superficie de área específica y tortuosidad del poro, son estimadas con un alto nivel de precisión en comparación con las mediciones en el volumen de las muestras enteras.

E. F. Médici and A. D. Otero (eds.), *Album of Porous Media*,
https://doi.org/10.1007/978-3-031-23800-0_66

SELF-ASSEMBLY OF DISCOTIC LIQUID CRYSTALS IN NANOPORES AS SEEN BY X-RAY DIFFRACTION AND MONTE CARLO COMPUTER SIMULATION

Zhuoqing Li[1], *Kathrin Sentker*[1], *Arne Zantop*[2], *Milena Lippmann*[3], *Oliver Seeck*[3], *Andreas Schönhals*[4], *Marco G. Mazza*[5], *and Patrick Huber*[1,3]

[1] Institute for Materials and X-Ray Physics, Hamburg University of Technology, Germany
[2] Institut für Theoretische Physik, Technische Universität Berlin
[3] Deutsches Elektronen-Synchrotron DESY
[4] Bundesanstaltt für Materialforschung und-prüfung, Berlin
[5] Max Planck Institute for Dynamics and Self-Organization, Göttingen and Loughborough University, UK

Discotic liquid crystals (DLCs), often driven by the $\pi-\pi$ interactions among their aromatic cores, may self-organize and stack up in columns, leading to a hexagonal ordered columnar liquid crystalline mesophase. Embedding of DLCs in nanoporous solids allows one to design hybrid materials with high charge carrier mobility on the single-nanopore scale and tailorable photonic properties[1, 2, 3]. (a) Self-assembly of DLCs into distinct columnar textures on the single-pore scale and the resulting polarisation state of light transmitted through the pore array. Self-assembly of radial aligned rings and axial aligned columns of discotic molecules on the single-pore scale, as evidenced by 2D X-ray diffraction, results in an either prolate or oblate ellipsoid of refractive indicatrix aligned to the pore axis direction. Thus, linear polarised light is split up by a parallel array of such pores into two beams with perpendicular polarisations and distinct propagation speeds. Their relative phase shift after passing the birefringent pore array is positive or negative, resulting in left- or right-hand elliptically polarised light, respectively, for the specific beam and membrane geometry. (b) Electron micrographs of porous anodized aluminium oxide membranes as host materials and radial Monte-Carlo simulation snapshots of DLCs confined in nanopores at decreasing temperature.

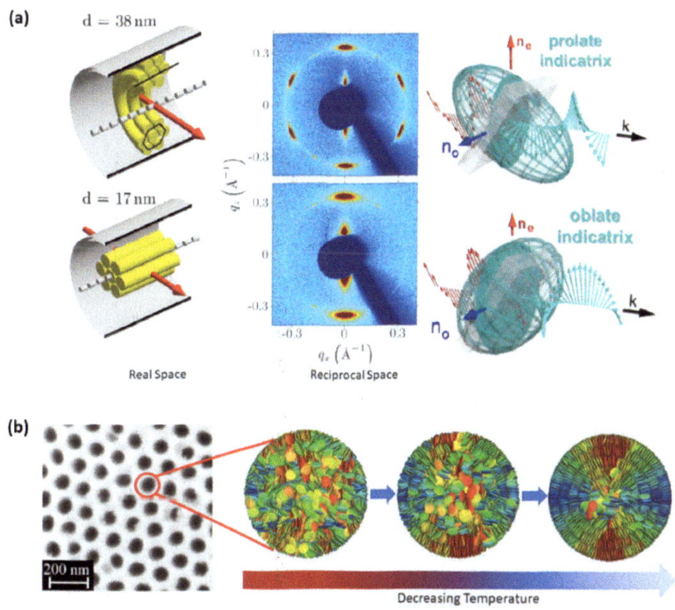

Los cristales líquidos discóticos (DLC, por sus siglas en inglés), a menudo guiados por las interacciones π-π de sus núcleos aromáticos, pueden autoorganizarse y apilarse en columnas, derivando en una mesofase de cristal líquido columnar hexagonal. La incrustación de DLC en sólidos nanoporosos permite el diseño de materiales híbridos con alta movilidad de portadores de carga a escala del nanoporo y propiedades fotónicas ajustables[1, 2, 3]. (a) Autoensamblado de DLC en distintas texturas columnares a escala del poro y el consiguiente estado de polarización de la luz transmitido a lo largo del conjunto de poros. El autoensamblado de anillos alineados radialmente y columnas de moléculas discóticas alineadas axialmente a escala del poro, tal y como lo evidencia la difracción de rayos X en 2D, resulta en un elipsoide oblongo o achatado de la indicatriz de refracción alineada con la dirección del eje del poro. Por lo tanto, la luz polarizada linealmente es dividida por un conjunto de poros paralelos en dos haces con polarizaciones perpendiculares y velocidades de propagación distintas. Su cambio de fase relativo después de pasar por el conjunto de poros birrefrigente es positivo o negativo, lo que resulta en una luz polarizada elíptica hacia la derecha o hacia la izquierda, respectivamente, específicos del haz y de la geometría de la membrana. (b) Micrografías electrónicas de membranas porosas de óxido de aluminio anodizado como materiales hospedantes e imágenes radiales de simulaciones Monte Carlo de DLC confinados en nanoporos a temperatura decreciente.

1. Sentker K. *et al.* (2019). *Nanoscale*, 11(48):23304.
2. Sentker K. *et al.* (2018). *Phys. Rev. Lett.*, 120(6).
3. Gang O. *et al.* (2020). *Soft Matter and Biomaterials on the Nanoscale*. World Scientific.

Contact: Zhuoqing Li <zqli0721@gmail.com>

E. F. Médici and A. D. Otero (eds.), *Album of Porous Media*,
https://doi.org/10.1007/978-3-031-23800-0_67

IMBIBITION OF WATER IN NANOPOROUS SILICA (MCM-41)

Lars Dammann[1,2], *Oliver Seeck*[1], *Robert Horst Meißner*[2,3], *and Patrick Huber*[1,2]

[1] Deutsches Elektronen-Synchrotron DESY
[2] Hamburg University of Technology
[3] Helmholtz-Zentrum Hereon

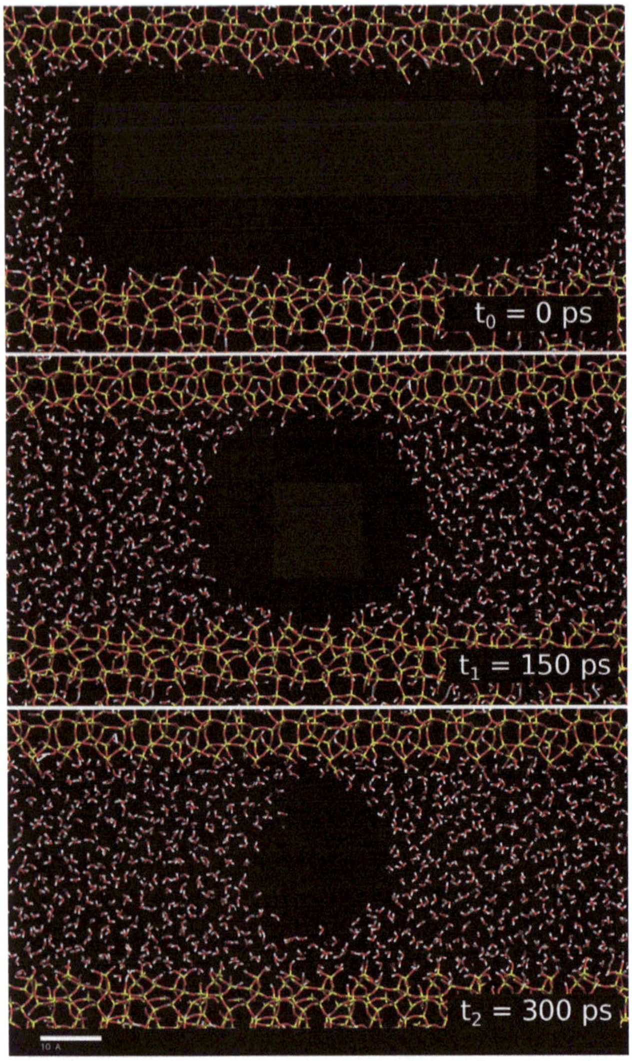

Se muestran los cortes transversales del resultado de una simulación de dinámica molecular en 3D de la dinámica de imbibición de moléculas de agua en sílice nanoporoso (MCM-41). Los tipos de átomo se representan con distintos colores. El blanco corresponde al hidrógeno, el rojo al oxígeno y el amarillo al silicio. La evolución en el tiempo se representa de arriba a abajo con t_0 = 0 ps, t_1 = 150 ps y t_2 = 300 ps. Los poros de MCM-41 tienen unos 109 Å de largo, 30 Å de diámetro y sus paredes tienen 10 Å de grosor. En los tiempos t_1 y t_2, se puede observar meniscos de agua en el poro de sílice causados por la naturaleza hidrofílica de la superficie de sílice hidrogenada. A nivel microscópico, la profundidad de la penetración de fluido L en un poro cilíndrico después de un tiempo t puede describirse por la ecuación de Lucas-Washburn, $L = \sqrt{v_i}\sqrt{t}$, en la que v_i es la llamada «velocidad de imbibición» que depende del cociente entre los parámetros del fluido, la interacción entre el fluido y la pared, el radio del poro y la condición de contorno hidrodinámica de deslizamiento libre[1]. El trabajo en cuestión muestra que esta relación no solo es válida en sistemas macroscópicos sino también en sistemas a escala nanoscópica. La simulación se realizó con LAMMPS[2]. Agradecemos la aportación de la estructura MCM-41[3]. La visualización fue creada con VMD[4].

Depicted are slices of a 3D molecular dynamics simulation of imbibition dynamics of water molecules in nanoporous silica (MCM-41). The atom types are represented by different colors. White corresponds to hydrogen, red to oxygen and yellow to silicon. The time evolution is presented from top to bottom at t_0 = 0 ps, t_1 = 150 ps and t_2 = 300 ps. The MCM-41 pores are around 109 Å in length, 30 Å in diameter and have a 10 Å wall thickness. At times t_1 and t_2 water menisci can be observed in the silica pore caused by the hydrophilic nature of the hydrogenated silica surface. Macroscopically the penetration depth of fluid L into a cylindrical pore after time t can be described by the Lucas-Washburn equation, $L = \sqrt{v_i}\sqrt{t}$, where v_i is the so-called “imbibition speed” that depends on the ratio of the fluid parameters, the fluid/wall interaction, the radius of the pore and the hydrodynamic slip-boundary condition[1]. The work at hand shows that this relation not only holds for macroscopic systems but also for systems of nanoscopic scale. The simulation was executed with LAMMPS[2]. We are grateful for the provided MCM-41 structure[3]. The visualization was created with VMD[4].

1. Gruener S. *et al.* (2009). *Phys. Rev. E*, 79(6).
2. Plimpton S. (1995). *J. Computational Physics*, 117(1):1.
3. Piane M.D. *et al.* (2014). *J. Phys. Chem. C*, 118(46):26737.
4. Humphrey W. *et al.* (1996). *J. Molecular Graphics*, 14(1):33.

Contact: Lars Dammann <lars.dammann@tuhh.de>

E. F. Médici and A. D. Otero (eds.), *Album of Porous Media*,
https://doi.org/10.1007/978-3-031-23800-0_68

DIRECT NUMERICAL SIMULATION OF FLOW THROUGH NANOSCALE SHALE PORES IN A MESOSCALE SAMPLE

Timothy Kneafsey, David Trebotich, and Terry Ligocki
Lawrence Berkeley National Laboratory

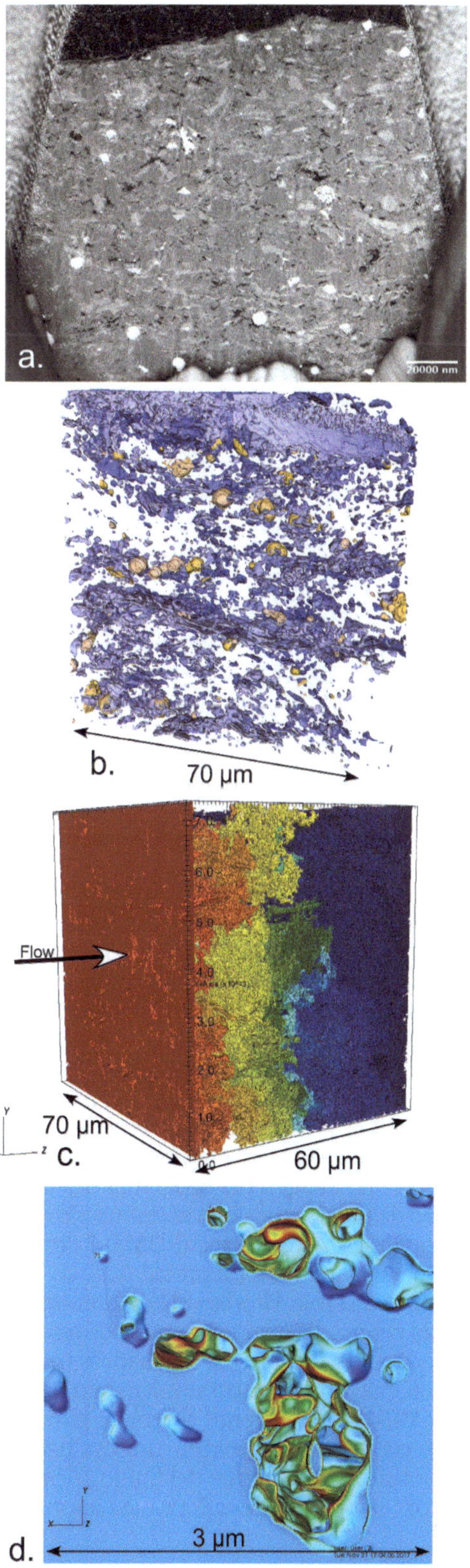

To better understand fluid flow through shale we performed direct numerical simulation (DNS) of flow in a pore geometry obtained from Marcellus shale image data[1]. The sample – a 100 μm cube, (a) – was imaged at 50 nm voxel size using a TESCAN FERA3 XMH focused ion beam/scanning electron microscope, milled with argon plasma by L. Chan, Tescan USA. Standard image processing techniques were used to filter the dataset and the gradient of the average value across a single image was normalized. FFTs were applied to directionally filter streaking in the images. The resulting pore-scale geometry (b: blue is pore, yellow is pyrite) is represented using implicit functions defined locally on a Cartesian grid. The simulation domain is $3072\times3072\times2688$ cells ($\sim$25 billion) for a 73.9 μm $\times$ 73.9 μm $\times$ 64.7 μm block of shale (24 nm resolution), and >150 billion degrees of freedom. The CFD solver in Chombo-Crunch was used to simulate incompressible viscous flow in the pore space (c: pressure gradient from high (red) to low (blue) and d: example of computed velocity into pores at a selected location).

Para comprender mejor el flujo de los fluidos a través de la lutita, realizamos una simulación numérica directa (DNS, por sus siglas en inglés) del flujo en una geometría porosa a partir de información de imágenes de lutita de Marcellus[1]. De la muestra, un cubo de 100 μm (a), molida con plasma de argón, se obtuvieron visualizaciones utilizando un microscopio electrónico de barrido con haz iones enfocados TESCAN FERA3 XMH con un tamaño vóxel de 50 nm gracias a L. Chan, Tescan EE.UU. Se utilizaron técnicas estándar de procesamiento de imágenes para filtrar el conjunto de datos y normalizar el gradiente del valor promedio por imagen. Se aplicó FFT para filtrar direccionalmente rayas de las imágenes. La geometría a escala de poro resultante (b, en azul los poros y en amarillo la pirita) es representada mediante funciones implícitas definidas localmente en una grilla cartesiana. El dominio de la simulación es de $3072\times3072\times2688$ celdas ($\sim$25 mil millones) para un bloque de lutita de 73,9 μm $\times$ 73,9 μm $\times$ 64,7 μm (resolución de 24 nm), y >150 mil millones de grados de libertad. Se utilizó el simulador CFD en el paquete Chombo-Crunch para simular flujo viscoso incomprensible en el espacio poroso: (c) gradiente de prcsión de mayor -rojo- a menor -azul- y (d) ejemplo de velocidad calculada dentro de los poros en una ubicación previamente seleccionada.

1. Trebotich D. and Graves D. (2015). *Commun. Appl. Math. Comput. Sci.*, 10(1):43.

Contact: Timothy Kneafsey <tjkneafsey@lbl.gov>

E. F. Médici and A. D. Otero (eds.), *Album of Porous Media*,
https://doi.org/10.1007/978-3-031-23800-0_69

IN SITU OPTICAL SPECTROSCOPY DURING LIQUID IMBIBITION INTO A NANOPOROUS PHOTONIC CRYSTAL RESOLVES PRECURSOR FILM SPREADING

Luisa G. Cencha[1], *Guido Dittrich*[2], *Claudio L.A. Berli*[3], *Patrick Huber*[2], *and Raul Urteaga*[4]

[1] Polymer Reaction Engineering Group, INTEC (Universidad Nacional del Litoral-CONICET)
[2] Hamburg University of Technology and Deutsches Elektronen-Synchrotron DESY
[3] INTEC (Universidad Nacional del Litoral-CONICET)
[4] IFIS-Litoral (Universidad Nacional del Litoral-CONICET)

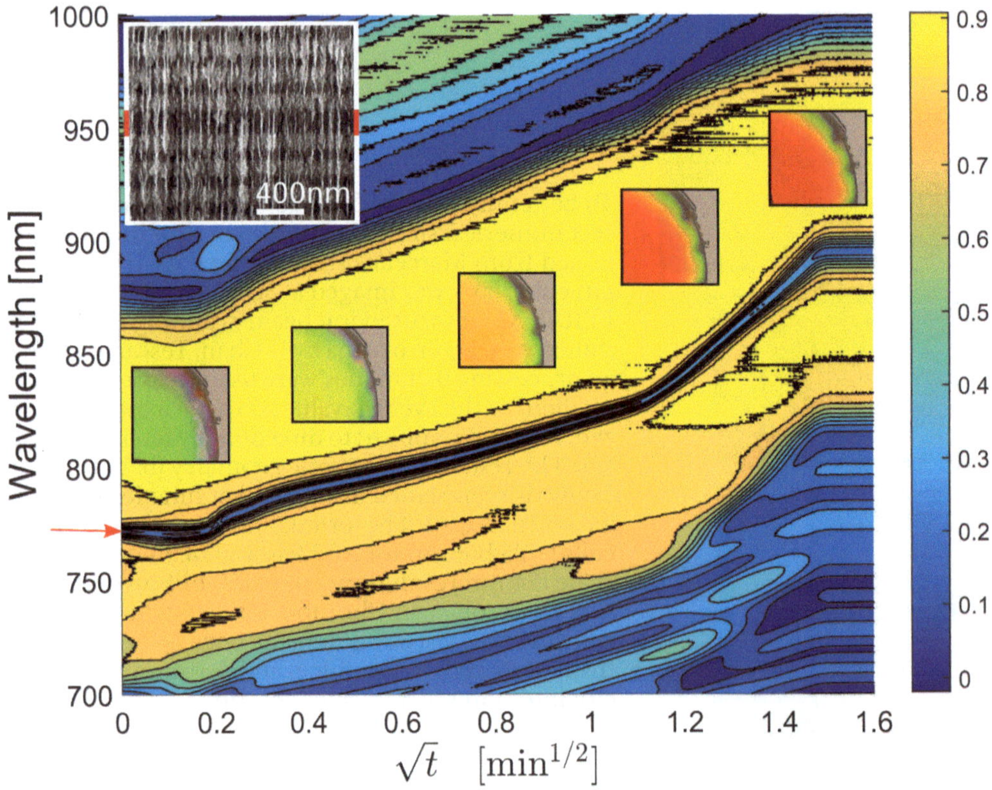

A photonic crystal (PC) can be integrated into nanoporous silicon by electrochemical etching. The inset in the upper left of the main image displays a SEM side view of the PC. Its resonance wavelength is determined by the size of a high porosity central defect layer, which is the biggest of the horizontally aligned dark regions highlighted by red edges. It is surrounded by smaller regions of alternating porosities acting as mirror layers. The main image displays the time evolution of the visible-to-near-infrared spectrum during the imbibition with polydimethylsiloxane (PDMS). The PC light reflection has a prominent local minimum at 775 nm indicated by a red arrow. Once liquid substitutes air in the central cavity, the resonance wavelength shifts proportionally. Insets show the color change of a similar sample tuned in the visible region in direct optical imaging. The technique allows high precision assessment of the fluid front's shape, including even a sub-nanometer sized precursor film[1].

1. Cencha L.G. *et al.* (2020). *Phys. Rev. Lett.*, 125(23).

Un cristal fotónico (CF) puede integrarse en silicio nanoporoso mediante grabado electroquímico. El recuadro en la parte superior izquierda de la imagen principal muestra una visualización lateral con MEB del CF. Su longitud de onda de resonancia se determina mediante el tamaño de una capa central defectuosa de alta porosidad, que es la mayor de las regiones oscuras alineadas horizontalmente resaltadas con bordes rojos. Esta capa está rodeada por regiones de menores porosidades alternantes que actúan como capas espejo. La imagen principal muestra la evolución en el tiempo del espectro visible casi infrarrojo durante la imbibición con polidimetilsiloxano (PDMS). El reflejo de la luz del CF tiene un mínimo local prominente de 775 nm indicado por una flecha roja. Una vez el líquido sustituye el aire en la cavidad central, la longitud de onda de resonancia se desplaza proporcionalmente. Los recuadros muestran el cambio de color de una muestra similar sintonizada en la region visible mediante imágenes obtenidas por microscopía óptica directa. La técnica permite una evaluación de alta precisión de la forma del frente del fluido, incluyendo la capa precursora de un tamaño inferior al nanómetro[1].

Contact: Guido Dittrich <guido.dittrich@tuhh.de>

E. F. Médici and A. D. Otero (eds.), *Album of Porous Media*,
https://doi.org/10.1007/978-3-031-23800-0_70

STUDY OF STRESS FIELD AND FRACTURE NETWORK DEVELOPMENT WITH ROCK ANALOGS

Bolivia Vega and Anthony Kovscek
Stanford University

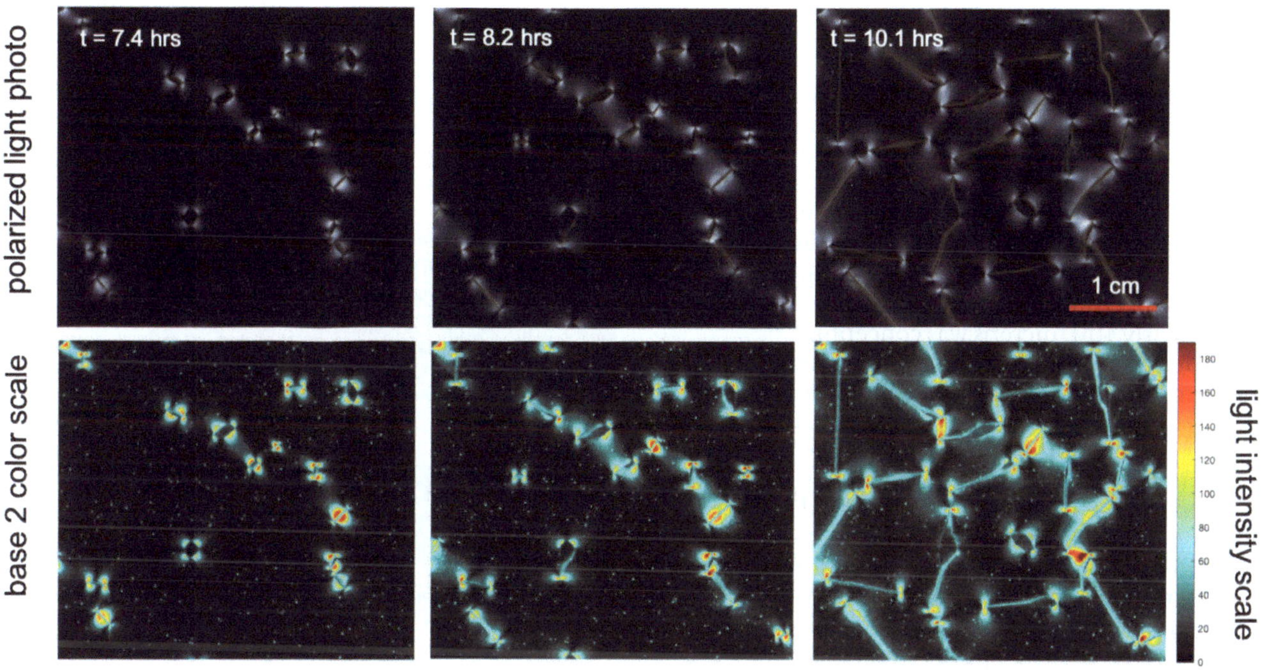

Analog materials are used to simplify a complex system while exhibiting comparable properties. Gelatin is a polypeptide that has been used in geosciences as a rock analog to mimic some responses and behaviors of the Earth's lithosphere[1]. Gelatin rheology and viscoelasticity have been exploited in experiments designed to study fracture network development during maturation of source rock[2, 3]. A gelatin solution containing sugar and yeast is placed between two glass plates, and a yeast-prompted sugar fermentation process is started. Gas is expelled as a byproduct, and as the increasing stress surpasses the gelatin tensile strength, fractures nucleate and grow. The process is imaged at fixed time intervals with high-resolution (18 MP) photographic equipment. One of these studies[4] included the use of gelatin photoelastic properties. Under polarized light, gelatin areas refract light proportional to the local stress (upper row time sequence). The row below shows the original images converted into a 0-256 colormap to enhance the qualitative assessment of the stress distribution through early (left), mid (center), and late (right) fracture growth.

Materiales análogos, que exhiben propiedades similares, son utilizados para simplificar sistemas complejos. La gelatina es un polipéptido empleado en la geociencia como análogo de roca para imitar algunas respuestas y comportamientos de la litosfera terrestre[1]. La reología y la viscoelasticidad de la gelatina han sido explotadas en experimentos diseñados para estudiar el desarrollo de redes de fracturas durante la maduración de la roca madre[2, 3]. Se coloca una solución de gelatina que contiene azúcar y levadura entre dos placas de vidrio, y se inicia un proceso de fermentación del azúcar provocada por la levadura. El gas es expulsado como subproducto y, a medida que la tensión en aumento sobrepasa la resistencia a la tracción de la gelatina, las fracturas nuclean y crecen. Se toman imágenes del proceso a intervalos fijos con un equipo fotográfico de alta resolución (18 MP). Uno de estos estudios[4] incluyó el uso de gelatina con propiedades fotoelásticas. Bajo luz polarizada, las áreas de gelatina refractan luz proporcionalmente a la tensión local (secuencia de la fila superior). La fila inferior muestra las imágenes originales convertidas en un mapa de color 0-256 para mejorar la evaluación cualitativa de la distribución de tensiones durante el crecimiento temprano (izquierda), intermedio (centro) y tardío (derecha) de la fractura.

This research was funded by the Center for Mechanistic Control of Unconventional Formations (CMC-UF), an Energy Frontier Research Center funded by the U.S. Department of Energy, Office of Science under DOE (BES) Award DE-SC0019165.

1. Kavanagh J.L. *et al.* (2006). *Earth and Planetary Sci. Letters*, 245(3-4):799.
2. Kobchenko M. *et al.* (2013). *Europhysics Letters*, 102(6):66002.
3. Vega B. and Kovscek A.R. (2019). *J. Petroleum Sci. and Eng.*, 173:209.
4. Vega B. *et al.* (2019). *Transp Porous Med*, 131(2):503.

Contact: Bolivia Vega <bolivia@stanford.edu>

E. F. Médici and A. D. Otero (eds.), *Album of Porous Media*,
https://doi.org/10.1007/978-3-031-23800-0_71

THERMAL AND OPTICAL IMAGING IN A MICROMODEL

Nikolaos Karadimitriou[1], *Philipp Nuske*, *S. Majid Hassanizadeh*[2], *and Rainer Helmig*[3]

[1] Institute of Mechanics (CE), Stuttgart University
[2] Utrecht University
[3] Institute for Modelling Hydraulic and Environmental Systems, Department of Hydromechanics and Modelling of Hydrosystems, Stuttgart University

In this work, a Poly-Di-Methyl-Siloxane (PDMS) artificial porous medium was used for the simultaneous thermal and optical visualization of two-phase displacement, for the first time in the literature. An elongated micromodel, with dimensions of 14.4×39 mm^2 and a constant depth of 100 microns, and a mean pore size of nearly 400 microns was conceptualized, designed, and manufactured. The choice of the mean pore size was dictated by the native resolution (30 microns) of the ICI-7640P thermal camera used. A novel setup for the simultaneous thermal and optical imaging was developed, enabling us to monitor and record the same field of view for both cameras at the same temporal and spatial resolution in a synchronous way, so that a direct comparison between the two approaches could be performed. The data obtained during dynamic drainage regarding the spatial distribution of fluids at any given time, with the non-wetting phase (water) represented in blue and the wetting one (Fluorinert FC-43) in red (Subfigure B), as well as the thermal signature of each phase (Subfigure C), and their average temperature across the length of the porous network (Subfigure D), were used in the study of non-equilibrium heat transfer in multiphase flow for the identification of the relevant parameters and constitutive relations, like the heat transfer coefficients between phases[1].

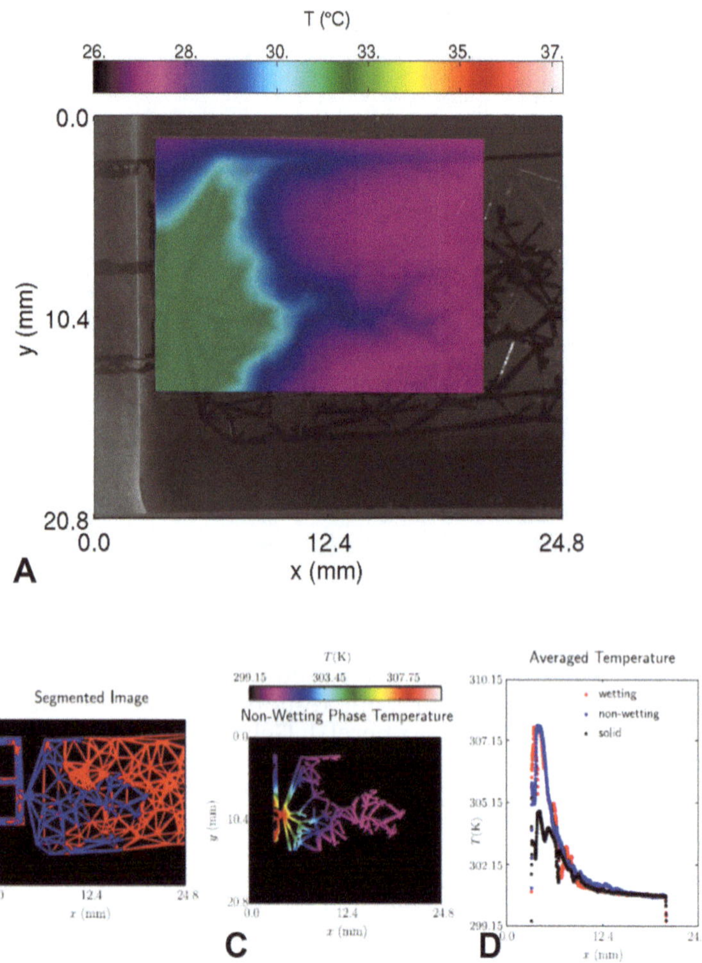

En este trabajo, se utilizó por primera vez en la literatura un medio poroso artificial de polidimetilsiloxano (PDMS) para una visualización térmica y óptica simultanea de un desplazamiento en dos fases. Se conceptualizó, diseñó y produjo un micromodelo elongado, con dimensiones de $14,4 \times 39$ mm^2 y una profundidad constante de 100 micrones, con un promedio de tamaño de poro de casi 400 micrones. La elección del tamaño de poro promedio fue dictada por la resolución intrínseca (30 micrones) de la cámara térmica ICI-7640P empleada. Se desarrolló una novedosa configuración para obtener simultáneamente imágenes térmicas y ópticas, que nos permitió monitorear y registrar el mismo campo de visión en ambas cámaras y la misma resolución temporal y espacial de una manera sincrónica, para poder realizar una comparación directa entre los dos métodos. Los datos obtenidos durante el drenado dinámico con respecto a la distribución espacial de fluidos en cualquier momento determinado, con fase no mojante (agua) representada en azul y la mojante (Fluorinert FC-43) en rojo (subimagen B), así como las características térmicas de cada fase (subimagen C) y sus temperaturas promedio a lo largo de la red porosa (subimagen D) fueron utilizados en el estudio de la transferencia de calor fuera de equilibrio en flujo multifásico para la identificación de los parámetros relevantes y relaciones constitutivas, como los coeficientes de transferencia de calor entre las fases[1].

1. Karadimitriou N.K. *et al.* (2014). *Lab Chip*, 14(14):2515.

Contact: Nikolaos Karadimitriou <nkaradimitriou@gmail.com>

E. F. Médici and A. D. Otero (eds.), *Album of Porous Media*,
https://doi.org/10.1007/978-3-031-23800-0_72

FLOW-DRIVEN CHANNELIZATION IN A PARTICLE-FILLED POROUS MEDIUM

Navid Bizmark and Sujit S. Datta
Princeton University, USA

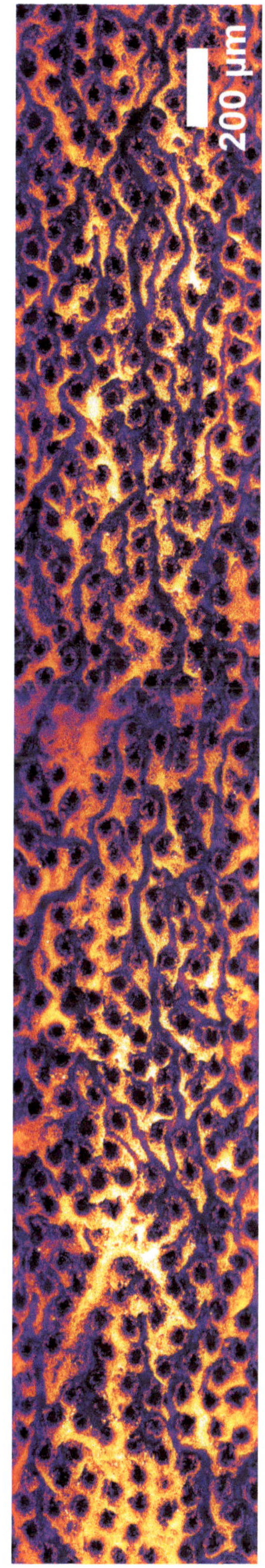

The coupling between fluid flow and solid deformability underlies many energy, environmental, geological, and industrial processes - in many cases, leading to beautiful multi-scale patterns. To investigate this coupling in the lab, we inject 1 μm-diameter colloidal particles into a rigid porous medium composed of densely-packed, sintered, 40 μm-diameter spherical glass beads (dark circles). The imposed flow direction is from left to right (with the page rotated 90° clockwise). The bead matrix is refractive index-matched and the particles are fluorescent, enabling them to be visualized using confocal microscopy. As the particles are transported, they attach to the bead surfaces[1, 2], eventually depositing in and completely filling the pore space (yellow speckles). As fluid continues to be injected, the viscous stresses cause this non-cohesive deposit to become continually eroded at specific locations[3] – reshaping it and eventually leading to the formation of meandering, branched channels (thick dark lines) that permeate through the medium and continue to support flow, much like a river network[4].

El acoplamiento entre el flujo del fluido y la deformabilidad del sólido constituye la base de muchos procesos energéticos, medioambientales, geológicos e industriales, y en muchos casos causan hermosos patrones multiescala. Para investigar este acoplamiento en el laboratorio, inyectamos partículas coloidales de 1 μm de diámetro en un medio poroso rígido compuesto por esferas de vidrio con un diámetro de 40 μm densamente empaquetadas (círculos oscuros). La dirección impuesta del flujo es de izquierda a derecha (rotando la página 90° en sentido horario). La matriz de esferas tiene un índice de refracción coincidente con el del fluido y las partículas son fluorescentes, lo que permite verlas utilizando microscopía confocal. A medida que las partículas son transportadas, se adhieren a las superficies de las esferas[1, 2], y eventualmente se depositan y llenan completamente el espacio del poro (brillo amarillo). Mientras el fluido sigue siendo inyectado, la tensión viscosa provoca que este depósito no cohesivo se erosione continuamente en lugares específicos[3], dando lugar a una nueva forma y eventualmente llevando a la formación de meandros, canales ramificados (líneas gruesas oscuras) que permean a través del medio y mantienen un flujo similar a una red fluvial[4].

1. Bizmark N. *et al.* (2020). *Sci. Adv.*, 6(46).
2. Schneider J. *et al.* (2021). *Phys. Rev. Fluids*, 6(1).
3. Jäger R. *et al.* (2017). *Phys. Rev. E*, 95(1).
4. Mahadevan A. *et al.* (2012). *Europhysics Letters*, 98(5):58003.

Contact: Sujit S. Datta <ssdatta@princeton.edu>

E. F. Médici and A. D. Otero (eds.), *Album of Porous Media*,
https://doi.org/10.1007/978-3-031-23800-0_73

STREAMLINES IN THE BULK OF A 3D POROUS MEDIA

Mathieu Souzy[1], *Henri Lhuissier*[2], *Yves Méheust*[1], *Tanguy Le Borgne*[1], *Bloen Metzger*[2]

[1] Aix Marseille Université, CNRS, IUSTI, UMR 7343, France
[2] Geosciences Rennes, UMR 6118, Université de Rennes, CNRS, France

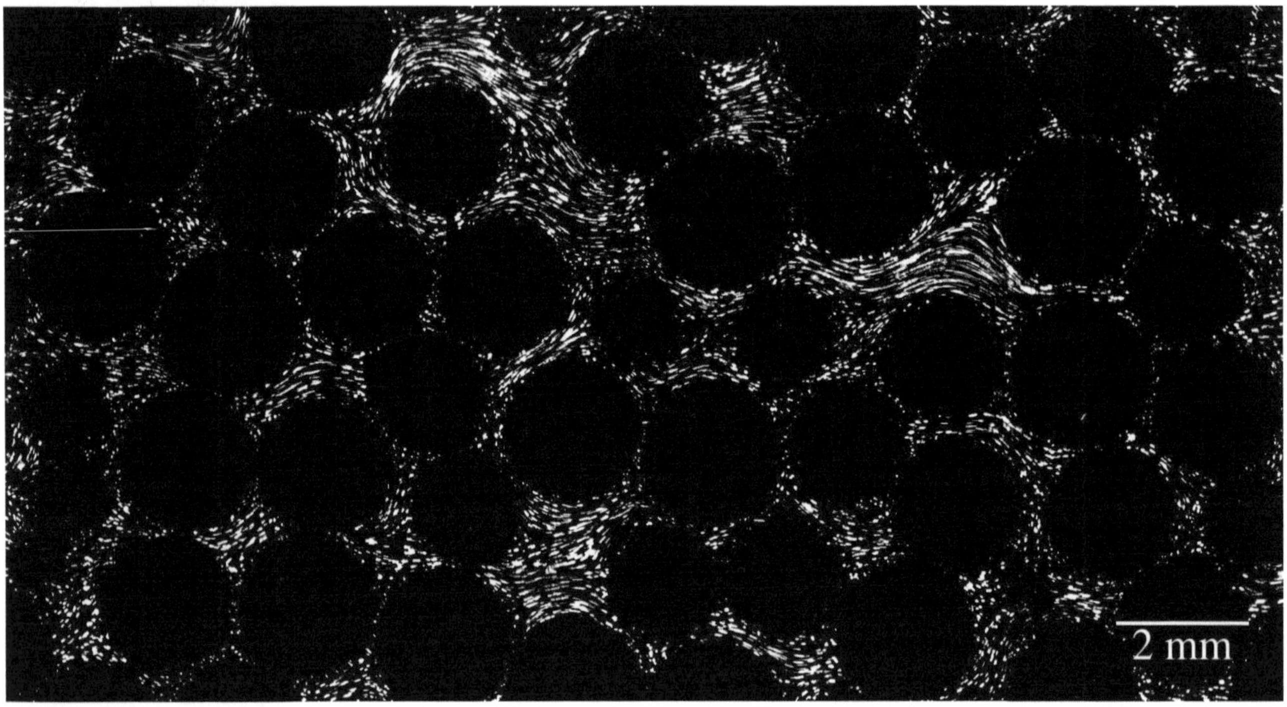

Overlay of successive PIV images illustrating the streamlines in the bulk of an index-matched 3D porous media made of a random bead-pack of monodisperse d = 2 mm diameter PMMA spheres. The total sample size is 100d × 12d × 12d, the interstitial fluid is Triton X-100 seeded with tiny passive non-diffusive fluorescent tracers (d_{tracer} = d/620), and the fluid is flowing from left to right. It reveals the strong flow velocity amplitude heterogeneities in porous media, which prescribe the mixing and dispersion process[1].

1. Souzy M. *et al.* (2020). *J. Fluid Mech.*, 891.

Contact: Mathieu Souzy <mathieu.souzy@inrae.fr>

Superposición de imágenes PIV (por sus siglas en ingles) que ilustran los flujos en el volumen de un medio poroso 3D con coincidencia del índice de refracción compuesto por un relleno aleatorio de esferas de PMMA monodispersas con un diámetro d = 2 mm. El tamaño de la muestra total es 100d × 12d × 12d, el fluido intersticial es Tritón X-100 con un esparcido de pequeños trazadores fluorescentes pasivos no difusivos (d_{tracer} = d/620), y el fluido fluye de izquierda a derecha. La imagen revela las fuertes heterogeneidades en la amplitud de la velocidad de flujo en los medios porosos que dicta el proceso de mezclado y dispersión[1].

E. F. Médici and A. D. Otero (eds.), *Album of Porous Media*,
https://doi.org/10.1007/978-3-031-23800-0_74

QUASI-TWO-DIMENSIONAL FOAM FLOW THROUGH AND AROUND A PERMEABLE OBSTACLE

Lauren Rose[1], *Natalia Shmakova*[2], *Natalya Penkovskaya*[2], *Benjamin Dollet*[3], *Christophe Raufaste*[4], *and Stéphane Santucci*[1]

[1] ENS de Lyon, CNRS, Laboratoire de Physique, Lyon, France
[2] Lavrentyev Institute of Hydrodynamics, Siberian Branch of the Russian Academy of Sciences, Novosibirsk, Russia
[3] Université Grenoble Alpes, CNRS, LIPhy, France
[4] Université Côte d'azur, CNRS, INPHYNI, France

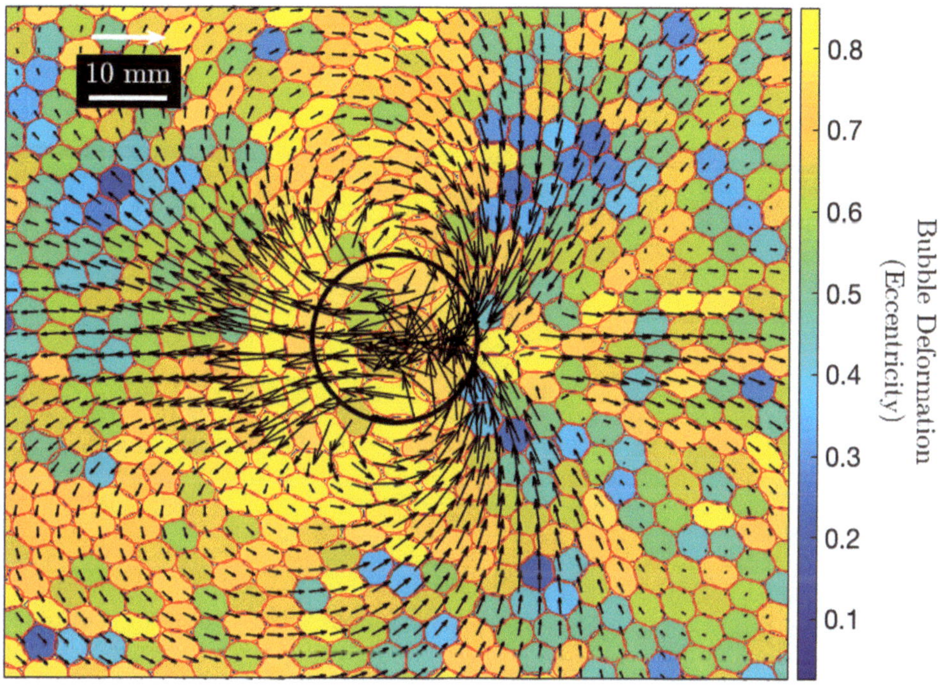

We quantify the motion and deformation of bubbles, elementary components of a 2D liquid foam, forced to invade (from left to right, as shown by the white arrow) a model porous medium: a Hele-Shaw cell, with a localized constriction (black circle) that reduces locally (by 50 %) its millimetric gap. The figure shows the instantaneous bubble deformation with a colormap giving the eccentricity measured for each bubbles, by fitting them as ellipses (shown in red). The superimposed black arrows give the average bubble velocity field in the frame of the flowing foam, computed by image correlation analysis (PIV)[1].

Foam flows in porous media are the subject of intense research, notably in the context of enhanced oil recovery. From a more fundamental perspective, our flow experiments highlight the elasto-plastic properties of foams, controlled by their liquid fraction, which are notably responsible of the symmetry breaking of the flow, with multi-polar deformation and velocity fields around the localized inhomogeneity. Such specific rheological properties are at the root of their use in numerous applications, with everyday-life products in food and cosmetic industries to large-scale industrial processes, as soil remediation.

1. Shmakova N. *et al.* (2020). *Phys. Rev. Fluids*, 5(9).

Contact: Stéphane Santucci <stephane.santucci@ens-lyon.fr>

Cuantificamos el movimiento y la deformación de burbujas, componente básico de una espuma líquida, forzada a entrar (de izquierda a derecha, tal y como lo muestra la flecha blanca) en un modelo de medio poroso: una celda de Hele-Shaw, con una constricción localizada (círculo negro) que reduce localmente (un 50 %) su brecha milimétrica. La figura muestra la deformación instantánea de las burbujas con un mapa de colores indicando la excentricidad medida para cada burbuja, al mostrarlas como elipses (en rojo). Las flechas negras superpuestas representan el promedio del campo de velocidad de las burbujas en el marco del flujo de la espuma, calculada mediante un análisis de correlación de la imagen (PIV)[1].

Los flujos de espuma en medios porosos son un tema de intensa investigación, especialmente en el contexto de la recuperación mejorada de petróleo. Desde una perspectiva más fundamental, nuestros experimentos de flujo resaltan las propiedades elastoplásticas de las espumas, controladas por su fracción líquida, que son particularmente responsables de la ruptura de la simetría del flujo, con una deformación multipolar y campos de velocidades alrededor de la inhomogeneidad localizada. Tales propiedades reológicas específicas son el motivo de sus numerosas aplicaciones, que van desde productos de uso diario en las industrias alimentaria y cosmética, hasta procesos a escala industrial como la remediación de suelos.

E. F. Médici and A. D. Otero (eds.), *Album of Porous Media*,
https://doi.org/10.1007/978-3-031-23800-0_75

THE FLOW OF SUPERCRITICAL CO_2 FOAM FOR MOBILITY CONTROL

Benyamine Benali
University of Bergen

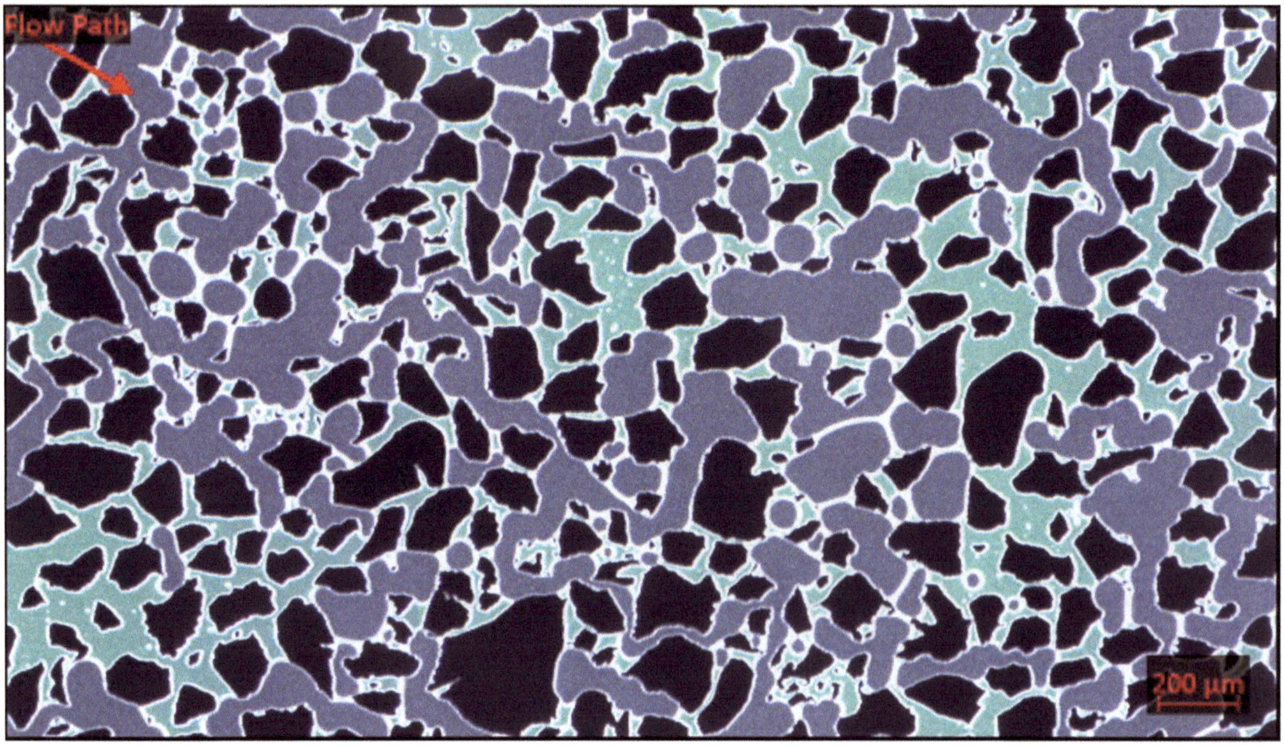

CO_2 bubbles (gray) are held separated temporarily using a 5000 ppm surfactant solution (green, reflection from a fluorophore) in a microfluidic pore network (black grains) at elevated pore pressure (100 bars). The bubbles are separated by thin surfactant-stabilized films called lamellae (white) which span several pores, orthogonally to the CO_2 flow direction indicated by the red arrow in the upper left corner of the image, thus reducing CO_2 mobility. The CO_2 foam visualization is composed by combining a brightfield and a fluorescence microscopy image.

Contact: Benyamine Benali <benjamin.benali20@gmail.com>

Burbujas de CO_2 (gris) se mantienen temporalmente separadas utilizando una solución surfactante de 5000 ppm (verde, reflejo de un fluorocromo) en una red de poros de microfluidos (gránulos negros) a una elevada presión de poro (100 bar). Las burbujas están separadas por finas películas de estabilizador surfactante llamadas láminas (blanco) que abarcan varios poros, de manera ortogonal a la dirección del flujo del CO_2, tal como indica la flecha roja en la esquina superior izquierda de la imagen, por lo tanto reduciendo la movilidad del CO_2. La visualización de la espuma de CO_2 está compuesta por la combinación de una imagen de microscopía de campo brillante y otra de fluorescencia.

E. F. Médici and A. D. Otero (eds.), *Album of Porous Media*,
https://doi.org/10.1007/978-3-031-23800-0_76

MULTIMODAL VISUALIZATION OF CALCIUM CARBONATE ($CaCO_3$) IN THE PORE SPACE OF A MICROMODEL

Malin Haugen, Bergit Brattekås, and Martin Fernø
Department of Physics and Technology, University of Bergen, Norway

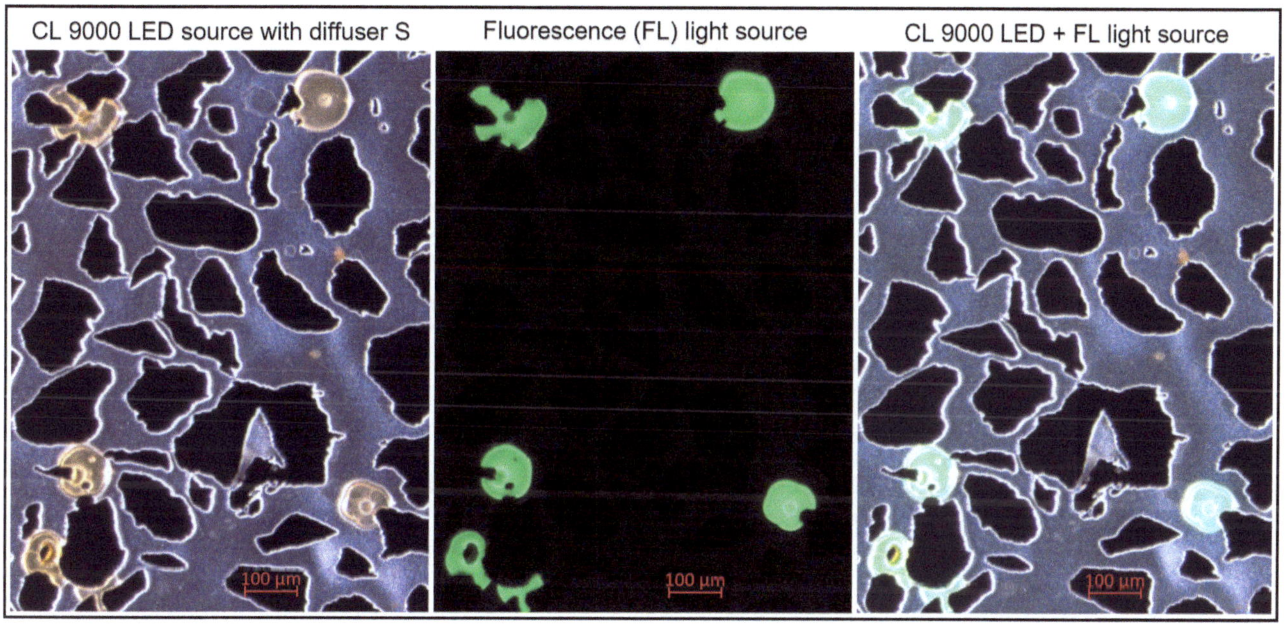

$CaCO_3$ was generated from a microbiological process where the bacteria *Sporosarcina pasteurii* enabled hydrolysis of urea ($CO(NH_2)_2$). A cementation solution with 7.35 wt% calcium chloride and 3.00 wt% urea in distilled water, with 0.10 wt% fluorescence (FL) tracer (Fluorescein sodium salt), was injected into the pore space to precipitate $CaCO_3$ minerals. The images are generated after the precipitation process, when the cementation solution (containing FL tracer) is displaced by deionized water (without FL tracer). The minerals appear orange in the cold-light (CL 9000 LED) source, and green in the FL light source (filter set 38 HR GFP). The presence of FL tracer in the cementation solution enabled molecules to be included in the mineralization, and thereby generating fluorescent minerals. Internal mineral structure and $CaCO_3$ quantification is possible through image analysis. This approach is of great value for pore-scale reactive flow studies, and an important tool to determine CO_2 storage security in carbonate formations.

Se generó $CaCO_3$ mediante un proceso microbiológico en el que la bacteria *Sporosarcina pasteurii* permitió hidrólisis de urea ($CO(NH_2)_2$). Una solución de cementación con cloruro de calcio al 7,35 % en peso, urea en agua destilada al 3,00 % en peso y un trazador (sal de fluoresceína sódica) fluorescente (FL) al 0,10 % en peso, fue inyectada en el espacio poroso para precipitar minerales de $CaCO_3$. Las imágenes se generaron después del proceso de precipitación, cuando la solución de cementación (que contenía el trazador FL) se desplazó con agua desionizada (sin trazador FL). Los minerales se observan de color naranja bajo la fuente de luz fría (CL 9000 LED), y verde bajo la fuente de luz FL (filtro a 38 HR GFP). La presencia del trazador FL en la solución de cementación permitió que las moléculas estuvieran incluidas en la mineralización, por lo que generó minerales fluorescentes. La estructura mineral interna y la cuantificación de $CaCO_3$ son posibles mediante el análisis de imágenes. Este enfoque es de gran valor para los estudios de flujo reactivo a escala de poro y una herramienta importante para determinar la seguridad del almacenamiento de CO_2 en formaciones de carbonato.

Contact: Martin A. Fernø <Martin.Ferno@uib.no>

E. F. Médici and A. D. Otero (eds.), *Album of Porous Media*,
https://doi.org/10.1007/978-3-031-23800-0_77

MICROFLUIDIC EXPERIMENTS ON CONSTRAINED OIL REMOBILIZATION INDUCED BY WATER TRANSPORT IN THE OIL PHASE

Lifei Yan[1], *Mohammad H. Golestan*[2], *Wenyu Zhou*[2], *S. Majid Hassanizadeh*[1], *Carl F. Berg*[2], *Amir Raoof*[1], *and Eirik Flekkøy*[3]

[1] Utrecht University
[2] Norwegian University of Science and Technology
[3] PoreLab, the Njord Centre, Department of Physics, University of Oslo, Norway

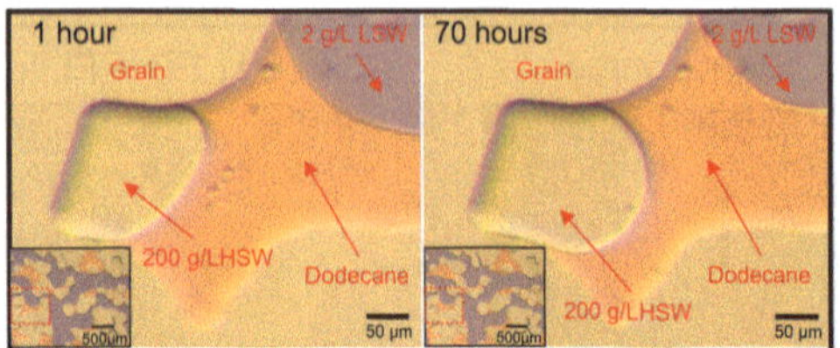

Figure 1: Images of an isolated LSW/dodecane/HSW system after one hour and 70 hours.

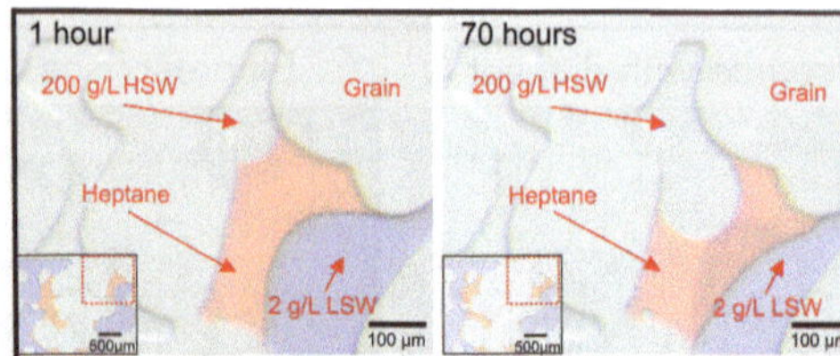

Figure 2: Images of an isolated LSW/heptane/HSW system after one hour and 70 hours.

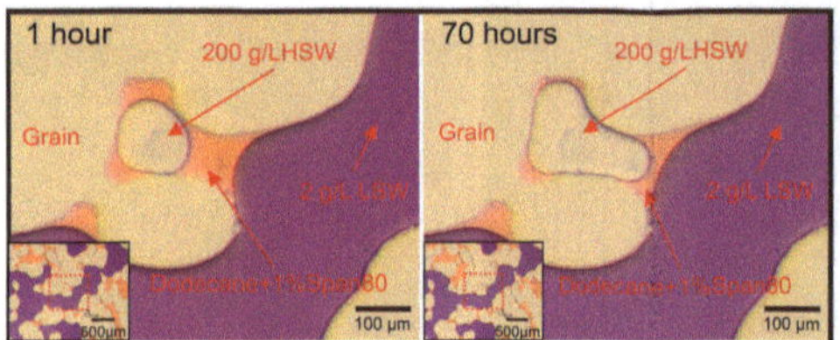

Figure 3: Images of an isolated LSW/dodecane+1% Span80 /HSW system after one hour and 70 hours.

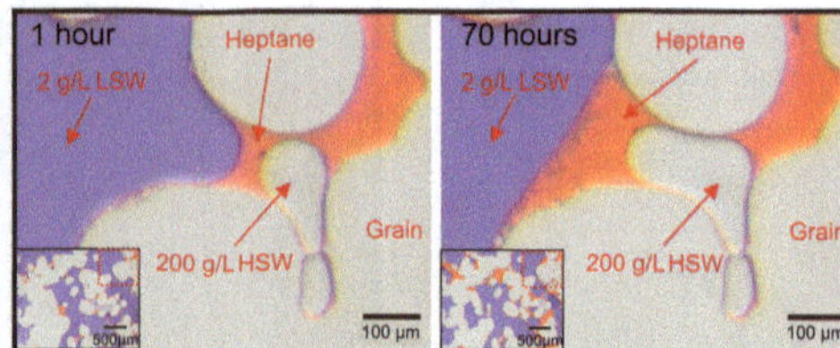

Figure 4: Images of an isolated LSW/heptane+1% Span80 /HSW system after one hour and 70 hours.

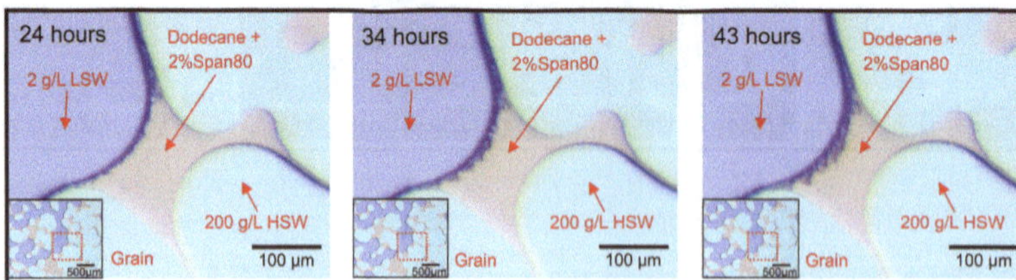

Figure 5: Images of dynamic spontaneous emulsification at LSW/alkane and HSW/alkane interfaces after 24 hours, 34 hours, and 43 hours.

The image illustrates the direct observation of trapped high-salinity water expansion at a microscopic scale and its contribution to the oil movements after 70 hours of continuous monitoring. The pure n-heptane phase has a higher water permeation than pure n-dodecane. With the addition of surfactant, water transport in both alkane phases is enhanced by water-in-oil emulsions to some extent, resulting in extra water flux and a non-linear increase in high-salinity water area. The dynamic process of spontaneous emulsification is captured as well. We were able to quantitatively estimate water transport in two alkane phases due to the solubility difference of water at water/oil interfaces and pore-scale emulsions, which allows us to get insight into its influence on the remobilization of constrained oil. The experiments were conducted with hydrophobic glass microfluidic devices, mimicking the natural low-salinity waterflooding process in oil reservoirs at an oil-wet condition[1].

La imagen ilustra la observación directa a escala microscópica de la expansión de agua con alta salinidad atrapada y su contribución a los movimientos del petróleo después de 70 horas de monitoreo continuo. La fase de n-heptano puro tiene una mayor permeabilidad al agua que el n-dodecano puro. Al añadir un surfactante, aumenta hasta cierto punto el transporte del agua en ambas fases alcanas mediante emulsiones de agua en petróleo, resultando en un flujo de agua extra y un aumento no lineal de la zona de agua con alta salinidad. También se capta el proceso dinámico de emulsificación espontánea. Pudimos estimar cuantitativamente el transporte en dos fases alcanas debido a la diferencia de solubilidad del agua en las interfaces entre el agua y el petróleo y en las emulsiones a escala de poro, lo cual nos permite entender su influencia en la removilización del petróleo atrapado. Los experimentos se realizaron con dispositivos microfluídicos hidrofóbicos de vidrio, imitando el proceso natural de inundación por agua con baja salinidad en yacimientos petrolíferos mojados con petróleo[1].

1. Yan L. *et al.* (2020). *Fuel*, 274:117798.

Contact: Lifei Yan <l.yan@uu.nl>

E. F. Médici and A. D. Otero (eds.), *Album of Porous Media*,
https://doi.org/10.1007/978-3-031-23800-0_78

OIL BLOBS SPIN IN A PORE FULL OF FERROFLUID IN A ROTATING MAGNETIC FIELD

Ningyu Wang[1], *Yifei Liu*[2], *Luming Cha*[2], *Matthew Balhoff*[1], *and Masa Prodanovic*[1]

[1] University of Texas at Austin
[2] China University of Petroleum

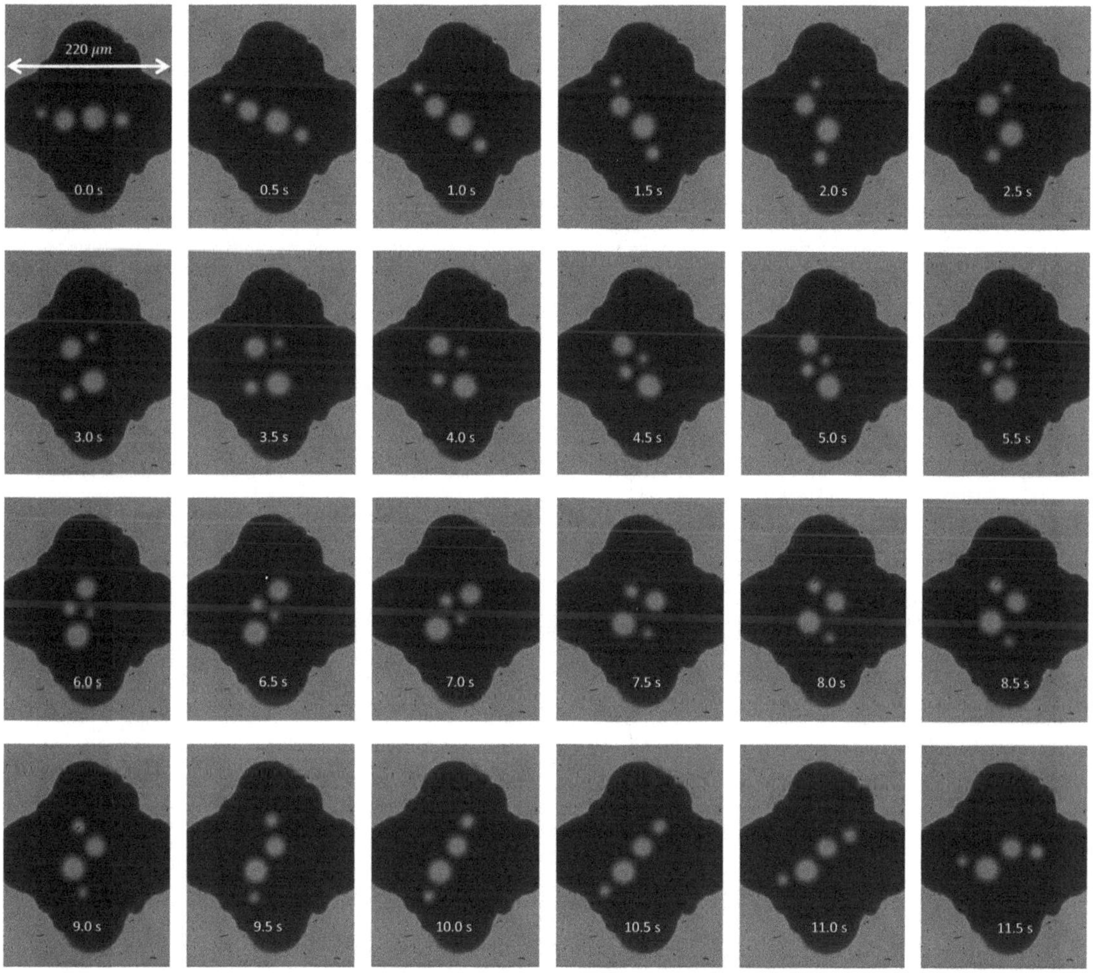

Four oil blobs (bright) are spinning in the EMG 700 water-based ferrofluid (dark) in a pore in a micromodel driven by a rotating magnetic field. It shows the immiscible fluid-ferrofluid two-phase flow in a pore under a rotating magnetic field. Ferrofluid is a potential flooding fluid for enhanced oil recovery and a potential tracer for reservoir characterization. Micromodels are used to visualize and study the flow in porous media at both pore-scale and core-scale. A pair of N52 magnets induce a 2.6 mT 8 rpm rotating magnetic field that provides the driving force. The pore length is 220 microns and the magnetic field is approximately uniform in the small pore. The experimental system is fully described in [1]. These images are selected from a stack of 3240 images taken by an Olympus optical microscope.

1. Wang N. *et al.* (2021). *SPE Journal*, 27(01):753.

Cuatro manchas de petróleo dan vueltas en el ferrofluido acuoso EMG 700 (oscuro) guiado por un campo magnético rotatorio en un micromodelo de un poro. La imagen muestra el flujo inmiscible bifásico de ferrofluido y fluido en un poro bajo un campo magnético rotatorio. El ferrofluido es un potencial fluido de inundación para la recuperación mejorada del petróleo y un posible trazador para la caracterización de yacimientos. Los micromodelos se utilizan para visualizar y estudiar el flujo en medios porosos, tanto a escala de poro como a escala de una muestra. Un par de imanes N52 inducen un campo magnético rotatorio de 2,6 mT a 8 rpm que proporciona la fuerza motriz. El largo del poro es de 220 micrones y el campo magnético es aproximadamente uniforme en el pequeño poro. El sistema experimental se describe en [1]. Las imágenes fueron seleccionadas de un grupo de 3240 imágenes tomadas por un microscopio óptico Olympus.

Contact: Ningyu Wang <wny001@gmail.com>

E. F. Médici and A. D. Otero (eds.), *Album of Porous Media*,
https://doi.org/10.1007/978-3-031-23800-0_79

BIOFILM ARCHITECTURE IN OPAQUE POROUS MEDIUM

Dorthe Wildenschild[1], *Gabriel Iltis*[1], *and Sassan Ostvar*[2]

[1] School of Chemical, Biological and Environmental Engineering, Oregon State University, USA
[2] Irving Medical Center, Columbia University in the City of New York

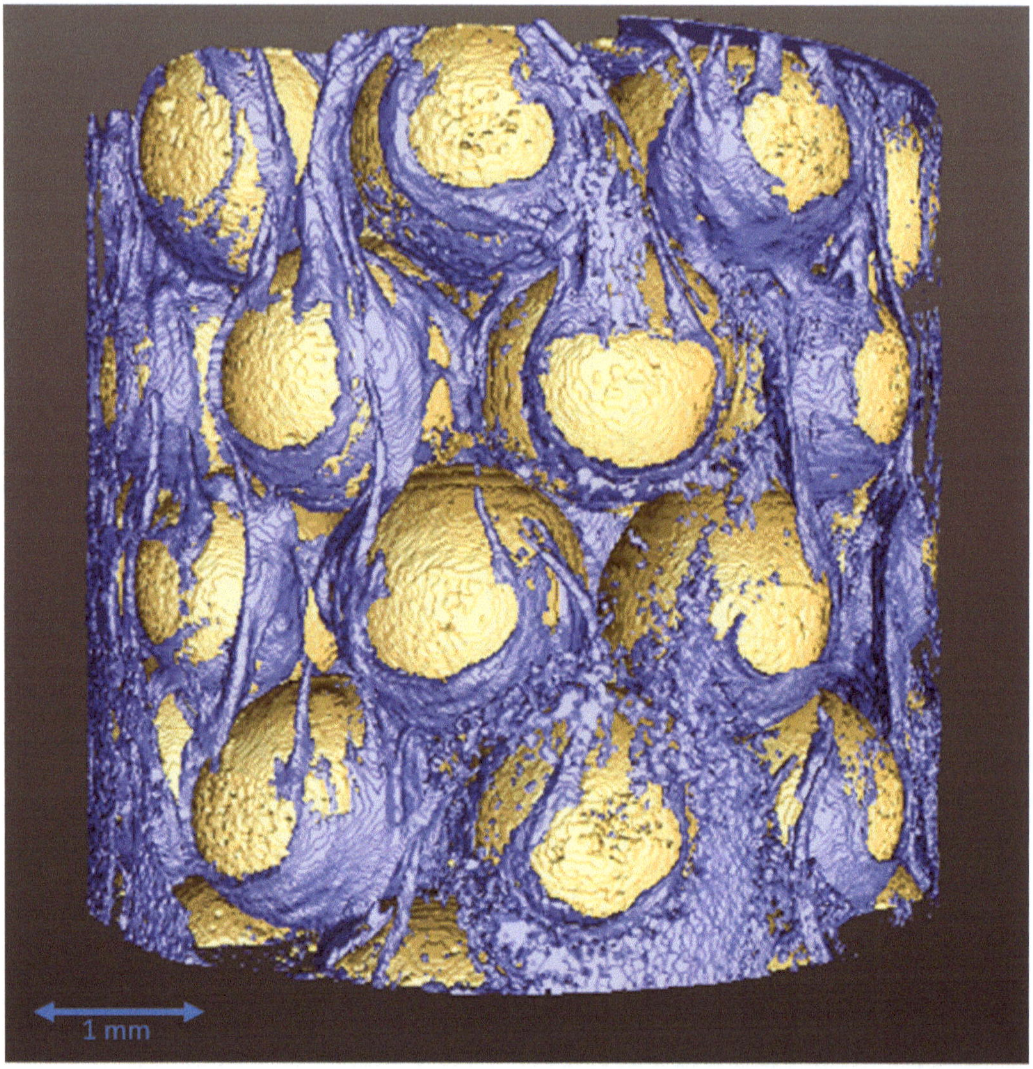

SShewanella oneidensis MR-1 biofilm (purple) grown in glass bead-pack (yellow), and imaged using X-ray microtomography (microCT) at GSECARS/beam-line 13-BMD at the Advanced Photon Source, Argonne National Laboratory. The image voxel size is 10.5 microns, and the field of view represents an approximately 5 mm tall section of the bead-pack. Flow is in the vertical direction (upwards). MicroCT imaging allows us to visualize biofilm architecture and structure in three dimensions as it exists inside an opaque porous medium, and also allows for quantification of a variety of metrics such as biofilm surface area, degree of clogging via porosity changes, and calculation of tortuosity changes due to biofilm growth[1].

1. Ostvar S. *et al.* (2018). *Advances in Water Resources*, 117:1.

Contact: Dorthe Wildenschild <dorthe@engr.orst.edu>

Biopelícula de *Shewanella oneidensis* MR-1 (violeta) cultivada en un empaquetamiento de esferas de vidrio (amarillo), y visualizada mediante microtomografía (microTC) de rayos X con el haz GSECARS de 13-BMD del Advanced Photon Source del Argonne National Laboratory. El tamaño de los vóxeles de la imagen es de 10,5 micrones y el campo de visión representa una sección de aproximadamente 5 mm del empaquetamiento de esferas. El flujo ocurre en dirección vertical (hacia arriba). La microTC nos permite visualizar la arquitectura y la estructura de la biopelicula en tres dimensiones a medida que sale de la parte interna de un medio poroso opaco y también permite la cuantificación de una variedad de métricas como el área de la superficie de la biopelícula, el grado de obstrucción mediante cambios en porosidad y el cálculo de los cambios de la tortuosidad debidos al crecimiento de la biopelícula[1].

E. F. Médici and A. D. Otero (eds.), *Album of Porous Media*,
https://doi.org/10.1007/978-3-031-23800-0_80

VI. TRANSPORT IN POROUS MEDIA

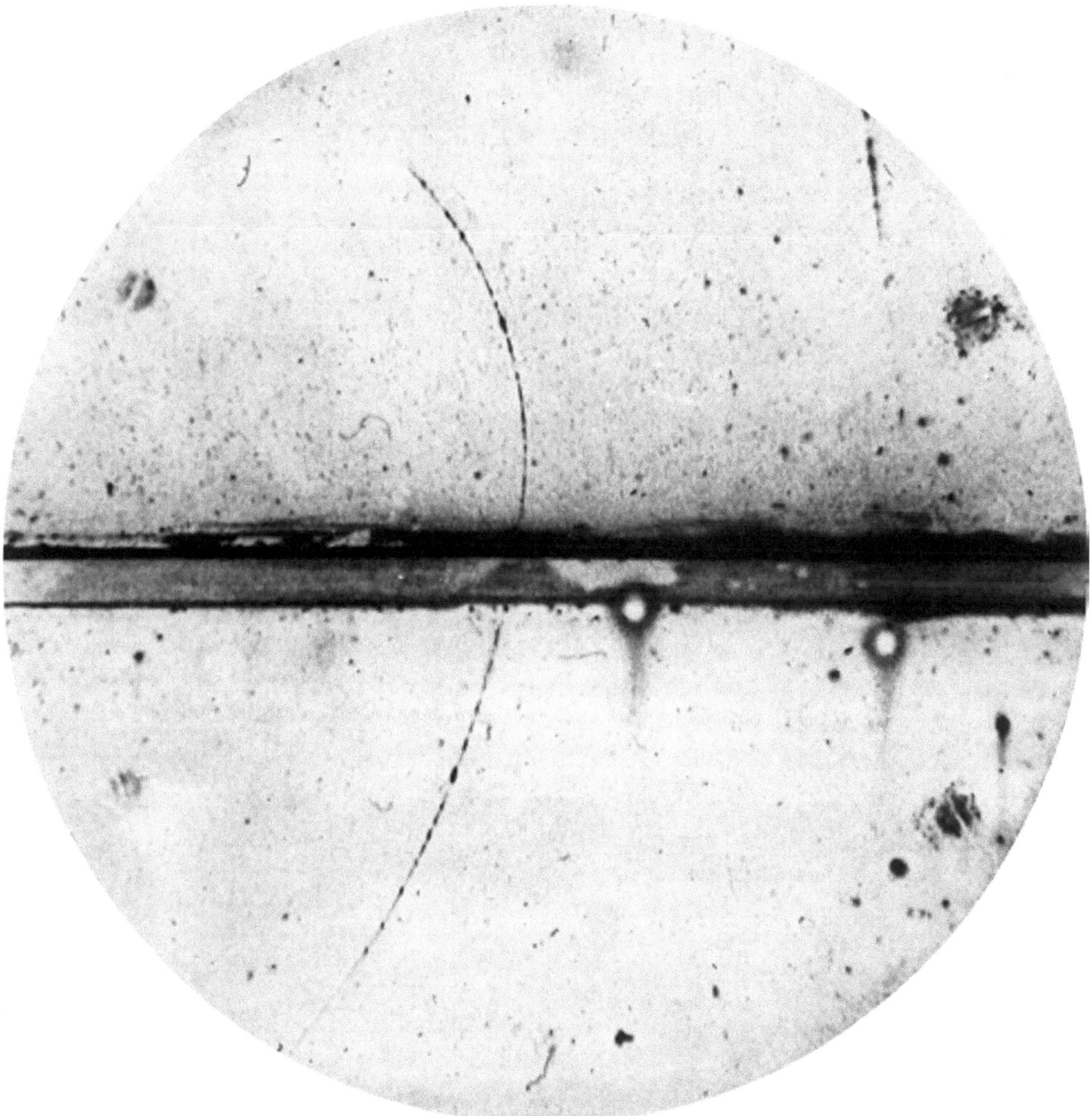

First image of a positron, a precursor discovery for the develpment of the positron emission tomography (PET).

Primera imagen de un positrón, descubrimiento precursor al desarrollo de la tomografía por emisión de positrones (PET, por sus siglas en inglés).

PORES IN WATER CONDUCTING CELLS OF PLANTS

Steven Jansen
Ulm University

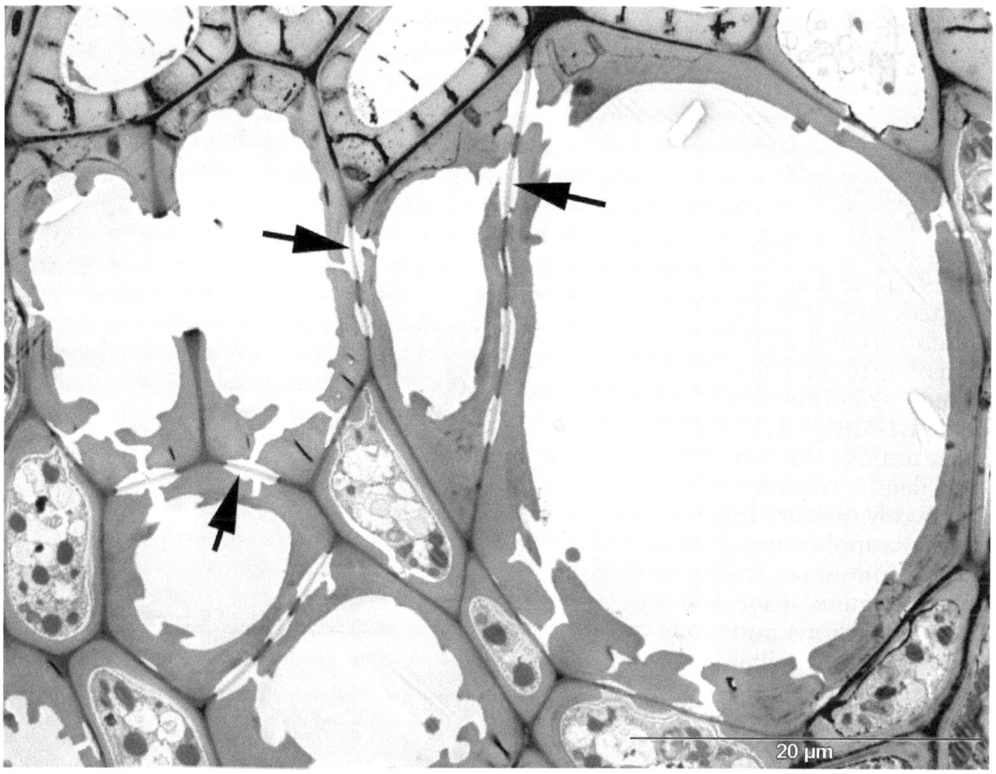

This figure shows a transverse section of wood tissue from *Hibiscus syriacus* observed with a transmission electron microscope. One of the secrets behind water transport under negative pressure, which plants are able to do effortlessly on a daily basis, relies on our understanding of gas-water-solid interactions in porous cell walls between water conducting cells (black arrows)[1, 2]. These mesoporous media have pore constrictions up to 20 nm, a porosity of ca. 80 %, and account for ca. 50 % of the total hydraulic resistivity between the roots and the leaves of a plant[3]. A challenge ahead represents the integration of porous media into evaporation-driven negative pressure devices, also called synthetic trees[4]. The same pores also play a crucial role in wood preservation and decomposition. The fresh wood sample was treated with a fixation solution, dehydrated, embedded in resin, and cut with an ultramicrotome. Observation was done with a transmission electron microscope (JEOL 1400). A total of 6 water conducing cells can be seen, with the large white areas representing the conduit lumina, which would be filled with sap under normal field conditions.

1. Jansen S. and Schenk H.J. (2015). *American J. Botany*, 102(10):1561.
2. Tyree M.T. and Zimmermann M.H. (2013). *Xylem structure and the ascent of sap*. Springer Science & Business Media.
3. Kaack L. *et al.* (2021). *New Phytol*, 230(5):1829.
4. Shi W. *et al.* (2020). *Sci Rep*, 10(1).

La figura muestra una sección transversal de un tejido de madera de *Hibiscus syriacus* observado mediante un microscopio electrónico de transmisión. Uno de los secretos tras el transporte de agua bajo presión negativa, que las plantas pueden hacer a diario sin dificultad, recae en nuestra comprensión de las interacciones entre el gas, el agua y el sólido en las paredes porosas de las celdas, entre celdas que conducen agua (flechas negras)[1, 2]. Estos medios mesoporosos tienen constricciones de poro de hasta 20 nm, una porosidad de aproximadamente el 80 %, y constituyen aproximadamente el 50 % de la resistividad hidráulica total entre las raíces y las hojas de la planta[3]. La integración de los medios porosos en los dispositivos de presión negativa inducidos por la evaporación, también llamados árboles sintéticos, representa un reto a futuro[4]. Estos mismos poros también tienen un papel crucial en la conservación y descomposición de la madera.

La muestra de madera fresca fue tratada con una solución fijadora, deshidratada, embebida en resina y cortada con un ultramicrótomo. La observación se realizó con un microscopio electrónico de transmisión (JEOL 1400). 6 celdas que conducen agua fueron observadas con amplias zonas blancas que representan los lúmenes de los conductos, que se llenarían de savia en condiciones normales.

Contact: Steven Jansen <steven.jansen@uni-ulm.de>

E. F. Médici and A. D. Otero (eds.), *Album of Porous Media*,
https://doi.org/10.1007/978-3-031-23800-0_81

IN SITU VISUALIZATION OF MULTIPHASE FLOW APPLYING PET-MR

Bergit Brattekås and Geir Ersland
Department of Physics and Technology, University of Bergen, Norway

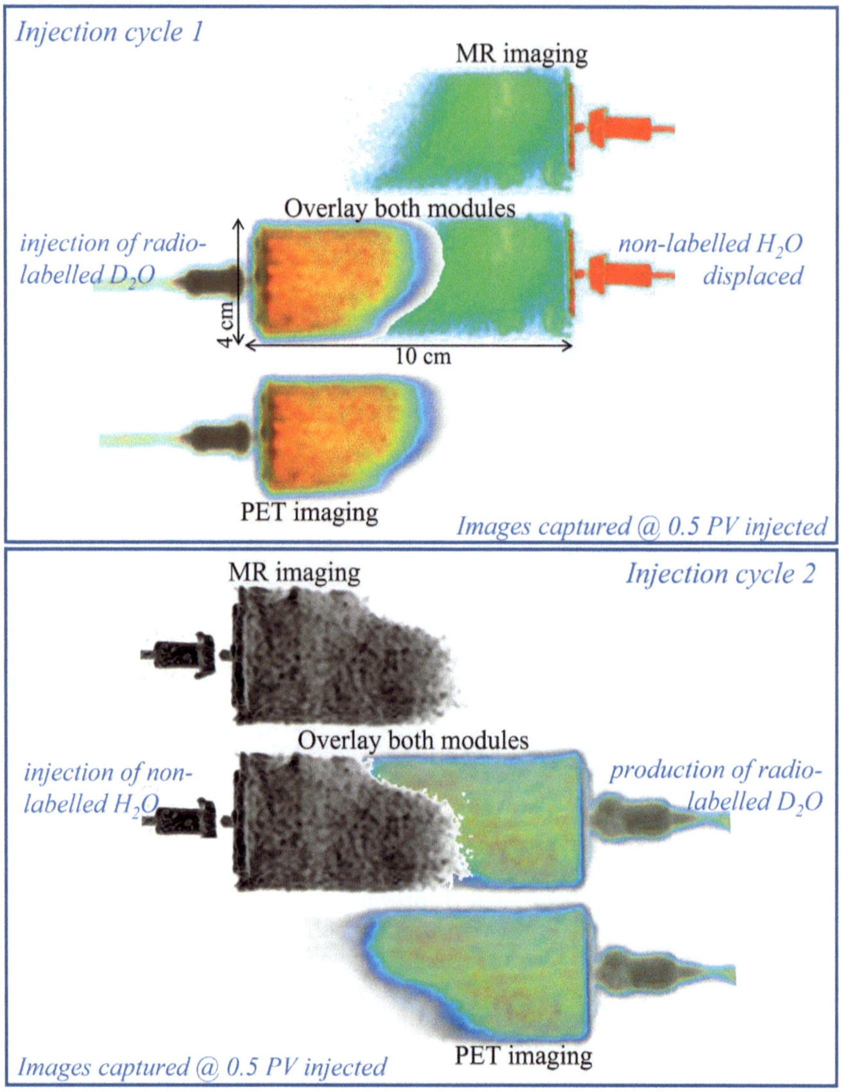

The recent combination of positron emission tomography (PET) with magnetic resonance (MR) imaging has high potential impact in porous media research: by enabling simultaneous and explicit quantification of up to three distinct fluids in the same porous system. Multimodal PET-MR quantify dynamic fluid saturations without the need for fluid density contrasts, and is therefore highly relevant in several porous media applications, including, but not limited to, porous media dispersion and adsorption, wettability alterations and carbon geo-sequestration. Miscible displacements imaged with simultaneous PET-MR are illustrated in this figure. In injection cycle 1, D_2O based 3.5 wt% NaCl brine labelled by positron-emitting radionuclide fluorodeoxyglucose 18F-FDG displaces H_2O based brine: rendering the displacing phase visible by PET and the displaced phase visible by MR imaging. In injection cycle 2 non-labelled H_2O brine (visible by MR) displaces the 18F-FDG labelled D_2O (visible by PET). Detailed procedures and analysis may be found in [1]; demonstrating the first application of PET-MR imaging in porous media research.

La reciente combinación de tomografía por emisión de positrones (TEP) con resonancia magnética (RM) tiene un alto potencial de impacto en la investigación de los medios porosos, al permitir la cuantificación simultánea y explícita de hasta tres fluidos distintos en el mismo sistema poroso. La combinación multimodal de TEP y RM cuantifica las saturaciones dinámicas de fluidos sin la necesidad de contrastes en la densidad de los fluidos, y por lo tanto esto es muy relevante para varias aplicaciones en medios porosos, incluyendo pero no limitado a dispersión y adsorción en medios porosos, alteraciones en la mojabilidad y geosecuestro de carbono. Esta figura muestra desplazamientos miscibles tomografiados con TEP/RM simultaneos. En el primer ciclo de inyección, D_2O basado en una solucion salina al 3.5% NaCl y etiquetado con fluorodeoxyglucosa 18F-FDG, radionúclido emisor de positrones, desplaza una solución salina de H_2O: representación de la fase desplazante visible con TEP y fase desplazada visible con RM. En el segundo ciclo de inyección, una solución salina de H_2O no etiquetada (visible con RM) desplaza el D_2O etiquetado con 18F-FDG (visible con TEP). Los procedimientos y análisis detallados pueden encontrarse en [1]; estos demuestran la primera aplicación de tomografia TEP/RM en la investigación de medios porosos.

1. Brattekås B. *et al.* (2020). *Advances in Water Resources*, 142:103641.

Contact: Bergit Brattekås <bergit.brattekas@uib.no>

E. F. Médici and A. D. Otero (eds.), *Album of Porous Media*,
https://doi.org/10.1007/978-3-031-23800-0_82

FLUID FLOW VISUALIZATION BY PET-CT

Bergit Brattekås[1], *Martin Fernø*[1], *and Heidi Espedal*[2]

[1] Department of Physics and Technology, University of Bergen, Norway
[2] Department of Clinical Medicine, University of Bergen, Norway

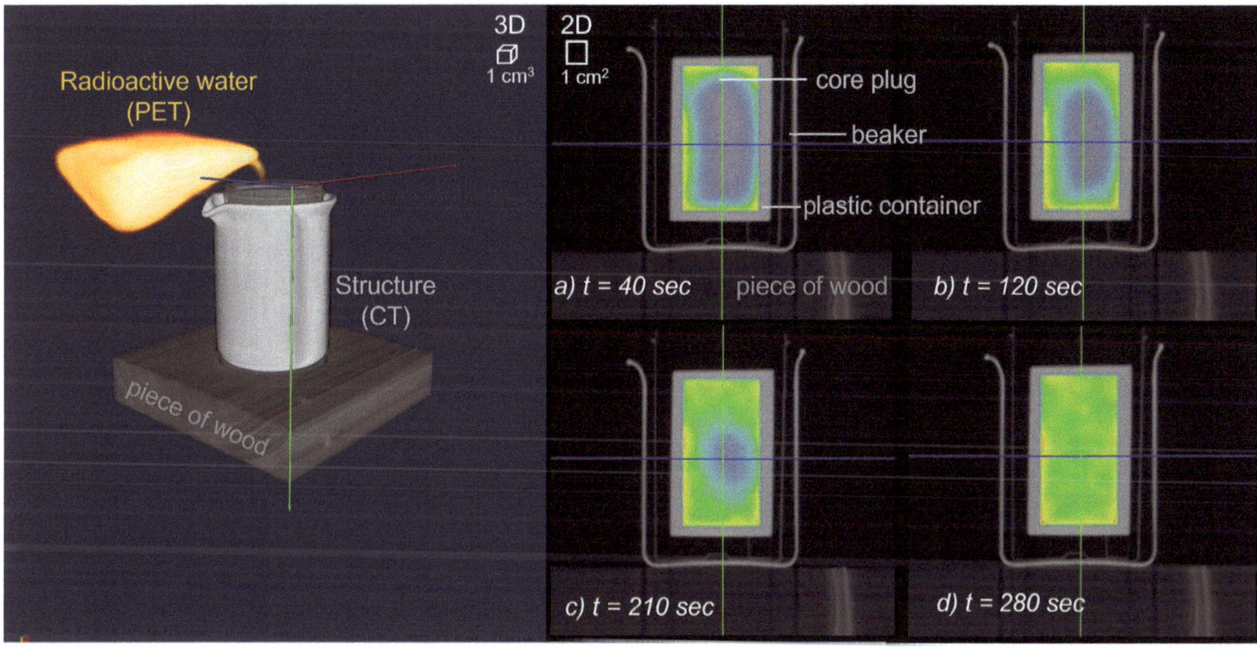

In situ imaging by combined positron emission tomography (PET) – computed tomography (CT) provides unique insight into porous media flow: by combining information about structure (visible by CT) with direct quantification of fluid flow paths (visible by PET). The 3D image (left) illustrates the concept: the greyscale CT image of a beaker resting on a piece of wood is overlaid by a colored PET image, showing another beaker filled with radioactive water. Upon entry into the container, radioactive water spontaneously imbibes into an air-filled chalk core plug under all faces open (AFO) boundary conditions (images a - d). PET imaging provides access to saturation development on voxel-by-voxel scale, at high temporal (seconds) and spatial (submillimeter scale) resolution. PET-CT is an emerging and versatile imaging technique to capture porous media flow, exemplified in the reference list [1, 2, 3, 4].

Las imágenes *in situ* mediante combinación de tomografía por emisión de positrones (TEP) y tomografía computarizada (TC) proporcionan una percepción única del flujo de los medios porosos: al combinar información sobre la estructura (visible mediante TC) con cuantificación directa de las trayectorias del flujo del fluido (visible con TEP). La visualización 3D (izquierda) ilustra el concepto: la imagen TC en gris de un vaso sobre un trozo de madera está sobrepuesta por una imagen TEP a color que muestra otro vaso lleno con agua radiactiva. Al entrar en el envase, el agua radiactiva embebe espontáneamente un núcleo de caliza lleno con aire bajo condiciones de contorno de tipo abierto en todas las caras (imágenes a - d). La imagen TEP permite evaluar la evolución de la saturación a una escala vóxel por vóxel, con una resolución temporal (segundos) y espacial (escala submilimétrica) altas. La TEP-TC es una técnica de imagen emergente y versátil para captar el flujo en medios porosos, tal y como se ejemplifica en la lista de referencias [1, 2, 3, 4].

1. Fernø M.A. *et al.* (2015). *Water Resour. Res.*, 51(9):7811.
2. Degueldre C. *et al.* (1996). *Earth and Planetary Sci. Letters*, 140(1-4):213.
3. Kulenkampff J. *et al.* (2008). *Physics and Chemistry of the Earth, Parts A/B/C*, 33(14-16):937.
4. Zahasky C. *et al.* (2019). *Advances in Water Resources*, 127:39.

Contact: Bergit Brattekås <bergit.brattekas@uib.no>

E. F. Médici and A. D. Otero (eds.), *Album of Porous Media*,
https://doi.org/10.1007/978-3-031-23800-0_83

ADVECTION AND DIFFUSION IN NANOPOROUS SHALE

Nirjhor Chakraborty and Zuleima Karpyn
Penn State University

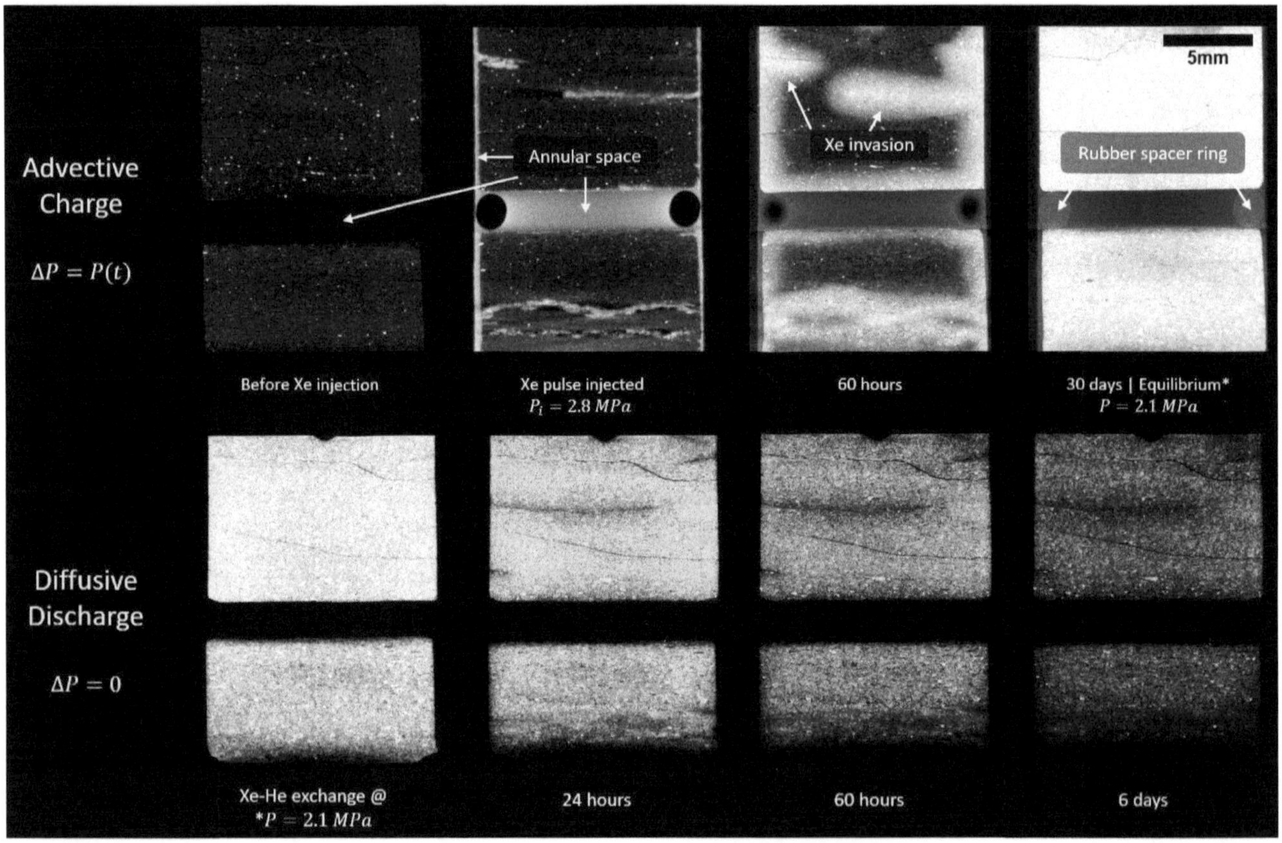

Pressure gradient-driven advective flow, commonly described by the Darcy equation, is the dominant mechanism driving gas flow in reservoir rocks such as gas-bearing sandstones. This is because the coefficient of advective flow-permeability, is high while the coefficient of concentration gradient driven diffusion is relatively small and negligible in such rocks with large pores. In nanoporous shales, however, permeability is really low. How significant then is the role of diffusion during gas transport in shales? Using X-ray CT imaging with radio-opaque xenon gas we monitor the advection-driven transport and uptake of gas in two ultra-low permeability (50 nanodarcy) 12 mm diameter shale plugs[1, 2]. We then expose the system to Helium at a constant pressure, the resulting concentration gradient causing xenon to be driven out of the sample via diffusion[3]. The duration of equilibrium, approximately a week in each case indicates clearly that diffusion cannot be ignored as a key transport mechanism in tight nanoporous shale.

1. Zhang M. *et al.* (2021). *Fuel*, 286:119300.
2. Chakraborty N. *et al.* (2020). *J. Natural Gas Sci. and Eng.*, 76:103120.
3. Chakraborty N. (2019). URL `https://etda.libraries.psu.edu/catalog/16253nic5064`.

El flujo advectivo guiado por el gradiente de presiones, comúnmente descrito por la ecuación de Darcy, es el mecanismo predominante que empuja el flujo de gas en las rocas de yacimientos como los de areniscas que contienen gas. Esto es porque el coeficiente del flujo advectivo (permeabilidad) es alto mientras que el coeficiente de difusión debido al gradiente de concentraciones es relativamente bajo y despreciable en las rocas con poros grandes. Sin embargo, en las lutitas nanoporosas la permeabilidad es realmente baja. Entonces, ¿cuán significativo es el papel de la difusión durante el desplazamiento del gas en las lutitas? Mediante el uso de imágenes TC de rayos X con gas xenón radioopaco, observamos el transporte advectivo y la absorción del gas en dos muestras cilíndricas de lutita con permeabilidad ultrabaja (50 nanodarcy) de 12 mm de diámetro[1, 2]. Después expusimos el sistema a helio a presión constante, y el gradiente de concentración resultante causó que el xenón saliera de la muestra por difusión[3]. El periodo de equilibrio, aproximadamente una semana en cada caso, indica claramente que no puede ignorarse la difusión como un mecanismo clave de transporte en lutita nanoporosa compacta.

Contact: Nirjhor Chakraborty <nirjhorpsu@gmail.com>

E. F. Médici and A. D. Otero (eds.), *Album of Porous Media*,
https://doi.org/10.1007/978-3-031-23800-0_84

ADSORPTION ENHANCED GAS UPTAKE IN NANODARCY PERMEABILITY SHALE

Nirjhor Chakraborty and Zuleima Karpyn
Penn State University

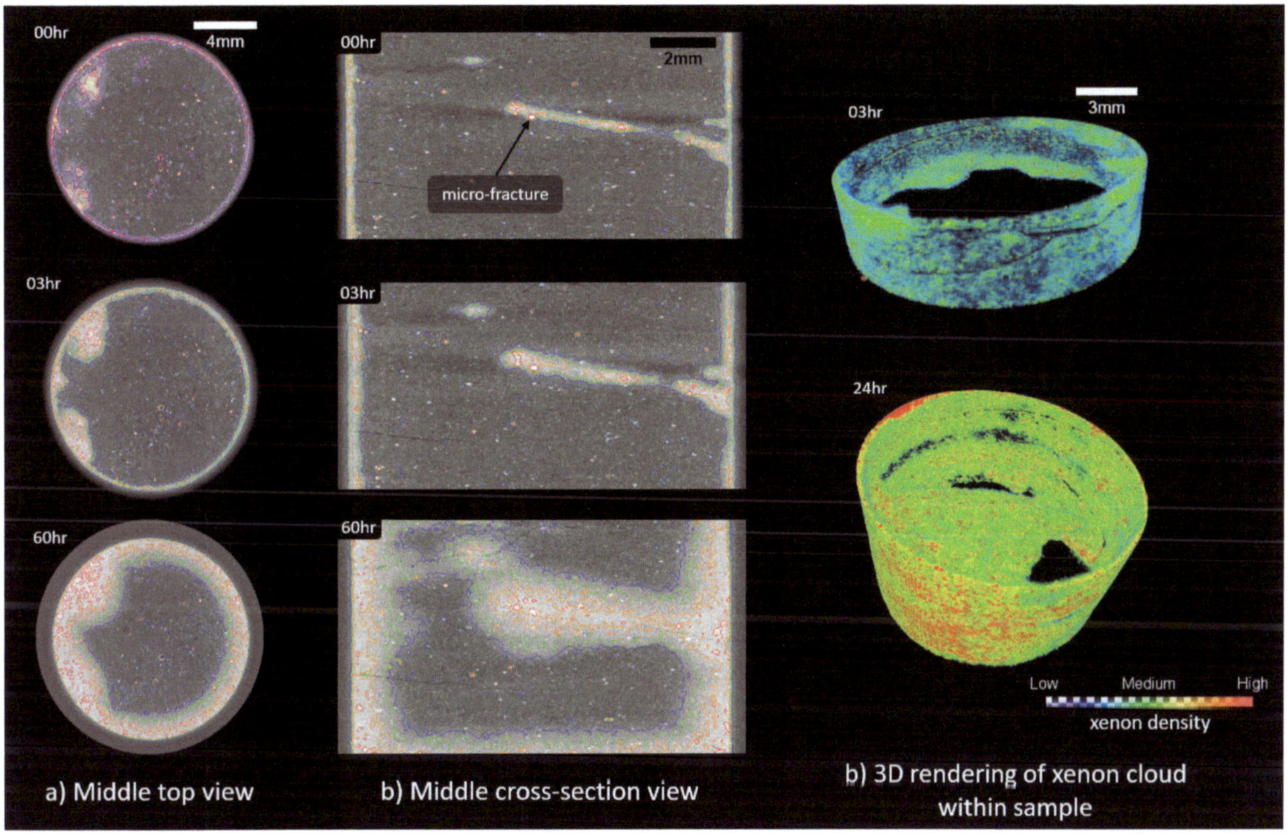

These images show the gas uptake into a cylindrical Marcellus shale rock sample 12 mm in diameter. Periodic 3D X-ray microCT imaging at a voxel resolution of 13.8 micron is used to track the progress of a fixed volume of radio-opaque xenon gas as it proceeds into the sample starting at an initial pressure of 2.8 MPa, dropping to 2.1 MPa by equilibrium (equilibrium state not shown)[1, 2]. Despite a large initial pressure-gradient of 2.8 MPa it takes approximately a week for xenon gas to fill up the sample. This slow pace is fundamentally due to the ultra-low nanodarcy permeability of the sample. Another exacerbating factor is gas adsorption at the fronts of the xenon cloud which further slows down progress. Despite xenon being a noble gas, significant amounts of gas uptake occur in the form of adsorption indicated by red contours in the 2D top- and cross-section- views and by the red-colored high-density regions in the 3D render of the xenon cloud within the sample.

Estas imágenes muestran la entrada de gas en una muestra de roca de lutita Marcellus cilíndrica de 12 mm de diámetro. Se utilizaron imágenes periódicas con microTC de rayos X en 3D con una resolución de vóxel de 13,8 micrones para seguir el progreso de un volumen fijo de gas xenón radio-opaco a medida que avanza en la muestra, empezando con una presión inicial de 2,8 MPa, bajando a 2,1 MPa por equilibrio (no se muestra el estado en equilibrio)[1, 2]. A pesar de un gradiente de presión inicial alto de 2,8 MPa, se necesita aproximadamente una semana para que el gas xenón llene la muestra. Este ritmo lento se debe fundamentalmente a la permeabilidad nanodarcy ultrabaja de la muestra. Otro factor exacerbante es la adsorción de gas en el frente de la nube de xenón, que ralentiza aún más el progreso. Pese a que el xenón es un gas noble, un monto significativo de gas es retenido debido a la adsorción que se indica por los contornos rojos de las vistas superiores (y transversales en 2D) y por las regiones de alta densidad en rojo de la imagen en 3D de la nube de xenón dentro de la muestra.

1. Zhang M. *et al.* (2021). *Fuel*, 286:119300.
2. Chakraborty N. *et al.* (2020). *J. Natural Gas Sci. and Eng.*, 76:103120.

Contact: Nirjhor Chakraborty <nirjhorpsu@gmail.com>

E. F. Médici and A. D. Otero (eds.), *Album of Porous Media*,
https://doi.org/10.1007/978-3-031-23800-0_85

TIME-LAPSE 3D IMAGING BY POSITRON EMISSION TOMOGRAPHY OF COPPER MOBILIZED IN SOIL BY THE HERBICIDE MCPA

Michael Kersten[1] *and Johannes Kulenkampff*[2]

[1] Geosciences Dept., Gutenberg-University
[2] Helmholtz Zentrum Dresden-Rossendorf (HZDR), Institute of Resource Ecology, Research Site Leipzig

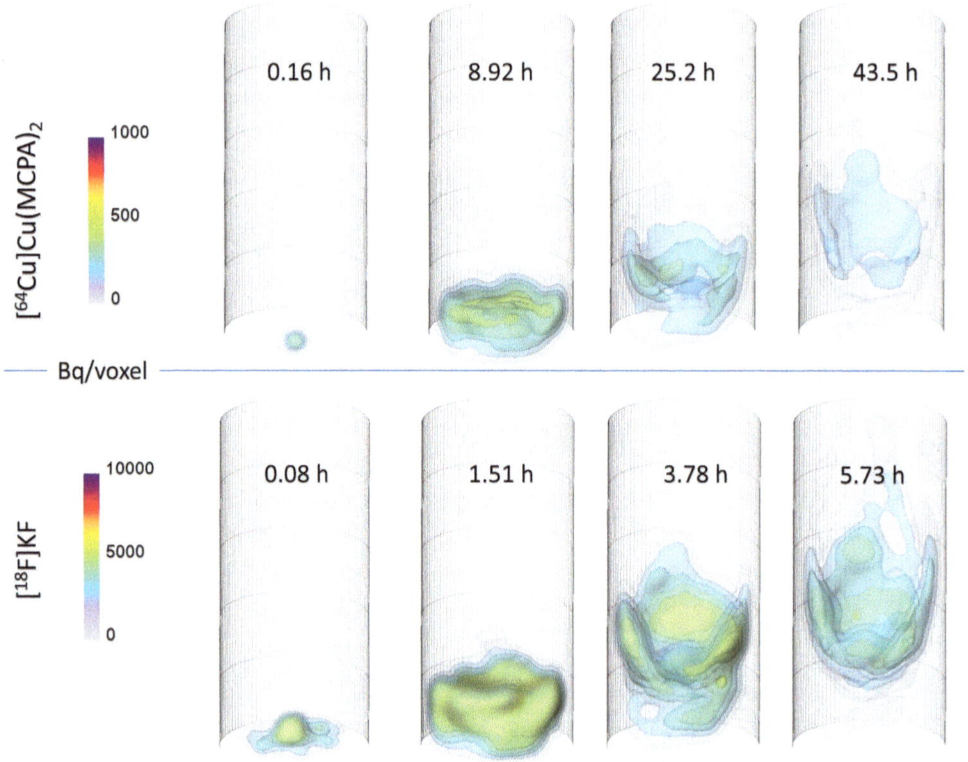

The MCPA herbicide mobilizes trace metals in soil by outcompeting adsorption by soil minerals[1]. Trace metal tracers are too immobile for the relatively short observation times available with positron emission tomography (PET). For the first time, however, time-lapse movies of Cu migration in columns filled with artificial soil (68 % sand, 26 % silt, 5 % illite clay, 1 % goethite) were recorded. Soil solution was supplied from bottom up at a flow rate of 100 μL/min, low enough to allow for equilibration while maintaining a residence time in the order of tracer half-life (12.7 h). Up to 5 mL or 250 MBq of Cu-64 tracer was applied in one pulse run. The figure show volume renderings at different times (voxel size = 1 mm^3, column high = 10 cm). The radiolabeled Cu-MCPA complex breakthrough occurred just about 10-fold slower than the inert solute tracer KF(F-18) used in control columns shown in the lower row of the figure. Imaging movies are accessible at [2].

1. Kulenkampff J. *et al.* (2018). *Sci. Rep.*, 8(7091). Open access CC BY 4.0.
2. URL `https://zenodo.org/record/852968`.

El herbicida MCPA moviliza los trazadores metálicos del suelo al adelantarse a la adsorción de los minerales[1]. Los rastros de los trazadores metálicos son demasiado inmóviles para los tiempos de observación relativamente cortos disponibles con las tomografías por emisión de positrones (TEP). No obstante, se realizaron por primera vez filmaciones de la migración de Cu con columnas llenas de suelo artificial (68 % de arena, 26 % de limo, 5 % de arcilla illita, 1 % de goethita). La solución de suelo se administró desde abajo hacia arriba con un caudal de 100 μL/min, lo suficientemente bajo como para permitir el equilibrio mientras se mantiene un tiempo de residencia del orden de la vida media de los trazadores (12,7 h). Se aplicaron hasta 5 mL o 250 MBq de trazador de Cu-64 en un solo pulso. La figura muestra representaciones volumétricas en distintos tiempos (tamaño del vóxel = 1 mm^3, alto de la columna = 10 cm). La compleja penetración de Cu-MCPA radioetiquetado ocurrió 10 veces más lento que el trazador soluto inerte KF(F-18) utilizado en las columnas de control mostradas en la fila inferior de la figura. Las películas de las imágenes se encuentran disponibles en [2].

Contact: Michael Kersten <kersten@uni-mainz.de>

E. F. Médici and A. D. Otero (eds.), *Album of Porous Media*,
https://doi.org/10.1007/978-3-031-23800-0_86

DOES THE INTERFACIAL TRANSITION ZONE INFLUENCE MOISTURE TRANSPORT IN CONCRETE?

Laura Dalton[1], Jacob LaManna[2], Scott Jones[3], and Mohammad Pour-Ghaz[1]

[1] North Carolina State University
[2] Physical Measurement Laboratory, National Institute of Standards and Technology
[3] Engineering Laboratory, National Institute of Standards and Technology

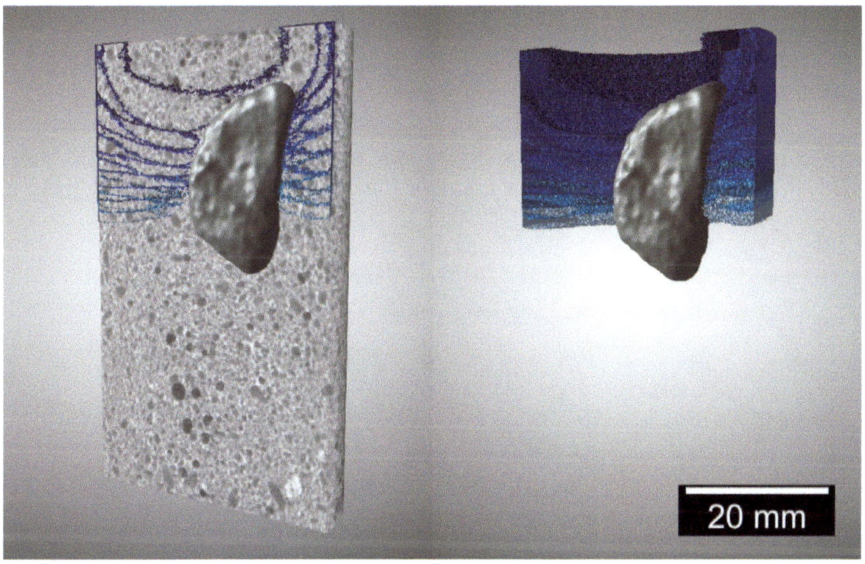

Several postulates exist to potentially explain the formation of the Interfacial Transition Zone (ITZ)[1, 2, 3]. ITZ is a region at the interface between aggregates and the binder in concrete that is believed to be more porous; the interconnected thickness of ITZ is reported to vary between 15 to 100 μm[3]. To better understand the contributions of ITZ during mass transport, a concrete cylinder containing a single coarse aggregate was scanned during water absorption using simultaneous Neutron and X-ray Tomography[4]. Volume subtractions between scans reveal the temporal absorption of water in the porous cement paste relative to the coarse aggregate. During water absorption, small pores (gel and capillary pores) fill with water prior to entrapped or entrained pores. Given the ITZ is an interconnected system of small pores, the region directly surrounding the aggregate is expected to fill with water faster than the bulk cement paste. However, no increase in water absorption is observed in the ITZ at the spatial (60 μm) and temporal (1 hour) resolutions in this study. These results suggest unsaturated moisture flow in the ITZ does not vary relative to the bulk cement paste.

Existen varios postulados para explicar potencialmente la formación de la Zona de Transición Interfacial (ZTI)[1, 2, 3]. La ZTI es una región en la interfaz entre los agregados y el aglutinante en el hormigón que se cree que es más porosa; y se reporta que el grosor interconectado de la ZTI varía entre 15 y 100 μm[3]. Para comprender mejor las contribuciones de la ZTI durante el transporte de masa, se escaneó un cilindro de hormigón que contenía un único agregado grueso durante la absorción de agua utilizando simultáneamente tomografía de rayos X y de neutrones[4]. Las sustracciones volumétricas entre escaneados revela la absorción temporal de agua en la pasta de cemento porosa en relación con el agregado grueso. Durante la absorción del agua, los poros pequeños (poros de gel y capilares) se llenan de agua antes que los poros con aire atrapado o encerrado. Dado que la ZTI es un sistema interconectado de poros pequeños, se espera que la región que rodea directamente el agregado se llene de agua más rápidamente que el volumen de la pasta de cemento. Sin embargo, no se observó un aumento de la absorción de agua en la ZTI en las resoluciones de espacio (60 μm) y tiempo (1 hora) de este estudio. Estos resultados sugieren que el flujo de humedad en la ZTI no saturada no varía en relación con la del volumen de la pasta de cemento.

1. Ollivier J. *et al.* (1995). *Advanced Cement Based Materials*, 2(1):30.
2. Diamond S. and Huang J. (2001). *Cement and Concrete Composites*, 23(2-3):179.
3. Scrivener K.L. and Nemati K.M. (1996). *Cement and Concrete Research*, 26(1):35.
4. LaManna J.M. *et al.* (2017). *Review of Scientific Instruments*, 88(11):113702.

Contact: Laura Dalton <laura.dalton@duke.edu>

E. F. Médici and A. D. Otero (eds.), *Album of Porous Media*,
https://doi.org/10.1007/978-3-031-23800-0_87

GAS TRAPPED IN THE PORE SPACE OF A SANDSTONE

Ying Gao, Ab Coorn, Niels Brussee, Hilbert van der Linde, and Steffen Berg
Shell Global Solutions International B.V.

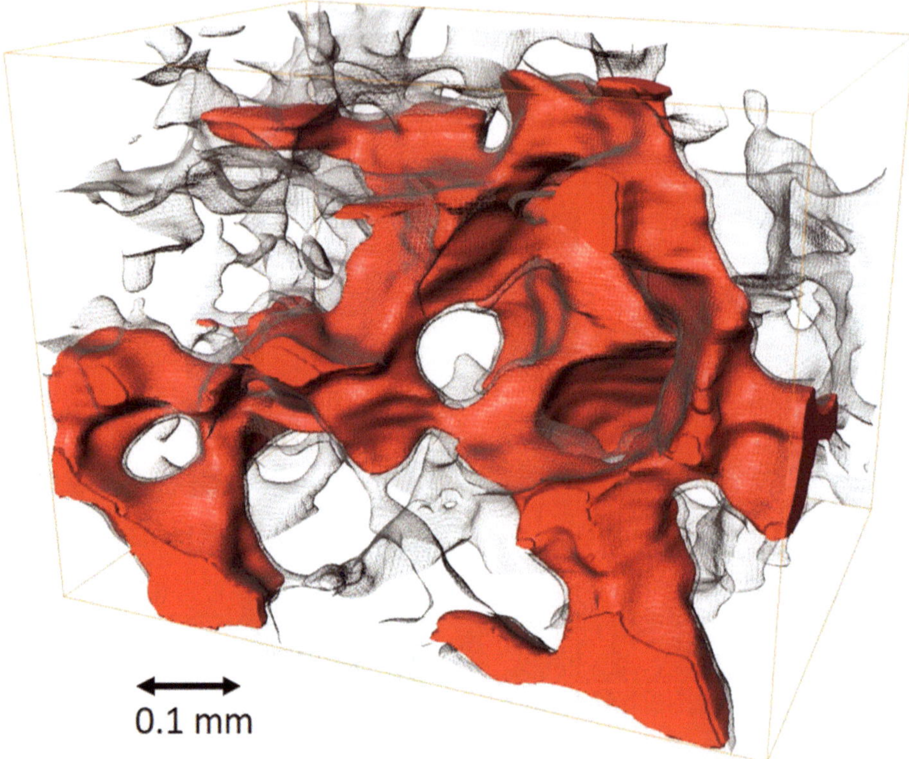

Hydrocarbon gas bubble (propane, red) trapped inside the pore space of Bentheimer sandstone[1, 2, 3]. The experimental study aims at determining the mobility of a gas phase in a gas-liquid system in porous media. It is relevant for many processes in subsurface engineering ranging from the recovery of oil and gas to CO_2 sequestration and underground hydrogen storage. The image shows a small region of interest of a larger rock sample (4 mm diameter and 10 mm length) during an experiment where the pore space of the rock was initially filled with a propane-decane liquid hydrocarbon mixture (gas is shown in red, liquid is transparent). Gas bubbles form when pressure is decreased below the bubble point pressure. The image was obtained by X-ray computed tomography *in situ* of a flow experiment where the cylindrical rock sample of 4 mm diameter and 10 mm length was enclosed in a miniature pressure vessel that allowed to control the fluid composition in the sample and the pressure. *In situ* 3D Imaging was performed using a benchtop micro-CT scanner (Zeiss Versa 520) at spatial resolution of 4 micrometers and a temporal resolution of 1 hour.

1. Berg S. *et al.* (2020). *Petrophysics–The SPWLA J. Formation Evaluation and Reservoir Description*, 61(2):133.
2. Gao Y. *et al.* (2021). *Oil Gas Sci. Tech. – Rev. IFP Energies nouvelles*, 76:43.
3. Xu K. *et al.* (2017). *Phys. Rev. Lett.*, 119(26).

Contact: Steffen Berg <steffen.berg@shell.com>

Burbuja de gas hidrocarburo (propano, en rojo) atrapada dentro del espacio poroso de arenisca de Bentheimer[1, 2, 3]. El objetivo del estudio experimental es determinar la movilidad de la fase gaseosa en un sistema de líquido y gas en un medio poroso. Este estudio es relevante en muchos procesos de ingeniería del subsuelo, que van desde la recuperación de gas y petróleo hasta el secuestro de CO_2 y el almacenamiento subterráneo de hidrógeno. En la imagen se observa la pequeña zona de interés de una muestra de roca grande (4 mm de diámetro y 10 mm de largo) durante un experimento en el que el espacio poroso de la roca se llenó inicialmente con una mezcla de hidrocarburo de propano-decano líquido (el gas se muestra en rojo; el líquido, transparente). Las burbujas de gas se forman cuando disminuye la presión por debajo del punto de presión de burbuja. La imagen fue obtenida mediante una tomografía computarizada de rayos X *in situ* de un experimento de flujo, en el que la muestra de roca cilíndrica de 4 mm de diámetro y 10 mm de largo fue puesta dentro de un recipiente de presión en miniatura, que permitió el control de la composición del fluido en la muestra, así como de la presión. Se tomaron imágenes 3D *in situ* utilizando un escáner de micro-TC de mesa (Zeiss Versa 520) con una resolución espacial de 4 micrómetros y una resolución temporal de 1 hora.

E. F. Médici and A. D. Otero (eds.), *Album of Porous Media*,
https://doi.org/10.1007/978-3-031-23800-0_88

SUBMICRON INSIGHT INTO THE SNAP-OFF PHENOMENON

Arjen Mascini, Veerle Cnudde, and Tom Bultreys
Ghent University

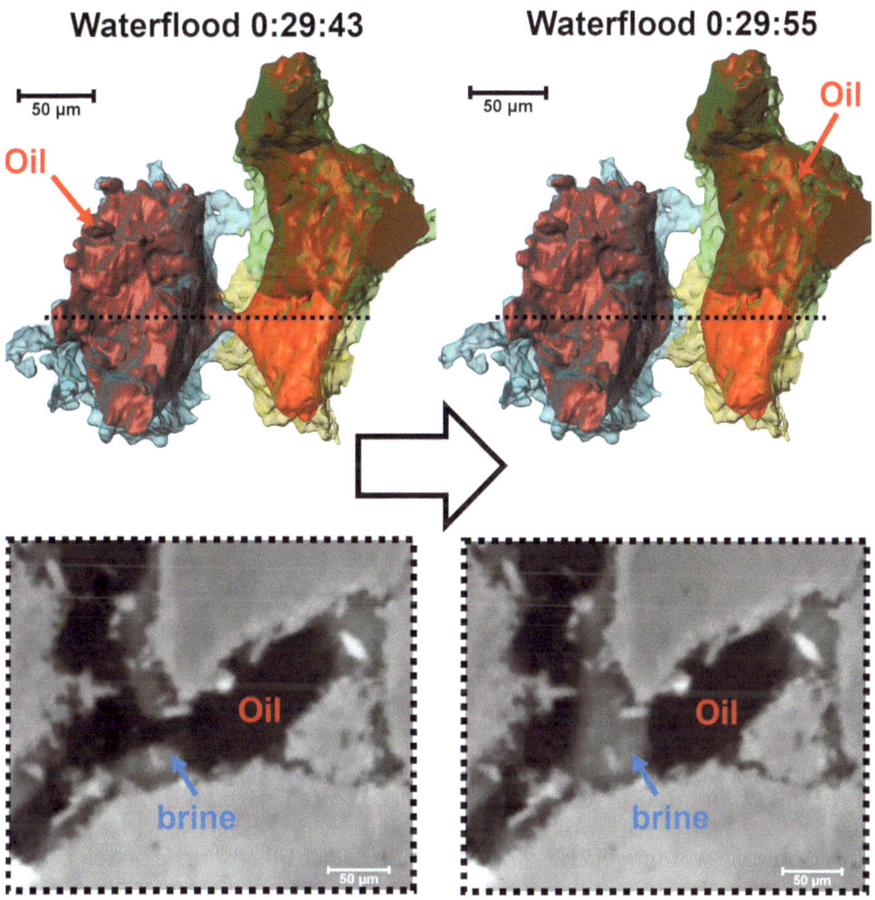

Snap-off is a multiphase displacement mechanism that can cause trapping of the non-wetting phase inside a porous medium. Understanding how, where and when snap-off occurs is of crucial importance to optimize subsurface storage of CO_2, hydrocarbon recovery and aquifer remediation operations. Here, we imaged snap-off inside a porous sandstone sample during a waterflooding experiment at very high spatial (0.8 μm/voxel) and temporal (12 s/image) resolution to investigate the snap-off phenomenon. In the top two figures renderings are shown of the non-wetting oil-phase (in red) inside the yellow, blue and green pores right before (left) and after the snap-off event. 2D slices through the tomogram in a perpendicular plane (dotted lines) for the same time steps are given below. Our imaging resolves the complex interplay between surface roughness, layer-swelling and pore geometry giving unprecedented insight into the pore-scale dynamics of snap-off. The X-ray imaging was performed at the TOMCAT beamline of the Swiss Light Source, Paul Scherrer Institute (Villigen, Switzerland).

Contact: Arjen Mascini <arjen.mascini@ugent.be>

El *snap-off* es un mecanismo de desplazamiento multifásico que puede causar retención de la fase no mojante dentro de un medio poroso. Comprender cómo, dónde y cuándo ocurre el *snap-off* es de crucial importancia para optimizar las operaciones de almacenamiento de CO_2 en el subsuelo, la recuperación de hidrocarburos y la remediación de acuíferos. Aquí, tomamos imágenes del *snap-off* dentro de una muestra de arenisca porosa durante un experimento de inundación de agua con una resolución espacial (0,8 μm/vóxel) y temporal (12 s/imagen) muy elevadas para investigar el fenómeno de *snap-off*. En las dos figuras superiores se muestra la fase de petróleo no mojante (en rojo) dentro de los poros amarillos, azules y verdes justo antes (izquierda) y después del evento de *snap-off*. Debajo, se observan cortes en 2D del tomograma en un plano perpendicular (líneas de puntos) para los mismos pasos de tiempo. Nuestra visualización resuelve la compleja interacción entre la rugosidad de la superficie, hinchamiento de la capa y la geometría del poro ofreciendo conocimientos sin precedentes sobre las dinámicas de *snap-off* a escala de poro. Las visualizaciones con rayos X se realizaron con un haz de luz TOMCAT del Swiss Ligth Source, en el Paul Scherrer Institute (Villigen, Switzerland).

E. F. Médici and A. D. Otero (eds.), *Album of Porous Media*,
https://doi.org/10.1007/978-3-031-23800-0_89

DISCONNECTED GAS FLOW IN HYDROPHOBIC POROUS MEDIA

Abdulla Alhosani, Branko Bijeljic, and Martin Blunt
Imperial College London

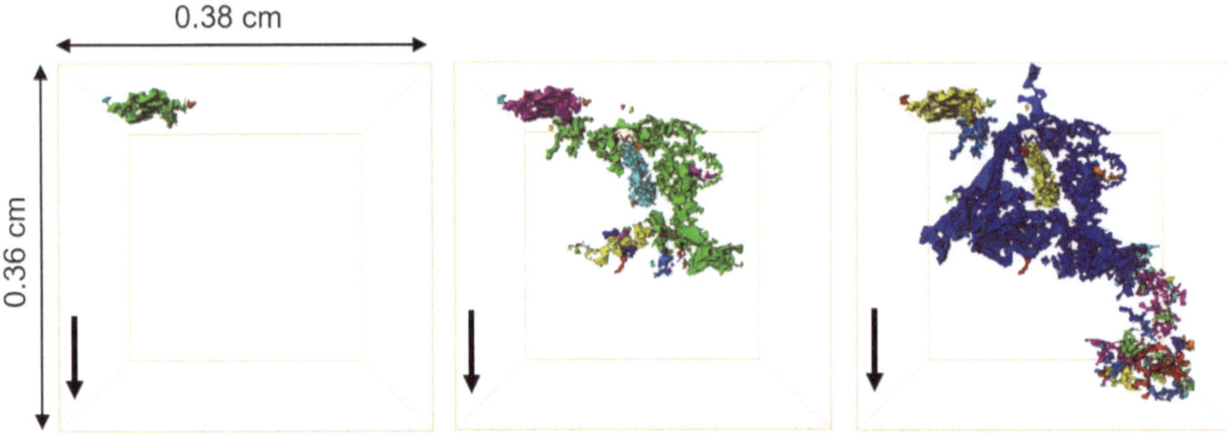

We show images of the gas configuration as it progresses through the pore space of a hydrophobic porous medium that contains oil and water – each color represents a disconnected gas cluster (rock, oil and water are rendered transparent) [the black arrows point to the flow direction]. The time-resolved images were acquired using a synchrotron X-ray light source. We observe a unique gas invasion pattern, gas progresses through the pore space in the form of disconnected clusters. This behaviour is attributed to the pore-scale events, made possible by double and multiple displacements, that govern the gas movement in the porous medium, which we name three-phase Haines jumps. As gas displaces either oil or water, it rapidly progresses to fill several pores, which causes it to retract from regions further away to enable this fast filling. This retraction leads to a permanent disconnection of gas ganglia, which fail to get reconnected as gas injection proceeds. The disconnected gas ganglia reach a new position of capillary equilibrium in the pore space and can only be displaced through double or multiple displacement events. This observation has huge implications on the design of secure CO_2 storage in oil reservoirs as it implies that CO_2 gets naturally trapped upon injection, and that the re-injection of water to trap CO_2 is not necessary and can have an adverse impact on the security of the storage.

Contact: Abdulla Alhosani <abdulla.alhosani17@imperial.ac.uk>

Mostramos imágenes de la configuración del gas a medida que avanza en el espacio de poro de un medio poroso hidrofóbico que contiene petróleo y agua. Cada color representa un cúmulo desconectado de gas (la roca, el petróleo y el agua son transparentes en la visualización; las flechas negras indican la dirección del flujo). Las imágenes resueltas en el tiempo fueron adquiridas mediante una fuente de rayos X de sincrotrón. Observamos un singular patrón de invasión de gas, el gas avanza entre el espacio poroso en forma de cúmulos desconectados. Este comportamiento se atribuye a los eventos a escala de poro, hechos posibles por los desplazamientos dobles y múltiples, que gobiernan el movimiento del gas en el medio poroso, denominados saltos de Haines de tres fases. A medida que el gas desplaza el petróleo o el agua, progresa rápidamente para llenar varios poros, lo que causa que se retracte de zonas más lejanas para permitir un llenado rápido. Esta retracción conduce a una desconexión permanente del ganglio de gas, que no se reconecta mientras que prosigue la inyección de gas. Este ganglio de gas desconectado alcanza una nueva posición de equilibrio capilar en el espacio poroso y solo puede ser desplazado mediante eventos de desplazamiento dobles o múltiples. Esta observación tiene inmensas implicaciones en el diseño del almacenamiento seguro de CO_2 en yacimientos de petróleo, puesto que implican que el CO_2 queda atrapado naturalmente tras la inyección, y que la reinyección de agua para retener el CO_2 no es necesaria y puede tener un impacto adverso en la seguridad del almacenamiento.

E. F. Médici and A. D. Otero (eds.), *Album of Porous Media*,
https://doi.org/10.1007/978-3-031-23800-0_90

FORMATION OF WATER MICRO-DISPERSION IN OIL AS A MECHANISM FOR EFFICIENT DISPLACEMENT IN LOW SALINITY WATERFLOODING

Ahmed M. Selem[1], *Nicolas Agenet*[2], *Martin J. Blunt*[1], *and Branko Bijeljic*[1]

[1] Imperial College London
[2] TotalEnergies

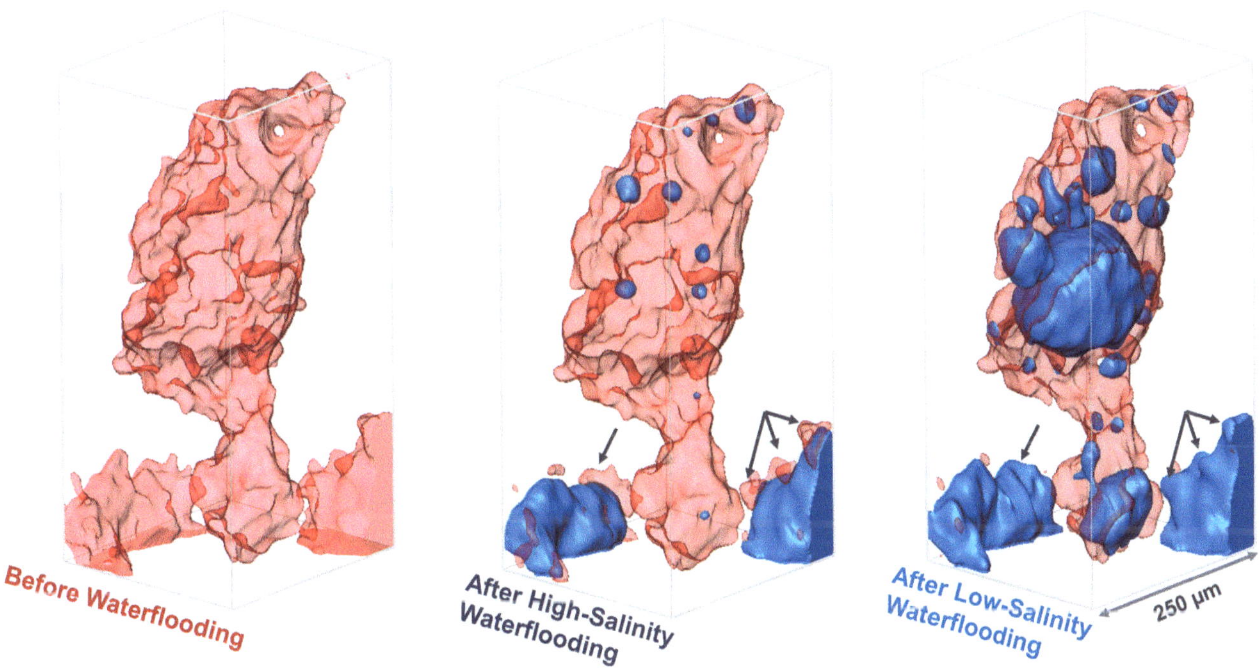

Visualization of water configuration, derived from micro-CT tomograms, in a carbonate rock sample aged in crude oil before and after tertiary waterflooding, i.e., high salinity followed by low salinity water injection. Oil and water are shown in red and blue, respectively, while the rock is rendered transparent. Before waterflooding pores were occupied by oil. Water micro-droplets slowly developed during high salinity brine, 31,092 ppm, injection within the oil phase. After the injection of 60 pore volumes of high salinity brine, only few water micro-droplets appear in the main pore, and oil was constrained to pore corners, as shown in the pores at the bottom. After low salinity brine, 1200 ppm, injection the water clusters grew and coalesced resulting in volume displacement of oil. In addition, oil was displaced from pore corners (as shown by the arrows) indicating a change from oil-wet towards more water-wet state. The development of water micro-dispersion is possibly one of the underlying mechanisms for the low salinity effect and wettability alteration resulting in an increase in oil recovery.

Contact: Ahmed Selem <a.selem18@imperial.ac.uk>

Visualización de la configuración del agua, derivada de microtomografías, en una muestra de roca de carbonato envejecida en petróleo antes y después de la inundación terciaria de agua, es decir, inyección de agua con alta salinidad seguida de una inyección de agua con baja salinidad. El petróleo y el agua se muestran en rojo y azul, respectivamente, mientras que la roca se visualiza transparente. Antes de la inundación por agua, los poros estaban ocupados por el petróleo. Lentamente surgieron microgotas de agua en la fase de petróleo durante la inyección de la solución de alta salinidad, 31092 ppm. Después de la inyección de 60 volúmenes porosos de la solución de alta salinidad, solo algunas microgotas de agua aparecieron en el poro principal, y el petróleo quedó restringido a las esquinas del poro, como se ve en los poros de la parte inferior. Después de la inyección de la solución de baja salinidad, 1200 ppm, los cúmulos de agua crecieron y se fusionaron, resultando en el volumen desplazado de petróleo. Además, el petróleo fue desplazado de las esquinas de los poros (como lo muestran las flechas), indicando un cambio de un estado mojado con petróleo hacia uno mojado por agua. El desarrollo de la microdispersión del agua es posiblemente uno de los mecanismos subyacientes del efecto de la baja salinidad y la alteración de la mojabilidad que resultan en un aumento de la recuperación del petróleo.

E. F. Médici and A. D. Otero (eds.), *Album of Porous Media*,
https://doi.org/10.1007/978-3-031-23800-0_91

PORE-SCALE TRAPPING AND FLOW FIELDS IN CARBONATE ROCKS

Kamaljit Singh[1] *and Martin Blunt*[2]

[1] Heriot-Watt University
[2] Imperial College London

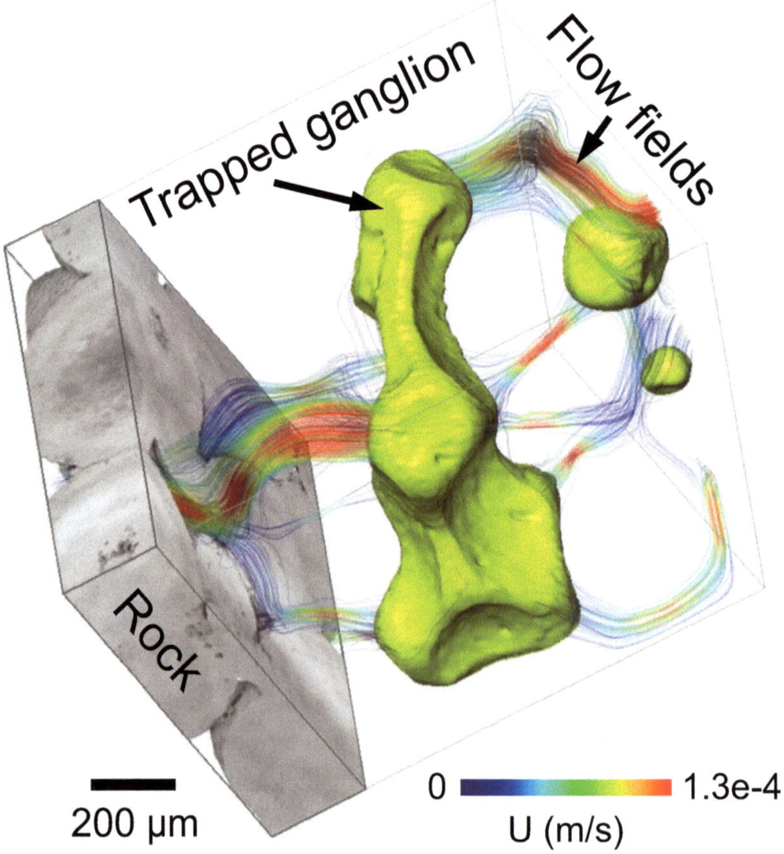

Understanding the pore-scale dynamics and trapping during multiphase flow in porous media is important in many processes such as water infiltration in soils, geo-sequestration of CO_2 and oil recovery. We investigate these pore-scale dynamic processes using time-resolved synchrotron X-ray micro-tomography. The image shows non-wetting (*nw*) phase ganglia or bubbles (green) trapped in a carbonate rock sample after a drainage-imbibition cycle. We also performed flow field simulations to track the paths of *nw* phase injection and its trapping. Colored lines show the streamlines of velocity fields in the pore space of the rock sample. These simulations were performed using an open-source computational fluid dynamics (CFD) software, OpenFOAM.

Contact: Kamaljit Singh <k.singh@hw.ac.uk>

Comprender la dinámica a escala del poro y la retención durante el flujo multifásico en medios porosos es importante en muchos procesos como la filtración de agua en suelos, el geosecuestro de CO_2 y la recuperación de petróleo. Investigamos estos procesos dinámicos de escala del poro utilizando microtomografías de rayos X de sincrotrón resuelta en el tiempo. La imagen muestra una fase con ganglios o burbujas no mojados (nm) (verde) atrapados en una muestra de roca carbonata después de un ciclo de drenado-imbibición. También realizamos simulaciones del campo de flujo para rastrear los caminos de inyección de la fase nm y su retención. Las líneas coloreadas muestran las líneas de corriente del flujo en el espacio poroso de la muestra de roca. Estas simulaciones se realizaron utilizando un software de código abierto de dinámica de fluidos computacional (CFD), OpenFOAM.

E. F. Médici and A. D. Otero (eds.), *Album of Porous Media*,
https://doi.org/10.1007/978-3-031-23800-0_92

NANO-IMAGING OF DIATOMITE WITH TRANSMISSION X-RAY MICROSCOPE

Bolivia Vega and Anthony Kovscek
Stanford University

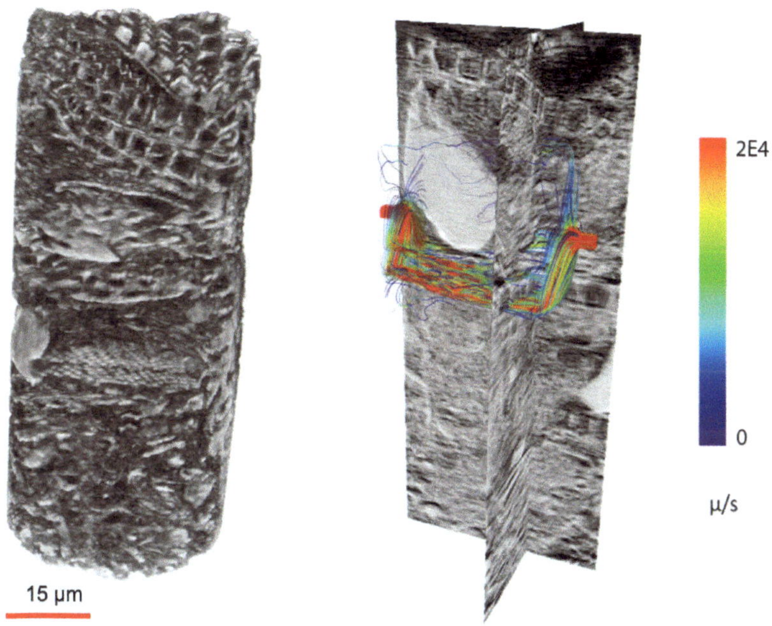

Diatomite is a porous (40–60 %) and low permeability (0.01–10 md) siliceous shale composed of diatom frustules and other detritus that have been converted into rock. Diatoms are microscopic algae that are tens to hundreds of microns in size with internal porosity. They are composed primarily of biogenic silica which is hydrous and amorphous[1]. Diatomaceous rock formations are mined, primarily to be used as filter media, and are hydrocarbon reservoirs estimated to hold over 10 billion of OOIP. Imaging-aided diatomite flow transport studies have been carried out at meso-scale[2, 3]. The image to the left shows a nano-CT volume reconstruction of a 30-micron diameter sandy diatomite sample featuring a silt particle and semi-intact fossilized algae structures such as frustules, porous plates, and fecal pellets. The image to the right features perpendicular slice projections of the volume with an overlay of simulated directional velocity streamlines. Significant decrease of flow at the top of the streamlines is observed due to the presence of a silt grain, while the bottom area shows continuous paths with larger velocities going through a frustule structure.

Work supported as part of the Center for Mechanistic Control of Unconventional Formations (CMC-UF), under DOE (BES) Award DE-SC0019165. Part of this research was carried out at the Stanford Synchrotron Radiation Lightsource, operated by the US Department of Energy Office of Science by Stanford University.

1. Ross C.M. *et al.* (2016). In *Western Regional Meeting*. SPE.
2. Vega B. and Kovscek A.R. (2014). In *Annual Technical Conference and Exhibition*. SPE.
3. Schembre J.M. *et al.* (2005). In *Western Regional Meeting*. SPE.

La diatomita es una lutita silícea compuesta por frústulas de diatomeas y otros detritus que se han convertido en roca porosa (40–60 %) con baja permeabilidad (0,01–10 md). Las diatomeas son algas microscópicas con porosidad interna que tienen un tamaño de decenas a cientos de micrones. Están compuestas principalmente de sílice biogénico que es hidratado y amorfo[1]. Las formaciones de roca de diatomea se minan, principalmente para utilizarse como filtro, y son reservas de hidrocarburo que se estima pueden contener hasta 10 mil millones de petróleo original *in situ* (OOIP, por sus siglas en inglés). Se realizaron estudios asistidos por visualización del transporte de flujo en diatomita en la mesoescala[2, 3]. La imagen de la izquierda muestra una reconstrucción volumétrica con nanotomografía computarizada de una muestra de diatomita arenosa de 30 micrones de diámetro caracterizada por una partícula de limo y estructuras fosilizadas semi intactas como frústulas, placas porosas y gránulos fecales. La imagen de la derecha muestra las proyecciones de corte perpendicular del volumen con una superposición de líneas de corriente de velocidad direccional simuladas. Se observa una disminución significativa del flujo de la parte superior de las líneas de corriente debido a la presencia del grano de limo, mientras que la parte inferior muestra caminos continuos con mayores velocidades a través de la estructura de frústula.

Contact: Bolivia Vega <bolivia@stanford.edu>

E. F. Médici and A. D. Otero (eds.), *Album of Porous Media*,
https://doi.org/10.1007/978-3-031-23800-0_93

PORE SCALE INVESTIGATION OF FREEZE-DRYING

Sebastian Gruber[1], *Petra Foerst*[1], *Maximilian Thomik*[2], *Evangelos Tsotsas*[2], *and Nicole Vorhauer-Huget*[2]

[1] Chair of Process Systems Engineering and Chair of Food Process Engineering, School of Life Sciences, Technical University of Munich, Germany

[2] Department of Thermal Process Engineering, Institute of Process Engineering, Otto-von-Guericke University Magdeburg, Germany

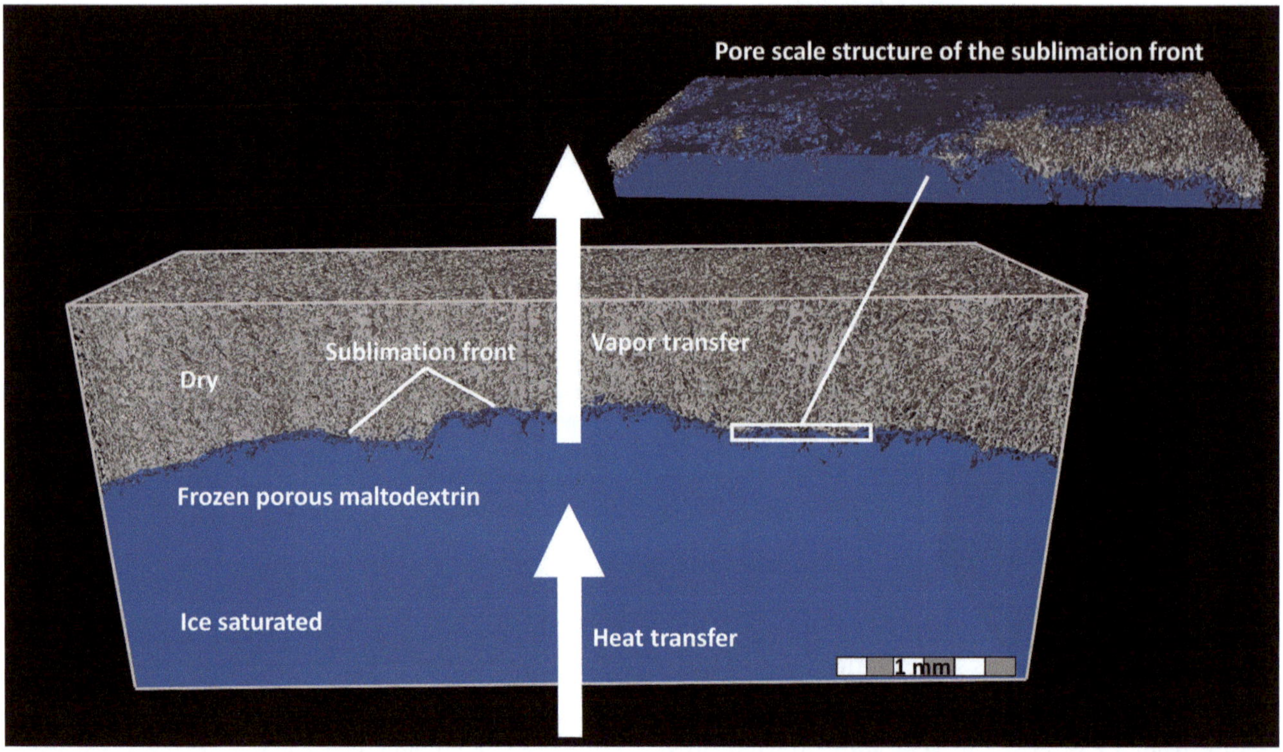

This image of frozen porous maltodextrin with size $5\times2.6\times1.2$ mm^3 (w$\times$h$\times$d) was obtained by X-ray micro tomography at a temperature of -15 °C and a pressure of 10 Pa. It shows the intermediate position of the sublimation front propagating into the material as a result of the phase change of water from frozen to vapor state in the course of freeze-drying[1]. The *in situ* visualization of the dynamically developing pore-scale structure of the sublimation front and its movement through the porous medium provides valuable insights into the process that would otherwise not be accessible. The ability of the method to access spatially and temporally resolved data about the water ice distribution builds the base for an enhanced process understanding. This way, it becomes possible to study the dependence of micro-scale heat and vapor transfer kinetics associated with the sublimation front structure. Such a deep knowledge is of high relevance especially for pharmaceutical and food production industries, where precious products are frequently preserved by freeze-drying, an expensive, time and energy consuming process.

1. Gruber S. *et al.* (2021). *Food Structure*, 29:100213.

Contact: Sebastian Gruber <sebi.gruber@tum.de>

Esta imagen de maltodextrina porosa congelada con un tamaño de $5\times2{,}6\times1{,}2$ mm^3 (w$\times$h$\times$d) fue obtenida mediante microtomografía de rayos X a una temperatura de -15 °C y a una presión de 10 Pa. La imagen muestra la posición intermedia del frente de sublimación propagándose en el material como resultado del cambio de fase del agua, de un estado congelado a uno de vapor, en el transcurso de liofilización[1]. La visualización *in situ* del desarrollo dinámico de la estructura a escala de poro del frente de sublimación y su movimiento a lo largo del medio poroso proporcionan información valiosa sobre el proceso que de otra manera no sería accesible. La habilidad del método para acceder a datos con resolución espacial y temporal sobre la distribución de agua y hielo constituye la base para una mayor comprensión del proceso. De esta manera, se posibilita estudiar la dependencia de la cinética de la transferencia de calor y vapor en la microescala asociada a la estructura del frente de sublimación. Tal conocimiento profundo es de gran importancia, especialmente para las industrias de producción farmacéutica y alimentaria, en las que con frecuencia se conservan preciados productos mediante procesos de liofilización que son costosos y demandan tiempo y energía.

E. F. Médici and A. D. Otero (eds.), *Album of Porous Media*,
https://doi.org/10.1007/978-3-031-23800-0_94

PLASTICITY DISTRIBUTION OF DIGITAL ROCK FROM ONE POINT OF THE YIELD ENVELOPE

Martin Lesueur
Delft University of Technology

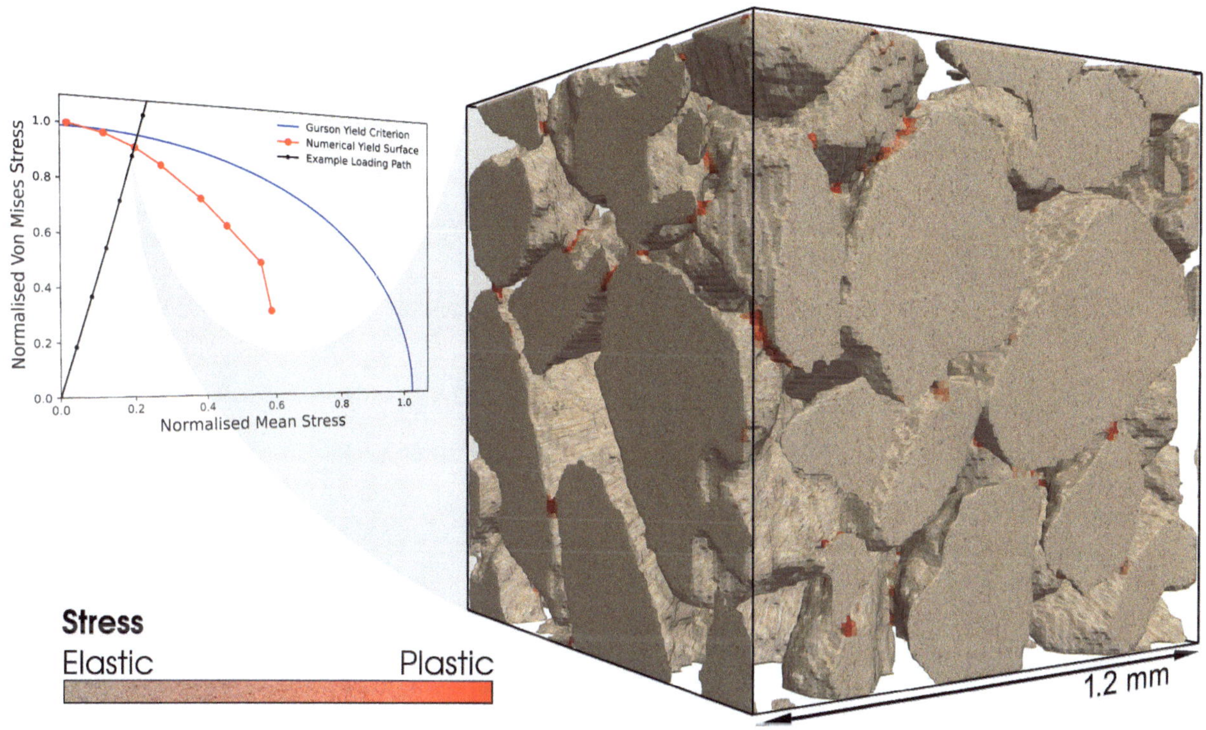

Color indicates stress concentration, with red values corresponding to elements having reached microscopic yield stress (i.e. plasticity). The sample is cropped to remove corner effects and the rock is textured for visualization. As expected, grain contacts are the zones under highest stress as they represent weak points in the chain force. Multiple compression experiments such as this one are simulated following different loading paths by varying the triaxial stress ratio (i.e. slope angle in p-q space shown here), using the framework from [1]. The yield points are computed using the classical offset method[2]. Grain contacts are assumed to follow a J2-plasticity to compare with the Gurson yield criterion[3]. For the same reason, the stresses are normalized with the cohesion value (i.e. Von Mises stress value at zero mean stress). Results on the left-hand side show that this sample (LV60A sandpack[4]) has a weaker compaction cap than would be expected from the Gurson yield criterion[3] computed for the same porosity, which illustrates the difference between the simple hollow sphere model and a complex rock microstructure.

1. Lesueur M. *et al.* (2017). *Geomechanics for Energy and the Environment*, 12:1.
2. Ross C.T.F. (1999). *Mechanics of Solids*. Elsevier.
3. Gurson A.L. (1977). *J. Eng. Materials and Tech.*, 99(1):2.
4. URL `https://doi.org/10.6084/m9.figshare.1153795`.

Contact: Martin Lesueur <m.lesueur@tudelft.nl>

El color indica la menor concentración de tensiones, con los valores en rojo que corresponden a elementos que han alcanzado la tensión de fluencia (es decir, plasticidad) a nivel microscópico. La muestra está cortada para evitar efectos de esquina y la roca está texturizada para poder visualizarla. Como se esperaba, en los contactos entre granos se encuentran las zonas de mayor tensión, ya que representan los puntos débiles en la cadena cinemática. Múltiples experimentos de compresión, como este, son simulados siguiendo distintos caminos de carga, variando la razón de tensiones triaxiales (es decir, el ángulo de inclinación en el espacio p-q que se muestra aquí), utilizando el marco de [1]. Los puntos de fluencia se calculan utilizando el método clásico de compensación[2]. Se asume que los contactos entre granos siguen una plasticidad J2 para poder compararlos con el criterio de fluencia de Gurson[3]. Por el mismo motivo, las tensiones están normalizadas por el valor de cohesión (es decir, el valor de la tensión de Von Mises con valor de tensión promedio cero). Los resultados de la izquierda indican que esta muestra (arena compactada LV60A[4]) tiene un límite de compactado más débil que el que se esperaría según el criterio de fluencia de Gurson[3] calculado con la misma porosidad. Esto ilustra la diferencia entre un modelo de esfera simple hueca y la compleja microestructura de las rocas.

E. F. Médici and A. D. Otero (eds.), *Album of Porous Media*,
https://doi.org/10.1007/978-3-031-23800-0_95

IMAGING THE PATH TO FAILURE IN ROCKS

François Renard
University of Oslo

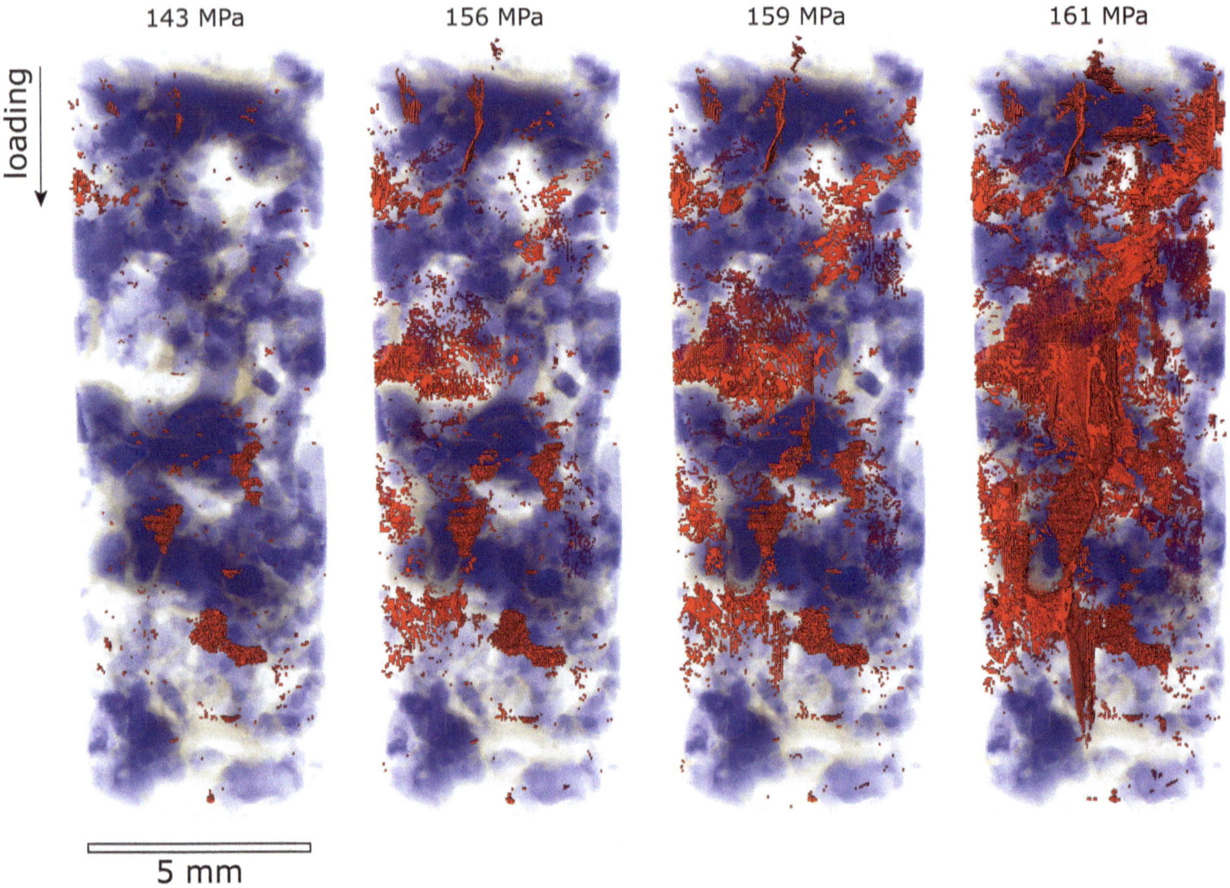

The mechanical evolution of rocks under increasing loading controls how earthquakes nucleate and brittle faults develop. Here, I have imaged the evolution of a rock sample toward system-size failure. Blue (pyroxene) and gray (quartz, feldspar) shades correspond to silicate minerals forming the monzonite rock. Microfractures that developed during loading are colored in red. This three-dimensional rendering of a rock progressing toward system-size failure was acquired under *in situ* conditions using synchrotron X-ray microtomography on beamline ID19 at the European Synchrotron Radiation Facility. The sample, 5 mm diameter and 10 mm height, was under a constant confining pressure of 25 MPa. The axial loading, indicated above each image, was increased by steps until the sample failed, producing a laboratory earthquake. The final failure was preceded by the nucleation, growth, and coalescence of microfractures in the sample that represent precursors to the final catastrophic failure.

La evolución mecánica de las rocas bajo una carga en aumento controla cómo ocurren la nucleación de los sismos y la propagación de fracturas frágiles. Aquí ilustro con imágenes la evolución de una muestra de roca hacia una falla total del sistema. Los colores azul (piroxeno) y gris (cuarzo, feldespato) corresponden a minerales de silicio que forman la monzonita. Las microfracturas que se desarrollan durante el aumento de la presión se ven de color rojo. Esta imagen tridimensional de una roca progresando hacia una falla total del sistema se realizó bajo condiciones *in situ* utilizando microtomografías de rayos X de sincrotrón en la línea de haz ID19 del Laboratorio Europeo de Radiación Sincrotrón (ESRF, por sus siglas in inglés). La muestra, de 5 mm de diámetro y 10 mm de altura, fue mantenida bajo una presión de 25 Mpa. La carga axial, indicada encima de cada imagen, fue aumentada paso a paso hasta que la muestra falló, y produjo un sismo de laboratorio. El fallo final fue precedido por una nucleación, crecimiento y coalescencia de microfracturas en la muestra que representan los precursores del fallo final catastrófico.

Contact: François Renard <francoisrenard99@gmail.com>

E. F. Médici and A. D. Otero (eds.), *Album of Porous Media*,
https://doi.org/10.1007/978-3-031-23800-0_96

VISUALIZATION OF THE UNIAXIAL COMPRESSION OF OPEN-CELL FOAMS

Matthias Ruf and Holger Steeb
University of Stuttgart - Institute of Applied Mechanics (CE)

The uniaxial compression of open-cell foam structures results typically in a stress-strain curve consisting of three regimes: linear-elastic, collapse, and densification regime. In this work, the elastic collapse of a cylindrical open-cell reticulated polyurethane (PUR) foam sample (diameter 52 mm, length 50 mm) with 10 pores per inch was investigated with *in situ* micro X-ray Computed Tomography (μ-XRCT) imaging. For this, an open, modular and flexible μXRCT system with an integrated universal testing machine was employed. After an initial *in situ* compression test up to 60 % compression strain, 5 snapshots of the deformed structure next to the initial state were performed. The red markers in the force-strain curve indicate the scanned deformation states. The images show the segmented sub-volume of the center region of the cylindrical foam sample and demonstrate the elastic collapse of the porous media. As the images illustrate, and is also known from the literature, the *plateau* of the compressive stress-strain is caused by distinct elastic buckling of the cell columns mainly oriented parallel to the loading direction[1, 2].

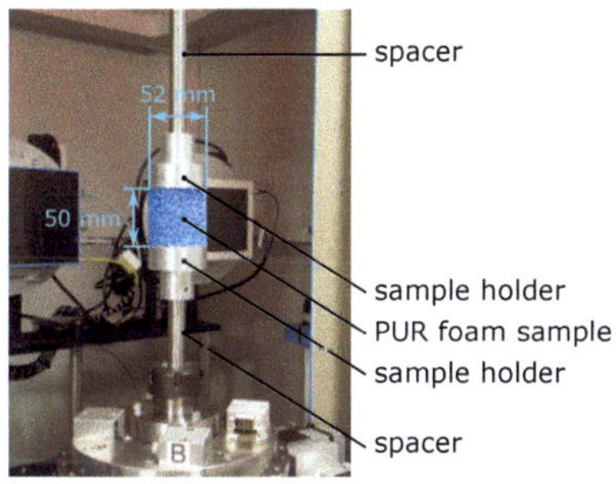

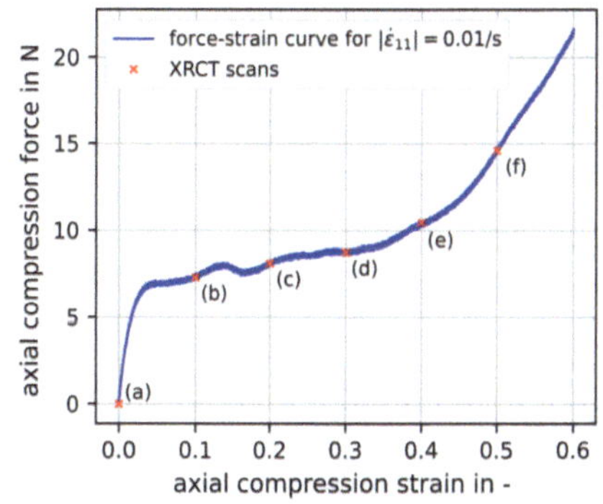

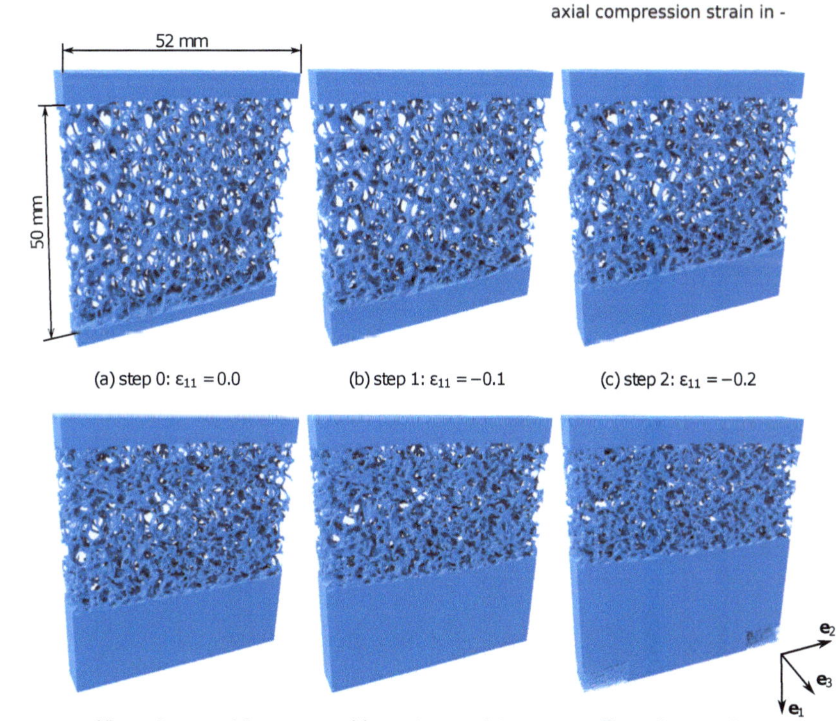

(a) step 0: $\varepsilon_{11} = 0.0$ (b) step 1: $\varepsilon_{11} = -0.1$ (c) step 2: $\varepsilon_{11} = -0.2$

(d) step 3: $\varepsilon_{11} = -0.3$ (e) step 4: $\varepsilon_{11} = -0.4$ (f) step 5: $\varepsilon_{11} = -0.5$

La compresión uniaxial de estructuras de espuma reticulada suele resultar en una curva de tensión-deformación compuesta por tres regímenes: elástico-lineal, colapso y densificado. En este trabajo, se investigó el colapso elástico de una muestra de espuma de poliuretano (PUR) reticulado cilíndrico de celda abierta (52 mm de diámetro, 50 mm de largo) con 10 poros por pulgada *in situ*, mediante imágenes con tomografía computarizada de rayos X (μ-TCRX). Para esto, se utilizó un sistema de μ-TCRX abierto, modular y flexible incorporado a una máquina de pruebas universal. Después de una prueba de compresión inicial *in situ* de hasta el 60 % de deformación por compresión, se tomaron 5 imágenes de la estructura deformada cerca del estado inicial. Los marcadores rojos en la curva de esfuerzo-deformación indican los estados de deformación escaneados. Las imágenes muestran el subvolumen segmentado de la región central de la muestra cilíndrica de espuma y demuestran el colapso elástico del medio poroso. Como muestra la imagen, y también es conocido en la literatura, el *plateau* de la tensión-deformación por compresión es causado por la distintiva deformación elástica de pandeo de las columnas de celda, principalmente orientadas en paralelo a la dirección de la carga[1, 2].

1. Ruf M. and Steeb H. (2020). *Review of Scientific Instruments*, 91(11):113102.
2. Gibson L. and Ashby M. (1999). *Cellular Solids*. Cambridge University Press.

Contact: Nikolaos Karadimitriou <nkaradimitriou@gmail.com>

E. F. Médici and A. D. Otero (eds.), *Album of Porous Media*,
https://doi.org/10.1007/978-3-031-23800-0_97

VII. COMPUTATIONAL MODELING

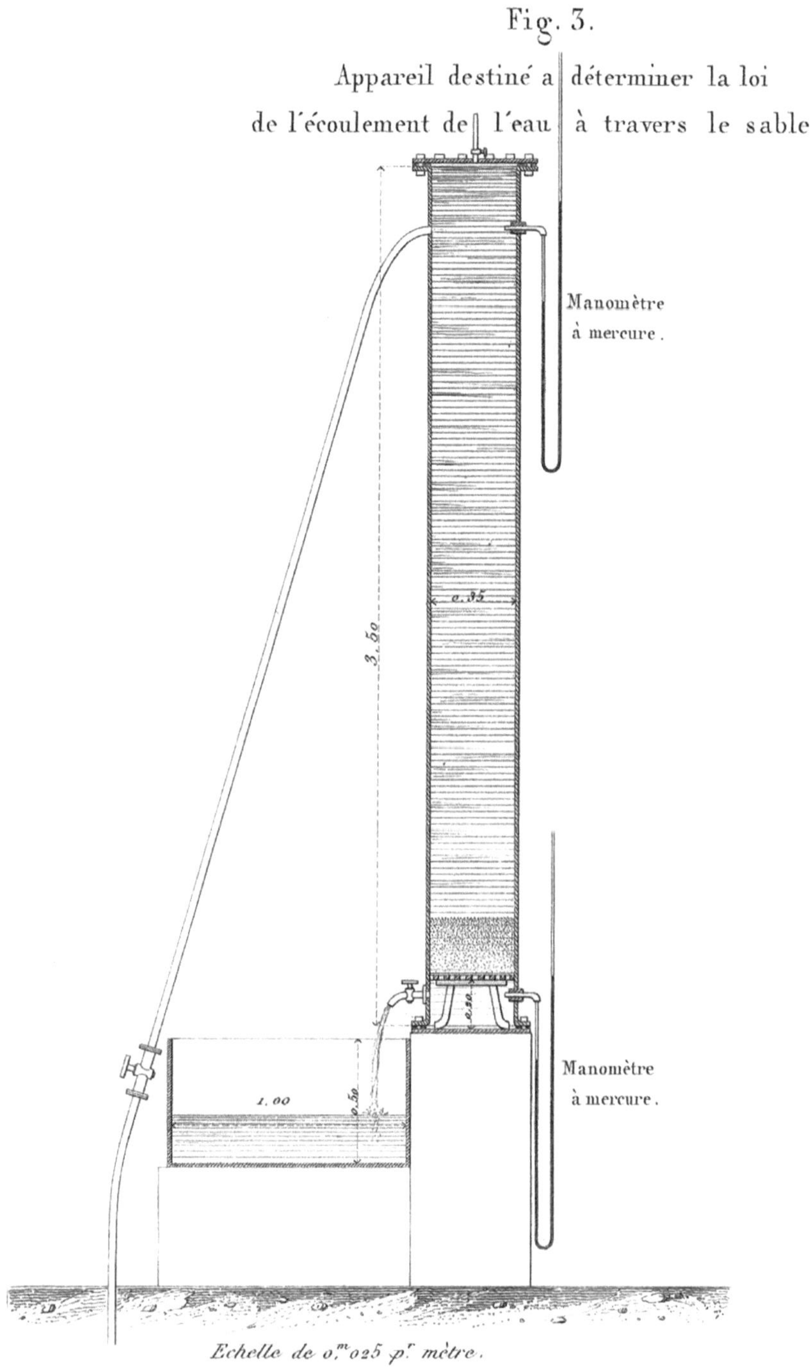

The work of Henry Darcy is considered to be one of the foundational works of the modern porous media[1]. Image courtesy of Dr. Glenn Brown, 2022.

La obra de Henry Darcy es considerada una de las obras fundacionales de los medios porosos modernos[1]. Imagen cortesía del Dr. Glenn Brown, 2022.

1. Darcy H. (1856). *Les fontaines publiques de la ville de Dijon exposition et application.* Dalmont, Paris.

OED-BASED METHOD FOR EXTRACTION OF PORE NETWORKS WITH HIGH ASPECT RATIO

Ninghua Zhan[1], *Rui Wu*[2], *and Abdolreza Kharaghani*[1]

[1] Otto von Guericke University Magdeburg
[2] Shanghai Jiao Tong University

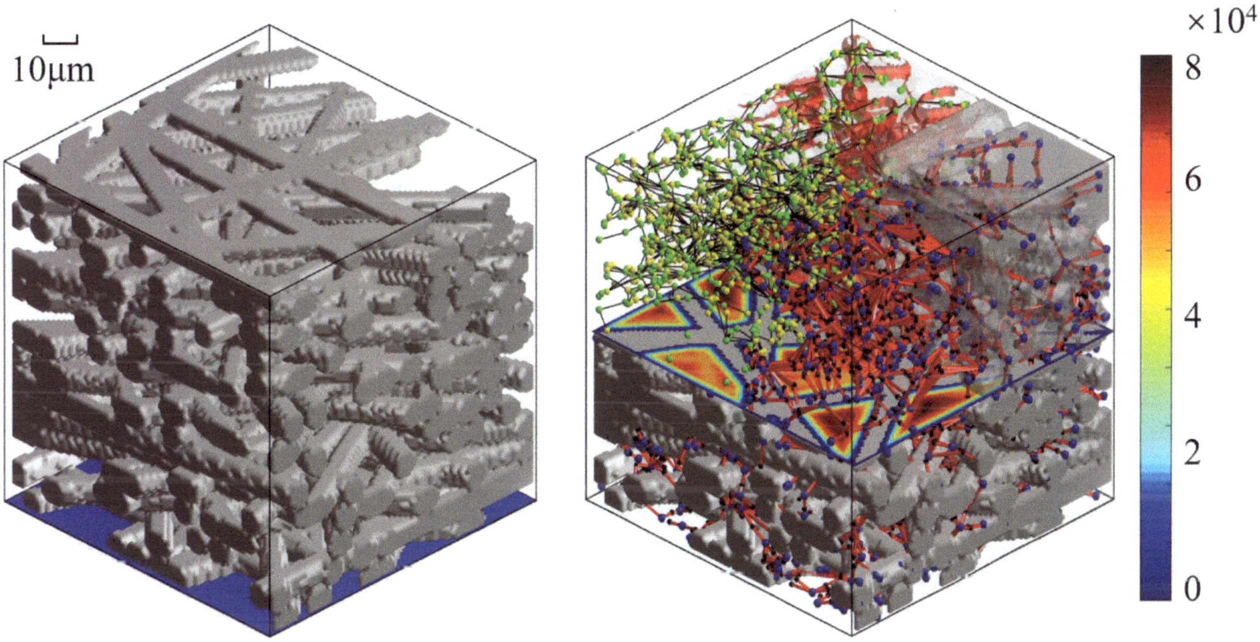

In existing extraction methods, Euclidean distance (ED) map is employed to discern the hierarchy of the void voxels so as to obtain the parameters required for the pore network extraction. However, for pores with high aspect ratio in porous media, many void voxels have the same hierarchy in ED map (i.e., ED map cannot distinguish pores with high aspect ratio). To address this issue, we propose an extraction method based on the concept of the Omnidirectional Euclidean distance (OED). The OED of a void voxel is a set of EDs from this void voxel to all the accessible solid boundary voxels. Any two void voxels usually have different OEDs. Hence, the hierarchy of the void voxels can clearly be discerned by using the OED map. The pore network including the pore bodies, pore throats, and corners extracted (right side) from a fibrous porous media generated by computer algorithm (left side) using the OED based method is illustrated in the figure. The hierarchy of the void voxels at the middle is characterized by various colors.

Contact: Ninghua Zhan <302883303@qq.com>

En los métodos de extracción existentes, el mapa de la distancia euclidiana (DE) se emplea para discernir la jerarquía de los vóxeles huecos y así obtener los parámetros requeridos en la extracción de la red de poros. Sin embargo, en poros con una elevada relación de aspecto en medios porosos, muchos vóxeles huecos tienen la misma jerarquía en los mapas DE (es decir, el mapa DE no puede distinguir poros con una elevada relación de aspecto). Para abordar esta cuestión, proponemos un método de extracción basado en el concepto de distancia euclidiana omnidireccional (DEO). La DEO de un vóxel hueco es un conjunto de distancias euclidianas desde este vóxel hueco hacia todos los contornos sólidos accesibles de los vóxeles. Cualquier conjunto de pares de vóxeles huecos suelen tener DEO distintos. Por lo tanto, la jerarquía de los vóxeles huecos puede discernirse claramente al utilizar el mapa DEO. En esta figura se ilustra la red de poros que incluye cuerpos y gargantas de los poros y esquinas extraídas (lado derecho) de un medio poroso fibroso generado mediante un algoritmo informático (lado izquierdo) utilizando el método basado en DEO. La jerarquía de los vóxeles huecos en el medio se caracteriza por varios colores.

E. F. Médici and A. D. Otero (eds.), *Album of Porous Media*,
https://doi.org/10.1007/978-3-031-23800-0_98

PORE TOPOLOGY METHOD (PTM): A COMPUTATIONALLY EFFICIENT APPROACH FOR PORE-SCALE MODELING AND DESIGN OF POROUS MATERIALS

M. Sadegh Riasi, Benjamin Paisley, and Lilit Yeghiazarian
University of Cincinnati

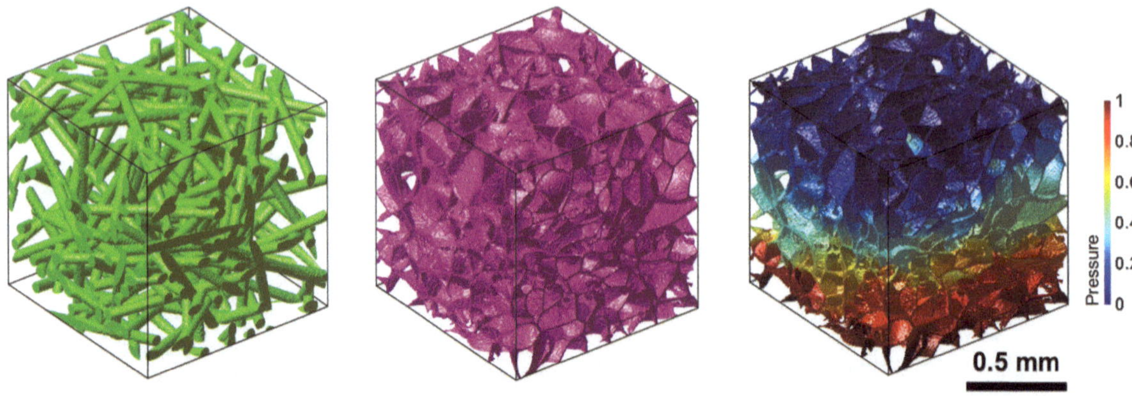

3D image of porous medium | Medial surface of void space | PDEs solved on medial surface

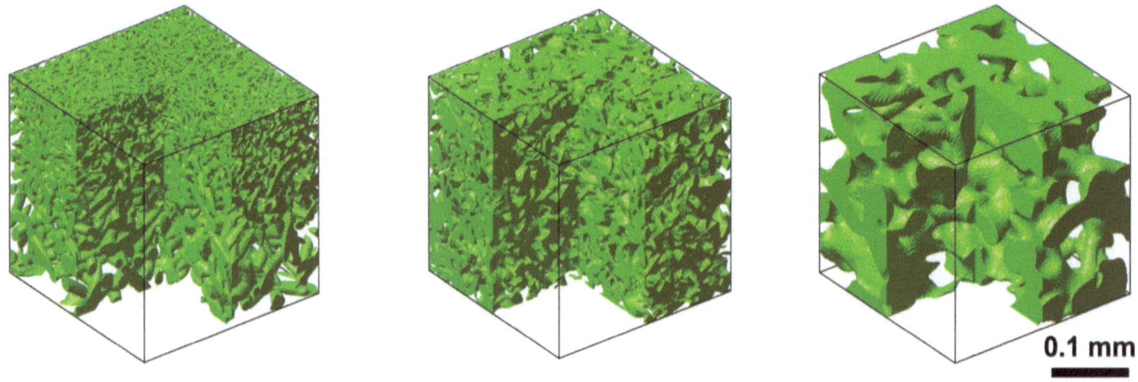

50% porosity with target vertical pore size gradient of 1:5 | 50% porosity with target pore size of 6 ± 3 μm | 50% porosity with mean pore size of 15 μm and maximal permeability

PTM FLOW (Top Row) is a pore-scale flow modeling methodology that reduces the computational cost by using the medial surface of the void space as the solution domain [1]. PTM can be used to characterize single-phase and multi-phase flow properties such as absolute permeability, relative permeability curves, as well as drainage and imbibition curves.

PTM DESIGN (Bottom Row) is a porous media design method that combines PTM FLOW with the Adjustable Level-Cut filtered Poisson Field (ALCPF) to generate 3D porous materials with target properties [2]. PTM DESIGN is capable of producing homogeneous and heterogeneous microstructures with target porosities, pore size distributions, pore size gradients and permeabilities.

Acknowledgments: National Science Foundation CBET-1248385 and CBET-1351361 awards.

1. Riasi M.S. *et al.* (2016). *Transp Porous Med*, 115(3):519.
2. Paisley B. *et al.* (2021). *J Por Media*, 24(1):41.

Contact: Lilit Yeghiazarian <yeghialt@ucmail.uc.edu>

FLUJO MTP (método de topología de poro, por sus siglas en inglés, fila superior) es una metodología para modelar el flujo a escala de poro que reduce el costo computacional al emplear la superficie media del espacio hueco como el dominio de la solución [1]. El MTP puede utilizarse para caracterizar las propiedades del flujo monofásico y multifásico, como la permeabilidad absoluta, las curvas de permeabilidad relativa, así como las curvas de drenado y de imbibición.

DISEÑO MTP (fila inferior) es un método de diseño del medio poroso que combina el flujo MTP con el campo de Poisson filtrado con corte nivelado ajustable (ALCPF, por sus siglas en inglés) para generar materiales porosos en 3D con propiedades determinadas [2]. El DISEÑO MTP puede producir microestructuras homogéneas y heterogéneas con determinadas porosidades, distribución de tamaño del poro, gradientes de tamaño de poro y permeabilidades.

E. F. Médici and A. D. Otero (eds.), *Album of Porous Media*,
https://doi.org/10.1007/978-3-031-23800-0_99

COUPLING OF PORE-NETWORK AND FINITE ELEMENT METHODS FOR RAPID QUANTIFICATION OF DEFORMATION

Pejman Tahmasebi
University of Wyoming

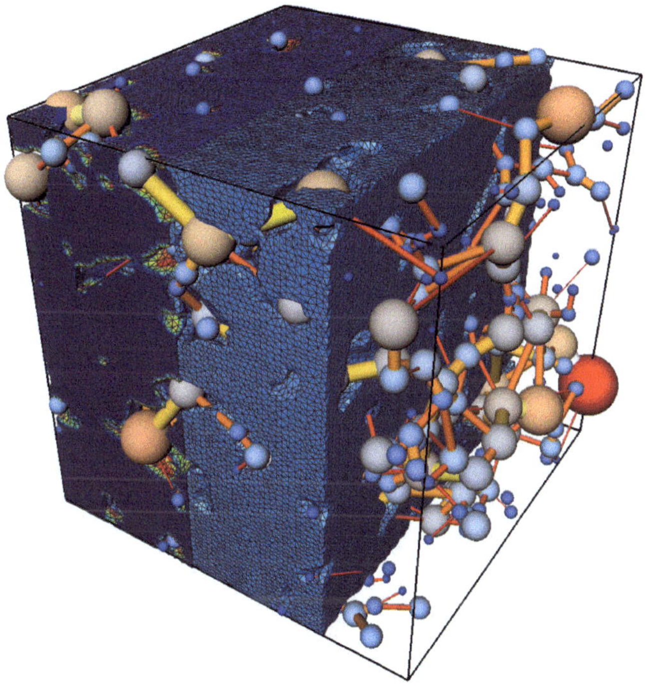

This image represents our recent study wherein we have proposed a new method for coupling the capability of pore-network (PN) models in flow simulation and finite element method (FEM) for solid deformation in complex porous media[1].
This is a Berea sandstone sample and the image shows the coupling of PN and FEM for quantifying deformation in porous media due to fluid injection and external stress. The PN, due to its high efficiency in simulating fluid flow, is used for conducting the flow simulation while FEM is used for the solid. These two methods, for the first time, are coupled and could result in an accurate prediction for deformation. The colors in the PN represent the radius of the pores and throats and, similarly, the color in the solid part, which is simulated using FEM, represents the amount of induced deformation in the solid.
The size of the sample is $(0.641 \text{ mm})^3$. In this simulation, the fluid (water) is injected from the left side. The color on the left side represents the deformation. To show three different aspects, the sample shows, from the left, solid with the containing deformation, solid mesh, and fluid domain shown by a pore-network model, respectively.

1. Fagbemi S. and Tahmasebi P. (2020). *J. Fluid Mech.*, 897.

Esta imagen representa nuestro reciente estudio en el que hemos propuesto un nuevo método que permite acoplar la capacidad de los modelos de red de poros (RP) para simular flujos y del método de los elementos finitos (FEM) para deformaciones sólidas en medios porosos complejos[1].
Esta es una muestra de arenisca de Berea y en la imagen se observa el acoplamiento entre los métodos de red de poros (RP) y de los elementos finitos (FEM) para cuantificar la deformación en medios porosos debido a la inyección de fluido y a tensiones externas. Debido a su alta eficiencia para simular el flujo de fluidos, la RP se utiliza para simular el flujo mientras que el FEM se utiliza para el sólido. Estos dos métodos se han acoplado por primera vez y podrían resultar en una predicción precisa para la deformación. Los colores en la RP representan los radios de los poros y gargantas, y similarmente, el color de la parte sólida, simulada con el FEM, representa la cantidad de deformación inducida en el sólido.
El tamaño de la muestra es de $(0{,}641 \text{ mm})^3$. En esta simulación, el fluido (agua) es inyectado por el lado izquierdo. El color de lado izquierdo representa la deformación. Para mostrar tres aspectos distintos, en la muestra se observa, desde la izquierda, el sólido conteniendo su deformación, la malla sólida y el dominio del fluido mostrado por el modelo de red de poros, respectivamente.

Contact: Pejman Tahmasebi <tahmasebi.pejman@gmail.com>

E. F. Médici and A. D. Otero (eds.), *Album of Porous Media*,
https://doi.org/10.1007/978-3-031-23800-0_100

HYBRID RANDOM WALKS FOR RADIATIVE/CONDUCTIVE HEAT TRANSFER IN POROUS MEDIA

Gerard Louis Vignoles
University of Bordeaux

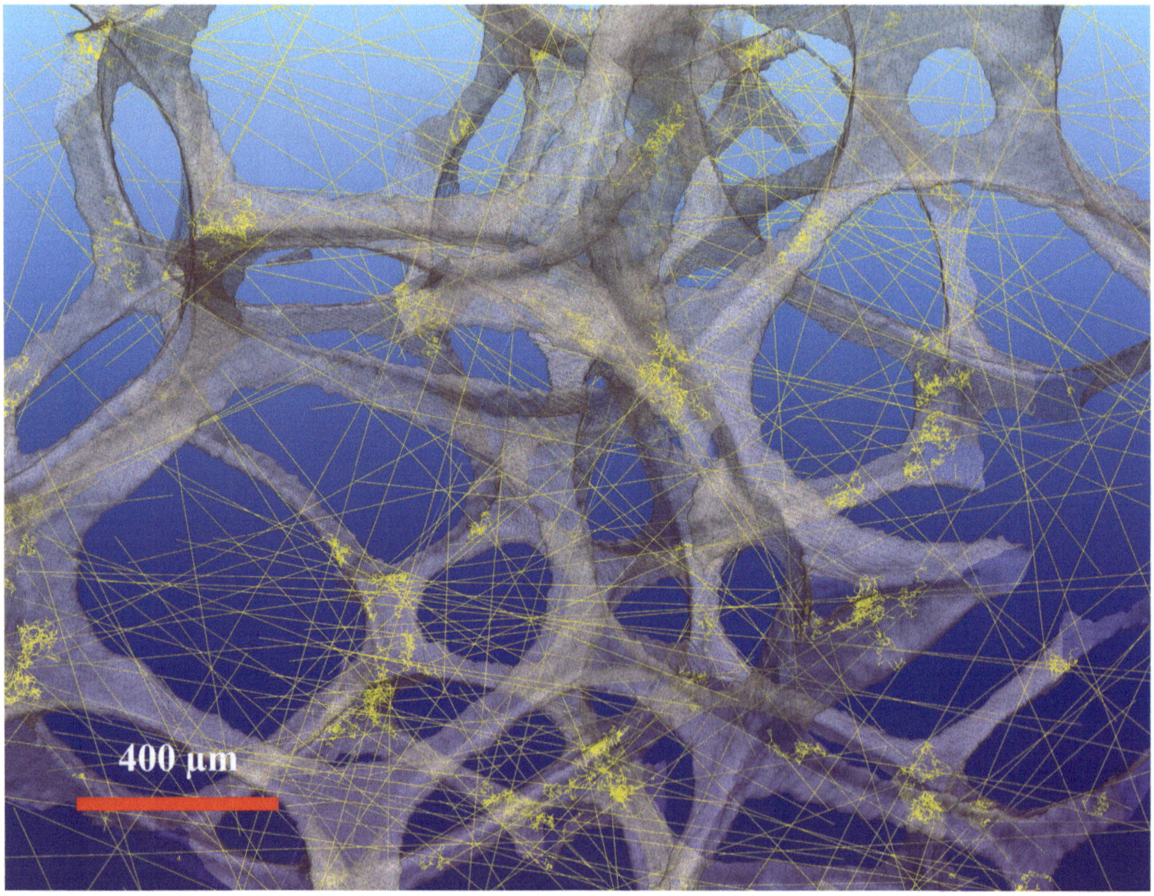

Coupled heat transfer by radiation and conduction is simulated by an original hybrid Monte-Carlo Random Walks scheme in a 3D representation of a porous medium discretized by a Simplified Marching Cube technique[1]. Here, the material is a ceramic open-cell foam (97 % porous, 60 pores per inch) used as a high-temperature insulator. The random walk (yellow lines) consists of straight paths in void space (blue background), representing radiative transfer, and sequences of uncorrelated short segments in the solid (semi-transparent gray zones), representing heat conduction. An average heat diffusivity is extracted from the statistics on thousands of random walks. The ratio between radiation and conduction is related to the probability, when a "conductive" walker hits the solid surface, to be "emitted" as a ray in the void. The main advantages of this algorithm are: (i) intimate coupling of radiation in pores and conduction in the solid, (ii) low memory requirements and (iii) rapid determination of averaged effective heat diffusivity.

1. Vignoles G.L. (2016). *International J. Heat and Mass Transfer*, 93:707.

Contact: Gerard Vignoles <vinhola@lcts.u-bordeaux.fr>

El efecto combinado de la transferencia de calor por radiación y conducción es simulado a través de un método de Monte Carlo híbrido original con un esquema de caminos aleatorios representando un medio poroso 3D discretizado con la técnica de Cubos de Marcha Simplificada[1]. En este caso, el material es una espuma reticulada de cerámica (97 % porosa, 6 poros por pulgada) utilizada como aislante a temperaturas elevadas. Los caminos aleatorios (líneas amarillas) consisten en segmentos rectos en el espacio vacío (fondo azul), que representan transferencia de calor por radiación, y secuencias de segmentos cortos desvinculados en la parte sólida (zonas grises medio transparentes) que representan la transferencia de calor por conducción. La difusividad térmica media se extrae de las estadísticas de miles de caminos aleatorios. La razón entre la radiación y la conducción está vinculada a la probabilidad, cuando un caminante «conductivo» alcanza la superficie sólida, para ser «emitido» en forma de rayo en el vacío. Las principales ventajas de este algoritmo son: (i) estrecho acoplamiento entre radiación en los poros y conducción en la parte sólida, (ii) baja capacidad de memoria necesaria y (iii) determinación rápida de la difusividad térmica efectiva promedio.

E. F. Médici and A. D. Otero (eds.), *Album of Porous Media*,
https://doi.org/10.1007/978-3-031-23800-0_101

FLOW IN BINARY POROUS MEDIA: VISCOUS DISSIPATION

Iván Colecchio[1], *Alejandro Boschan*[1], *Benoît Noetinger*[2], *and Alejandro Otero*[1,3]

[1] Facultad de Ingeniería, Universidad de Buenos Aires, Argentina
[2] IFP Energies Nouvelles, France
[3] Centro de Simulacion Computacional, CSC - CONICET, Argentina

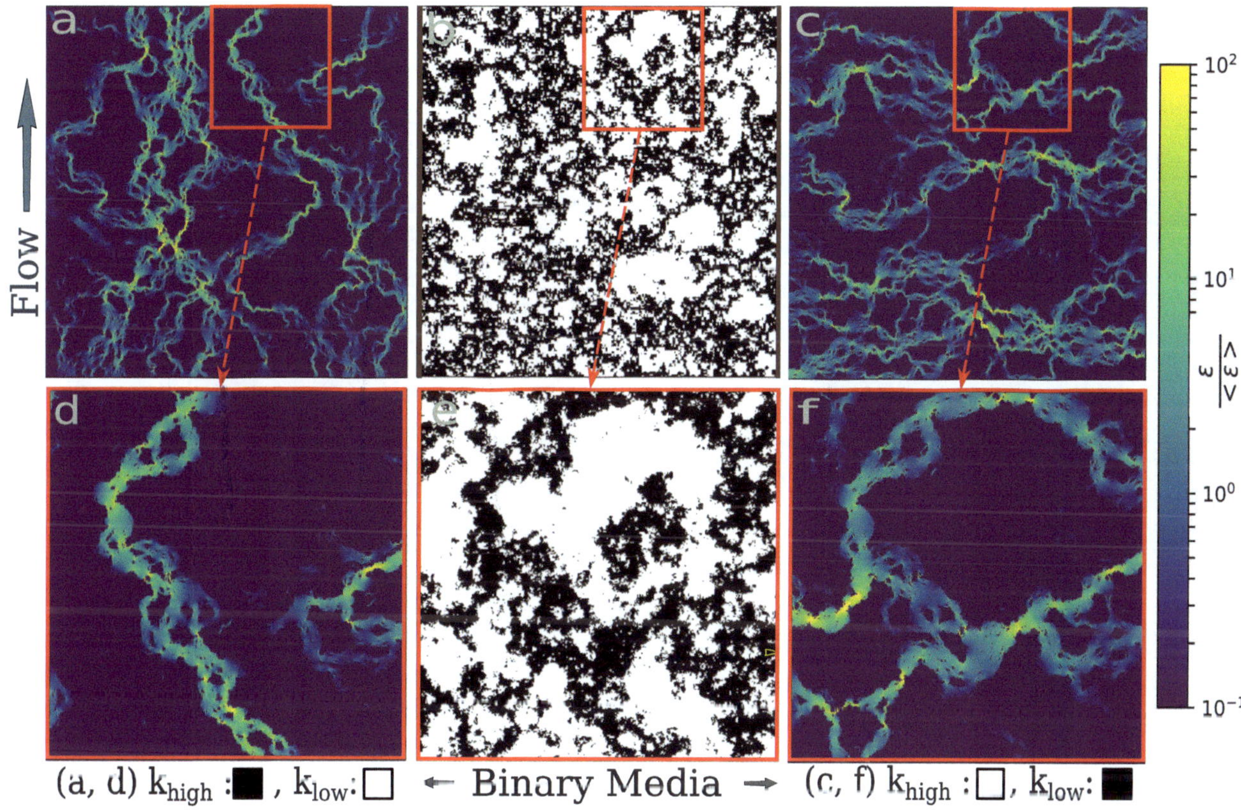

Flow and transport in porous media strongly depend on the connectivity of the high permeability component. Frequently, small critical regions determine the overall flow behavior. Energy dissipation approaches make it possible to detect these regions, yielding more accurate effective properties[1].
We've simulated flow in a synthetic binary medium (b) with a high conductivity (k_{high} = 100m/day) and a low one (k_{low} = 0.01m/day). No-flow boundary conditions are applied in the vertical boundaries while a pressure gradient exists between the inlet (bottom) and the outlet (top). The resulting energy dissipations maps are shown in (a) (with k_{high}: ■, k_{low} : □), and in (c) (with k_{high}: □, k_{low} : ■). Zooms over a small critical region are shown in the bottom row (d, e, f).
When k_{high} paths connect inlet and outlet (i.e. when percolation of the k_{high} component occurs) energy dissipation is distributed mostly along flow channels (a, d). Otherwise, channelization is absent and energy dissipation is distributed along barriers (c, f).

1. Colecchio I. *et al.* (2020). *Advances in Water Resources*, 140:103594.

El flujo y el transporte en medios porosos depende en gran parte de la conectividad del componente de alta permeabilidad. Con frecuencia, pequeñas regiones críticas determinan el comportamiento del flujo en general. Los enfoques basados en la disipación de energía permiten detectar estas regiones, dando propiedades efectivas más precisas[1].
Hemos simulado el flujo en un medio binario sintético (b), con una conductividad alta (k_{high} = 100/da) y otra baja (k_{low} = 0, 01m/da). Se aplicaron condiciones de contorno de flujo nulo en los lados verticales y un gradiente de presión entre la entrada (inferior) y la salida (superior). Los mapas de disipación de energía resultantes se muestran en (a) (con k_{high}: ■, k_{low} : □), y en (c) (con k_{high}: □, k_{low} : ■). La fila inferior (d, e, f) son ampliaciones de una región crítica pequeña.
Cuando los caminos de k_{high} conectan la entrada y la salida (por ejemplo cuando ocurre percolación en el componente k_{high}) la disipación de energía se distribuye principalmente a lo largo de canales de flujo (a, d). En el caso contrario, no ocurre canalización y la disipación de energía se distribuye a lo largo de barreras (c, f).

Contact: Iván Colecchio <icolecchio@fi.uba.ar>

E. F. Médici and A. D. Otero (eds.), *Album of Porous Media*,
https://doi.org/10.1007/978-3-031-23800-0_102

A SPHEREPACK AND ITS NAVIER-STOKES SOLUTION APPROXIMATION

Javier E. Santos and Masa Prodanovic
University of Texas at Austin

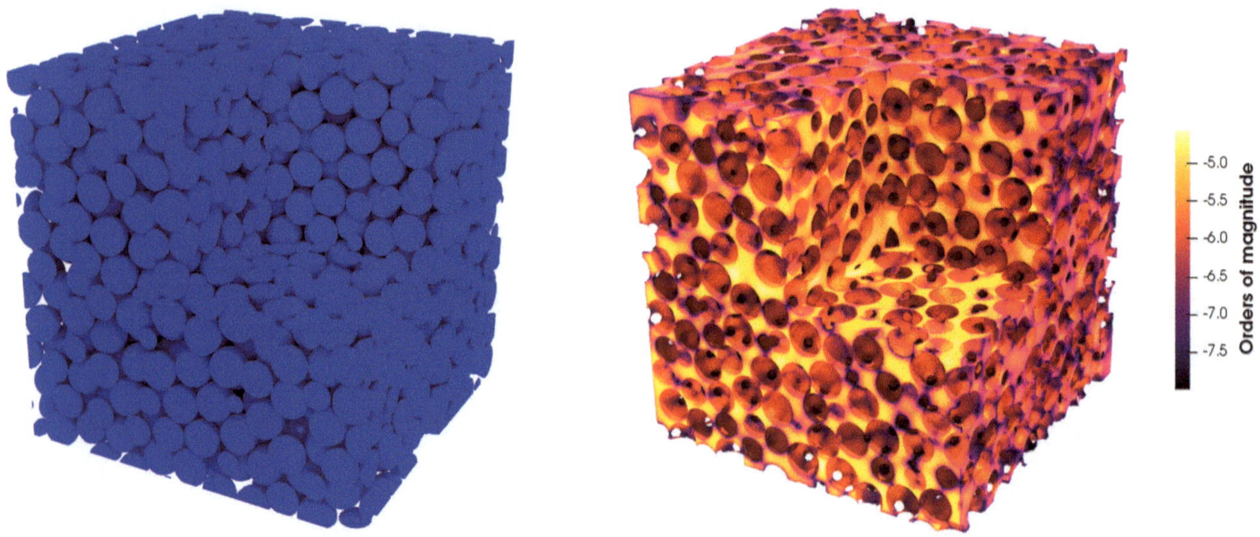

Understanding how fluids travel through subsurface rocks is crucial to design ground water management, hydrocarbon extraction, CO_2 sequestration, and contaminant remediation projects. Expressions that describe fluids traveling through simple domains exist, but they fail to forecast the behavior of fluids traveling through complex media. On the other hand, numerical methods like lattice Boltzmann provide a pathway to obtain the velocity field in complex geometries.

In this figure, the geometry, and the velocity field are shown. We used a small pressure gradient, implemented with constant pressure at the inlet (plane at the front of the image) and outlet (plane at the back of the image) to simulate laminar flow, which is used to compute permeability. The computational size of the domain is 512^3 and it can be scaled using the resolution of interest. This is valuable to assess flow channels in the subsurface (contaminant tracing, hydrocarbon movement in an oil reservoir) and bottlenecks for flow. The behavior of the fluids at the smaller scales is key to make inferences of larger domains; therefore, it is relevant to understand how fluids behave at the microscale.

Comprender cómo se dezplazan los fluidos a traves de la matriz porosa de yacimientos en el subsuelo es crucial para diseñar proyectos de sistemas de manejo de agua subterránea, extracción de hidrocarburos, secuestro de CO_2 y remediación de contaminación. Existen expresiones para describir el desplazamiento de fluidos en dominios simples, pero éstas no logran predecir el comportamiento de los fluidos que se desplazan a través de medios complejos. Por otro lado, los métodos numéricos como el lattice Boltzmann proporcionan una estrategia para obtener el campo de velocidades en geometrías complejas.

En esta figura se muestran la geometría y el campo de velocidades. Utilizamos un pequeño gradiente de presiones, implementado con presión constante en la entrada (plano frontal de la imagen) y en la salida (plano posterior de la imagen) para simular un flujo laminar, que se utiliza para calcular la permeabilidad. El tamaño computacional del dominio es de 512^3 y se puede escalar utilizando la resolución de interés. Esto es valioso para evaluar los canales de flujo preferenciales en el subsuelo (rastreo de contaminantes, movimiento de hidrocarburo en un yacimiento de petróleo) y restricciones en el flujo. El comportamiento de los fluidos en las escalas más pequeñas es clave para realizar inferencias en dominios mayores; por lo tanto, es relevante comprender cómo los fluidos se comportan en la microescala.

Contact: Javier E. Santos <jesantos@utexas.edu>

E. F. Médici and A. D. Otero (eds.), *Album of Porous Media*,
https://doi.org/10.1007/978-3-031-23800-0_103

FLOW IN TIGHT POROUS MEDIA

Pejman Tahmasebi
University of Wyoming

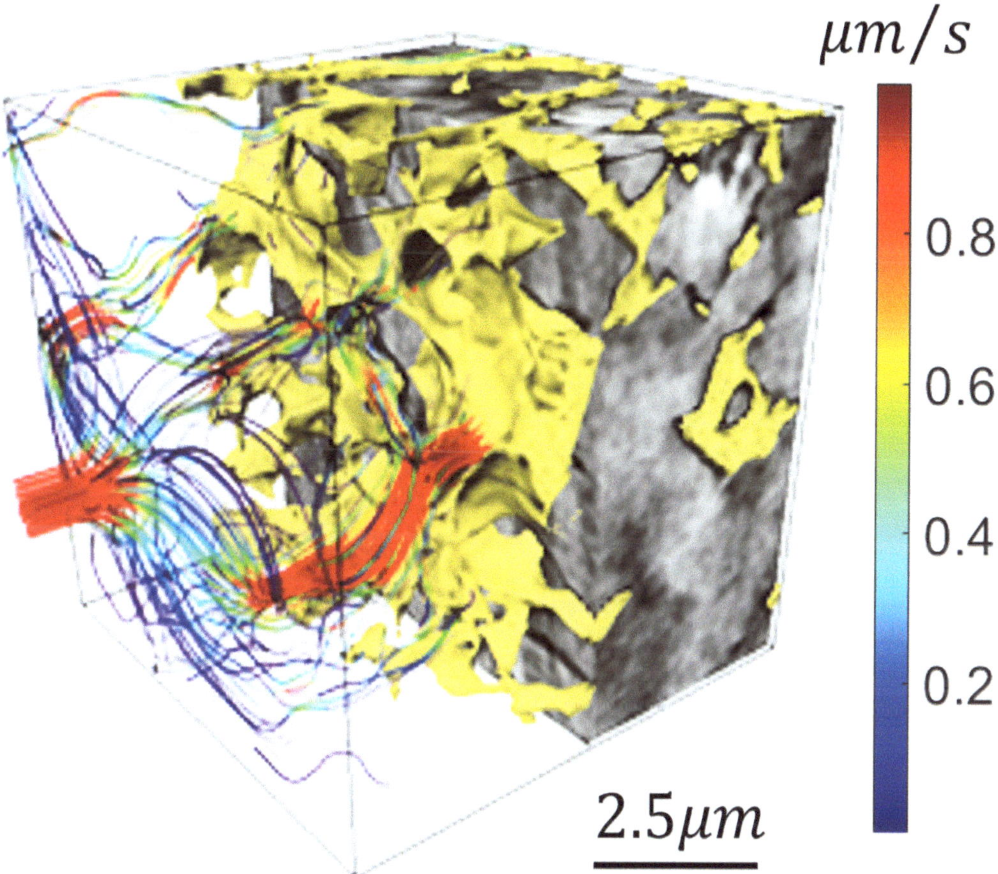

This image is the result of a direct flow simulation conducted in OpenFOAM and visualized in Paraview and it shows a shale sample accompanying flow streamlines for a gas. The flow is injected from one side and the outlet is in opposite direction, which is shown using streamlines and the velocity is normalized between 0 and 1. The yellow sections show the pore space and the color of the streamlines indicates the velocity which is higher in smaller pores. The size of the sample is $9\times9\times9\ \mu\text{m}^3$. The gray color represents the solid and the pores are shown in light yellow. The image is produced using SEM.

Esta imagen es el resultado de una simulación directa de flujo realizada en OpenFOAM y visualizada en Paraview, y en ella se observa una muestra de lutita acompañada por las líneas de corriente del flujo de un gas. El flujo es inyectado por un lado y la salida se encuentra en el lado opuesto, lo que se muestra utilizando las líneas de corriente y la velocidad que está normalizada entre 0 y 1. Las secciones amarillas muestran el espacio del poro y el color de las líneas de corriente indican la velocidad que es mayor en poros más pequeños. El tamaño de la muestra es de $9\times9\times9\ \mu\text{m}^3$. El color gris representa el sólido y los poros se muestran en amarillo claro. La imagen es producida con un MEB.

Contact: Pejman Tahmasebi <tahmasebi.pejman@gmail.com>

E. F. Médici and A. D. Otero (eds.), *Album of Porous Media*,
https://doi.org/10.1007/978-3-031-23800-0_104

SPATIAL VARIANCE OF PORE-SCALE FLUID VELOCITIES

Timo Koch[1], *Kilian Weishaupt*[2], *and Johannes Müller*[2]

[1] University of Oslo
[2] University of Stuttgart

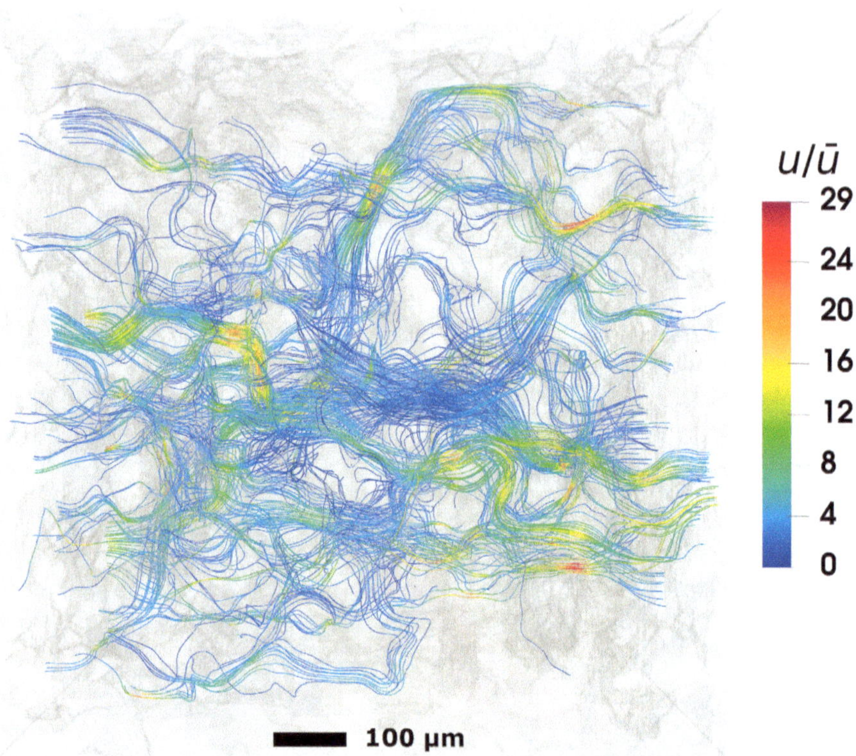

Simulation of fluid flow from left to right driven by a global pressure gradient through a Berea sandstone sample. The tortuous streamlines reflect the highly heterogeneous pore space structure. Local fluid velocity magnitudes (u) can exceed the average fluid velocity (ū) by orders of magnitude. Peak local Reynolds numbers in the narrowest pore throats estimated at 100 coincide with the appearance of recirculation structures.

Local velocities strongly affect transport and evolution processes. Narrow channels lead to a nonlinear increase of resistance in the non-Darcy flow regime. The local velocity field affects other processes such as the local convective heat transfer between fluid and solid, and pore space evolution by influencing mineral precipitation or biofilm growth.

Open-source software and data pipeline: the velocity field is computed with OpenFOAM[1] solving the Navier-Stokes equations on a computational grid based on binarized μCT images of Berea sandstone[2]. Streamline computation and rendering is performed in ParaView[3]. The image is based on Fig. 10 from [4] (CC-BY 4.0).

1. Jasak H. (2009). *Int. J. Naval Architecture and Ocean Eng.*, 1(2):89.
2. Dong H. and Blunt M.J. (2009). *Phys. Rev. E*, 80(3).
3. Ahrens J. *et al.* (2005). In *Visualization Handbook*, (717–731). Elsevier.
4. Koch T. *et al.* (2021). *Transp Porous Med*, 140(1):107.

Contact: Timo Koch <timokoch@math.uio.no>

Simulación del flujo de fluido guiado de izquierda a derecha por un gradiente de presión global a través de una muestra de arenisca de Berea. Las líneas de corriente tortuosas reflejan la estructura muy heterogénea del espacio poroso. Las magnitudes de la velocidad local del fluido (u) pueden exceder el promedio de la velocidad del fluido (ū) por varios órdenes de magnitud. Los números de Reynolds locales pico en la parte más estrecha de las gargantas de los poros que se estiman en 100 coinciden con la aparición de estructuras de recirculación.

Las velocidades locales afectan profundamente los procesos de transporte y evolución. Los canales estrechos conducen a un aumento no lineal de la resistencia en el régimen de flujo no-Darcy. La velocidad de flujo local afecta otros procesos como la transferencia de calor por convección local entre fluido y sólido, y la evolución del espacio poroso, al influenciar la precipitación mineral o el crecimiento de biopelículas.

Software de código abierto y secuencia de procesamiento de datos: la velocidad de flujo está calculada con OpenFOAM[1] resolviendo las ecuaciones Navier-Stokes en una grilla computacional basada en imágenes de μTC binarizadas de arenisca de Berea[2]. El cálculo y la visualización de las líneas de corriente se realizó con Paraview[3]. La imagen está basada en la Fig. 10 de [4] (CC-BY 4.0).

E. F. Médici and A. D. Otero (eds.), *Album of Porous Media*,
https://doi.org/10.1007/978-3-031-23800-0_105

FLUID FLOW IN SANDSTONE DIGITAL CORE MODEL

Kambiz Vafai[1] *and Jianchao Cai*[2]

[1] University of California, Riverside
[2] China University of Petroleum, Beijing, China

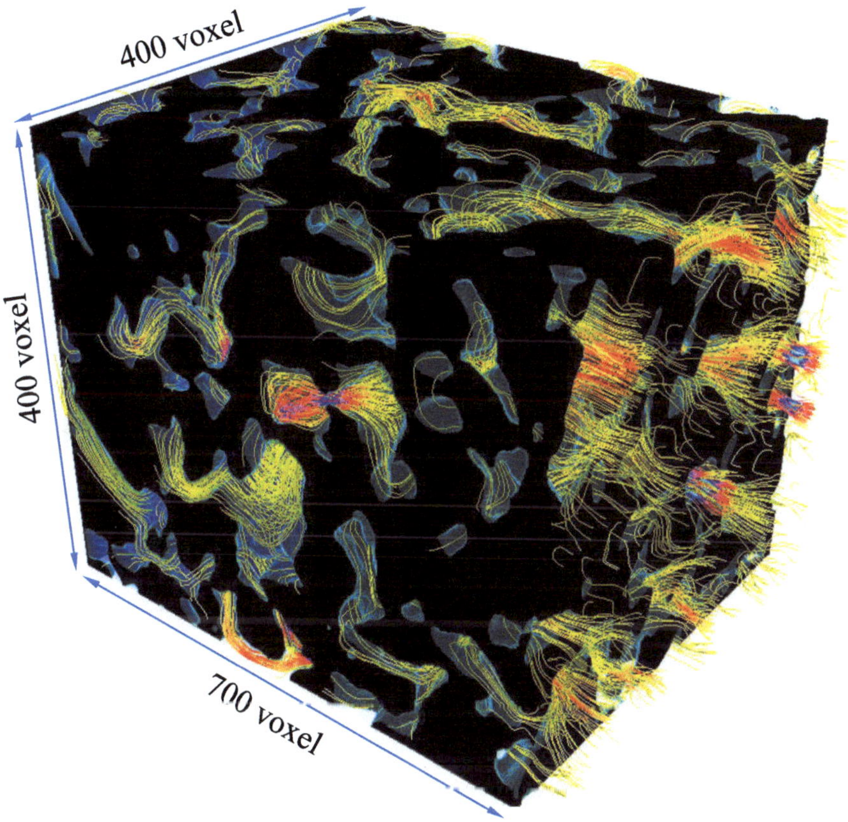

This figure presents fluid flow within the pore space of a saturated sandstone driven by a pressure difference. Flow path and flow rate are indicated by streamlines and different colors respectively. Low and high flow rates are mapped to yellow and red, respectively. Therefore, the velocity field distribution in the pore space can be clearly observed by this figure. The digital core model is acquired and reconstructed by CT scanning. The simulation of fluid flow in sandstone pore is obtained by using lattice Boltzmann method. The specific simulation settings are as follows: 1) Pressure boundaries are applied to the left and right boundaries to act as inlet and outlet boundaries. 2) Periodic boundaries are applied around the model, including the upper/lower and the front/back boundaries of the model. Under these conditions, the fluid flows from the left end to the right of the sandstone model. The black color represents the rock skeleton. The pore space is rendered in blue with different degrees of transparency. Streamlines with different colors represent the flow path of fluid in the pore.

Contact: Kambiz Vafai <vafai@engr.ucr.edu>

La figura presenta el flujo del fluido dentro del espacio poroso de arenisca saturada guiado por el gradiente de presión. El camino del flujo y el caudal se encuentran indicados por las líneas de corriente y los distintos colores respectivamente. Caudales altos y bajos están señalizados en amarillo y rojo, respectivamente. Por lo tanto, la distribución del campo de velocidades en el espacio poroso puede observarse claramente en esta figura. El modelo del núcleo digital es adquirido y reconstruido mediante una TC. La simulación del flujo del fluido en poros de arenisca se obtiene utilizando el método Lattice Boltzmann. Los ajustes específicos de la simulación son los siguientes: 1) Se aplican condiciones de borde de presión en los contornos izquierdo y derecho para actuar como fronteras de entrada y salida. 2) Se aplican fronteras periódicas alrededor del modelo, incluyendo los contornos superior e inferior, frontal y trasero del modelo. Bajo estas condiciones, el fluido fluye del contorno izquierdo al derecho del modelo de arenisca. El color negro representa el esqueleto de roca. El espacio poroso se visualiza en azul con distintos grados de transparencia. Las líneas de corriente con distintos colores representan el camino del flujo del fluido en el poro.

E. F. Médici and A. D. Otero (eds.), *Album of Porous Media*,
https://doi.org/10.1007/978-3-031-23800-0_106

FLOW THROUGH THE MICROSTRUCTURE OF NONWOVEN AIR FILTER MEDIA

Mehdi Azimian
Math2Market GmbH

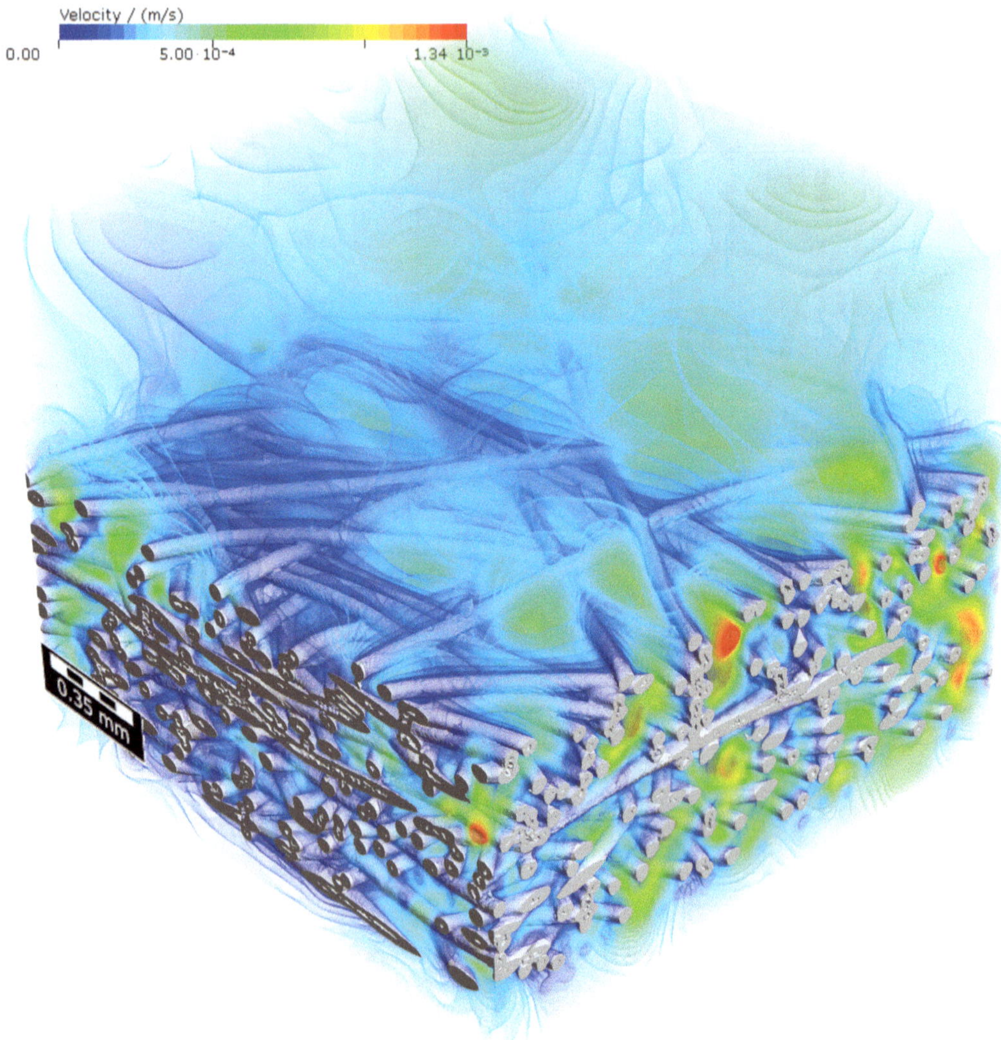

Image shows the flow field computed for the microstructure of a nonwoven air filter media. An air filter media consisting of hollow and solid circular fibers was scanned by μCT and is shown in gray. The flow field was computed for a pressure drop between the bottom (inlet) and top (outlet) of the filter media by solving the Navier-Stokes equations using the GeoDict software package. Flow direction is from top to bottom of the image. We compute creeping flow based on the steady Stokes equation and the Reynolds number is zero. The velocity is rendered with colors ranging from low (blue) to high (red) in the pore space before, after, and between the fibers. The flow velocity also governs the transparency, so that one can see through low velocity areas, while high velocity (red) areas appear rather solid.

Contact: Barbara Planas <barbara.planas@math2market.de>

La imagen muestra el campo de flujo calculado en la microestructura de un medio no entretejido para el filtrado de aire. Un medio para filtrado de aire compuesto por fibras circulares sólidas y huecas fue escaneado mediante μTC y se muestra en gris. Se calculó el campo de flujo para una caída de presión entre la parte inferior (entrada) y la superior (salida) del filtro, resolviendo las ecuaciones de Navier–Stokes utilizando el paquete de software GeoDict. En la imagen, la dirección de flujo va desde arriba hacia abajo. Calculamos el flujo de arrastre según la ecuación de Stokes con un número de Reynolds cero. Se visualiza la velocidad con colores que muestran de la más baja (azul) a la más alta (rojo) en el espacio poroso antes, después y entre las fibras. La velocidad del flujo también determina la transparencia, por lo que se puede ver a través de las zonas de baja velocidad, mientras que las de alta velocidad (rojo) son opacas.

E. F. Médici and A. D. Otero (eds.), *Album of Porous Media*,
https://doi.org/10.1007/978-3-031-23800-0_107

TORTUOSITY OF THE FLOW IN POROUS MEDIA

Maciej Matyka[1], *Zbigniew Koza*[1], *and Arzhang Khalili*[2]

[1] Institute for Theoretical Physics, Faculty of Physics and Astronomy, University of Wrocław, Poland
[2] Max Planck Institute for Marine Microbiology, Germany

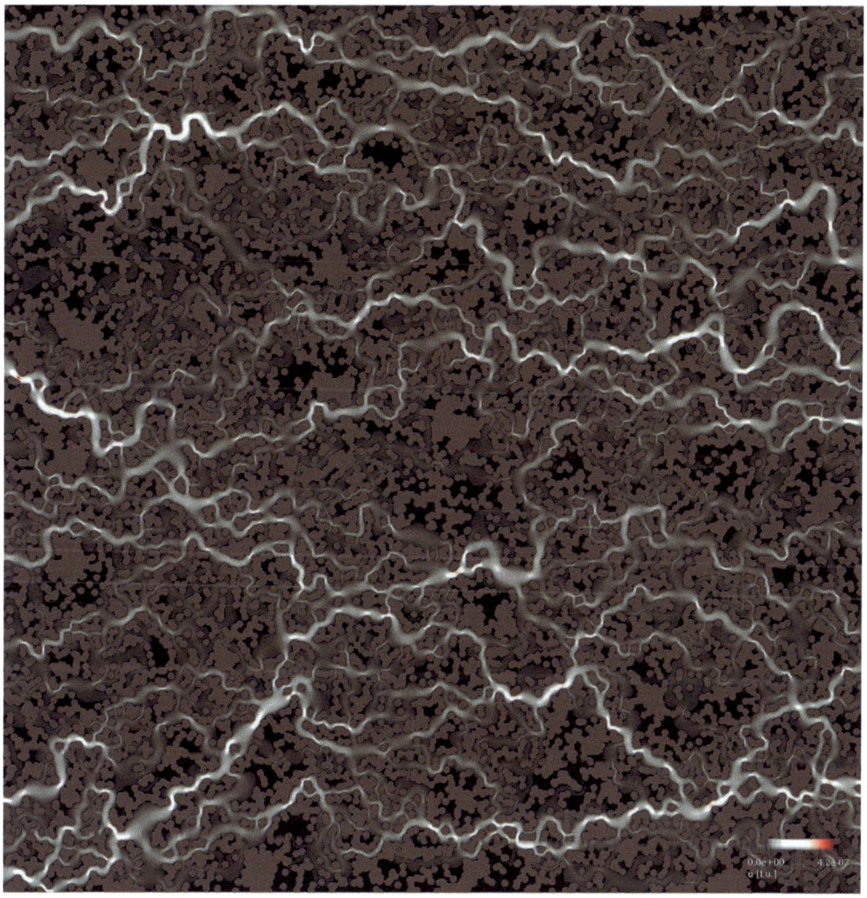

Irrespective of the type of the solid matrix shaping a porous medium, the interconnected voids appear as highly complex and tortuous paths. Here, we visualize the velocity field of a fluid flowing from left to right through such a tortuous path within a digital model of a thin layer of porous media. A porous matrix is immersed in the fluid. The deviation from a straight path line of the flow is a direct indication of the phenomena called tortuosity, an important physical property of a porous medium[1].

The absolute momentum visualized as a color gradient conveys information on the paths chosen by the transport mechanism. Brighter sections correspond to larger fluid velocities (with top peak as red color). It wouldn't be visible without proper adjustment of color/brightness/contrast.

The sample has 1000000 (one million) of computational cells. To find the solution to the fluid flow we used the Palabos code which implements the Lattice Boltzmann method[2].

1. Matyka M. *et al.* (2008). *Phys. Rev. E*, 78(2).
2. Latt J. *et al.* (2021). *Computers & Mathematics with Applications*, 81:334.

Independientemente del tipo de matriz sólida que da forma al medio poroso, la interconexión de los huecos aparecen como caminos sumamente complejos y tortuosos. Aquí, visualizamos la velocidad de flujo de un líquido que fluye de izquierda a derecha a través de esos caminos en un modelo digital de medio poroso delgado. Una matriz porosa es sumergida en el fluido. La desviación del flujo con respecto al camino recto del flujo es una indicación directa del fenómeno llamado tortuosidad, una importante propiedad física de un medio poroso[1].

La cantidad de movimiento absoluta, visualizada como un gradiente de colores, provee información sobre los caminos elegidos por el mecanismo de transporte. Las secciones más brillantes corresponden a mayores velocidades del fluido (con un pico en el color rojo). Esto no sería visible sin un ajuste adecuado de color, brillo y contraste.

La muestra tiene 1000000 (un millón) de celdas computacionales. Para encontrar la solución del flujo del fluido utilizamos el código Palabos que implementa el método Lattice Boltzmann[2].

Contact: Maciej Matyka <maciej.matyka@uwr.edu.pl>

E. F. Médici and A. D. Otero (eds.), *Album of Porous Media*,
https://doi.org/10.1007/978-3-031-23800-0_108

HIGH RESOLUTION SIMULATION OF FLUID FLOW IN PRESS FELTS USED IN PAPER MANUFACTURING

David Trebotich and Terry Ligocki
Lawrence Berkeley National Laboratory

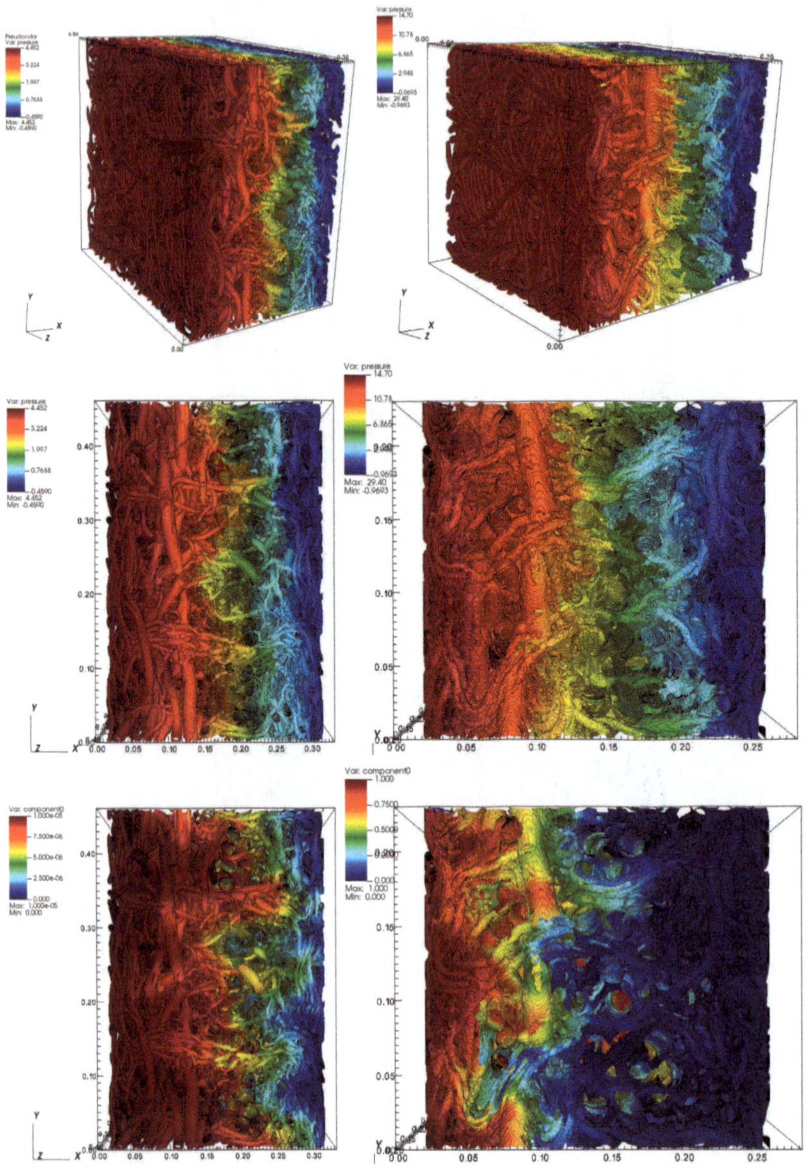

Se ha desarrollado un sistema computacional para simular con precisión el flujo de humedad desde o hacia la pulpa porosa (deshidratación/mojado) durante el proceso de prensado en la producción de papel. Una comprensión predictiva de la física de este proceso de absorción conllevará al desarrollo de medidas de prevención para la reabsorción del agua por parte del papel o pulpa comprimidos, y por lo tanto, reducirá la energía de secado. Las simulaciones pueden utilizarse para analizar la influencia de la heterogeneidad a escala del poro en el flujo de agua a través de las complejas estructuras porosas de fieltro, y para calibrar parámetros macroscópicos utilizados en modelos continuos como los de porosidad y permeabilidad absolutas, así como correlaciones de porosidad-permeabilidad para distintas calidades de diseños de fieltro. Presentamos los resultados de la simulación del flujo de agua en estructuras porosas complejas de fieltros de prensado para secar papel sin comprimir (imágenes de la izquierda, 0,1 MPa, grilla de 832×1152×1152) y comprimidas (imágenes de la derecha, 5,0 MPa, grilla de 1408×1152×1152). El panel de la imagen muestra la presión del fluido en 3D (superior), un corte transversal en 2D de las presiones (centro) y un corte transversal en 2D del frente del trazador (inferior). Datos de la imagen de fieltro comprimido obtenidos de J.M. Voith SE & Co. KG a través de Agenda 2020 Technology Alliance.

A computational framework has been developed to accurately simulate flow of moisture from/to the porous pulp medium (dewatering/rewetting) during the pressing process in paper making. Predictive understanding of the physics of this wicking process will lead to development of preventative measures for the reabsorption of water by the compressed paper/pulp, thereby reducing drying energy. Simulation can be used to analyze the influence of pore scale heterogeneity on water flow through complex pore structures in felts, and further calibrate macroscopic parameters used in continuum models such as bulk porosity and permeability, as well as porosity-permeability correlations for different grades of felt designs. We present simulation results of water flow in complex pore structures of paper drying press felts under uncompressed (left images, 0.1 MPa, 832×1152×1152 grid) and compressed (right images, 5.0 MPa, 1408×1152×1152 grid) conditions. Image panel shows fluid pressure in 3D (top), with 2D slice of pressure (middle), and tracer front in 2D slice (bottom). Press felt image data obtained from J.M. Voith SE & Co. KG through the Agenda 2020 Technology Alliance.

Contact: David Trebotich <dptrebotich@lbl.gov>

E. F. Médici and A. D. Otero (eds.), *Album of Porous Media*,
https://doi.org/10.1007/978-3-031-23800-0_109

VISUALIZATION OF WATER ADSORPTION IN CLAY WITH COMPLEX SURFACE WETTABILITY

Rui Xu and Masa Prodanovic
The University of Texas at Austin

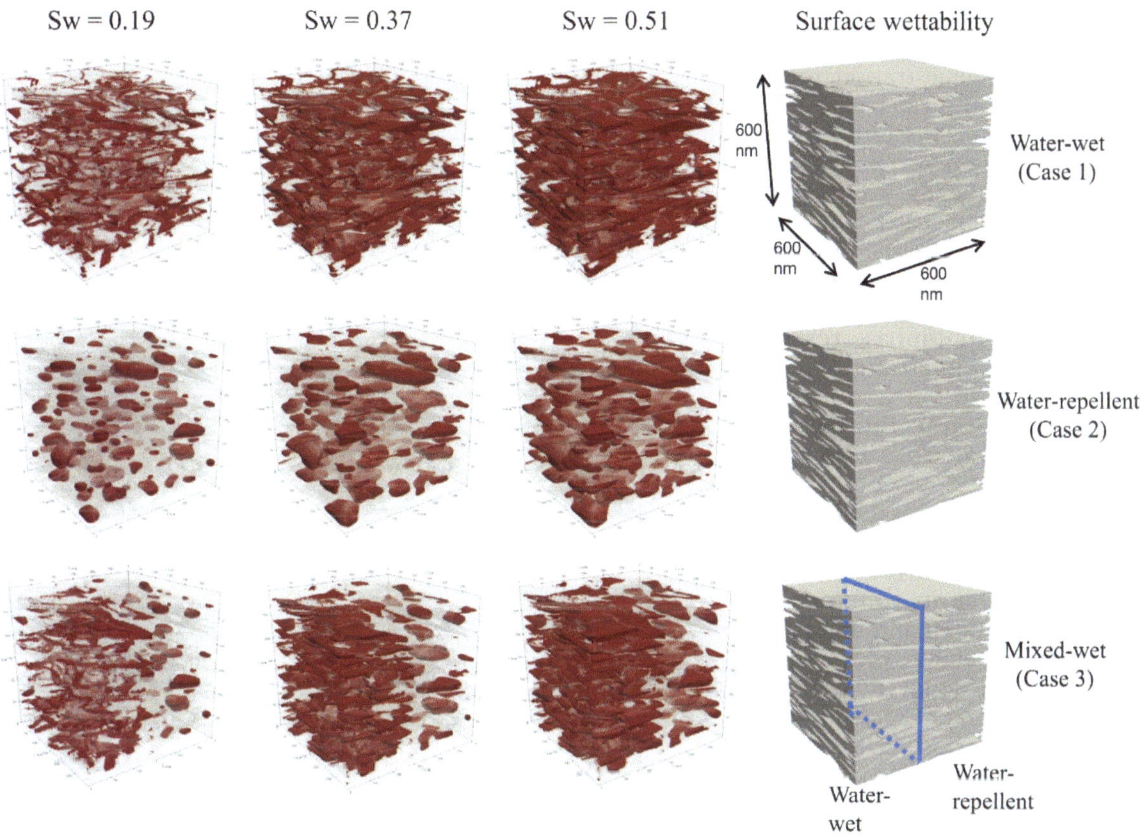

In our previous work[1], we used lattice Boltzmann method (LBM) to simulate water adsorption in a 3D reconstructed clay structure (a 600 nm^3 cube with a voxel resolution of 2 nm). Shown in the figure is the equilibrium 3D distributions of condensed liquid water phase at 400 Kelvin in the clay pore system for three wettability cases (Case 1: water-wet, Case 2: water-repellent, and Case 3: mixed-wet) at three water saturations Sw = 0.18, 0.37, and 0.51. Red indicates the condensed liquid water, and the clay particles are shown in semitransparent gray. The surface wettability of the three cases is shown in the last vertical track.

Results show that water tends to accumulate at the corners and surfaces of the hydrophilic (i.e., water-wet) pore walls, while for hydrophobic (i.e., water-repellent) surfaces, water condenses in the center of the pore space, forming quasispherical vapor/liquid interfaces as a result of the force balance among surface tensions. A mixed behavior is observed for the mixed-wet case.

1. Xu R. *et al.* (2020). *Water Resour. Res.*, 56(10)

Contact: Rui Xu <rui_xu@utexas.edu>

En nuestro trabajo previo[1], utilizamos el método Lattice Boltzmann (LBM) para simular la adsorción del agua en una estructura de arcilla reconstruida en 3D (un cubo de 600 nm^3 con una resolución de vóxel de 2 nm). En las figuras se observan las distribuciones 3D en equilibrio de la fase de agua líquida condensada a 400 Kelvin en el sistema de poros de la arcilla en tres casos de mojabilidad (caso 1: mojable por agua, caso 2: repelente de agua y caso 3: mojabilidad intermedia) para tres saturaciones de agua Sw = 0,18, 0,37 y 0,51. El rojo indica el agua líquida condensada y las partículas de arcilla se muestran en gris semitransparente. La mojabilidad de la superficie de los tres casos se indica en la última columna vertical.

Los resultados muestran que el agua tiende a acumularse en las esquinas y en las superficies de los contornos porosos hidrofílicos (es decir, mojable por agua); mientras que en las superficies hidrofóbicas (es decir, repelente de agua), el agua se condensa en el centro del espacio poroso, formando interfaces de vapor y líquido casi esféricas como resultado del equilibrio de fuerzas entre las tensiones superficiales. Se observa un comportamiento mixto en el caso de mojabilidad intermedia.

E. F. Médici and A. D. Otero (eds.), *Album of Porous Media*,
https://doi.org/10.1007/978-3-031-23800-0_110

CAPROCK TOPOGRAPHY AND INJECTION POSITION INFLUENCE ON THE SOLUBILITY TRAPPING PHENOMENA DURING CO_2 GEOLOGICAL SEQUESTRATION

Pradeep Reddy Punnam, Dasika Prabhat Sourya, and Vikranth Kumar Surasani
Birla Institute of Technology and Science, Pilani, Hyderabad Campus

In CO_2 geological sequestration, multiphase flow of CO_2-water and solubility trapping i.e., dissolution of CO_2 in water, triggers series of reactions between aqueous CO_2 and minerals[1]. Injection location and caprock topography of the geological formation controls the CO_2 migration and solubility trapping[2]. To elucidate this effect, a multiphase reactive transport simulation using PFLOTRAN is conducted on a domain (10 km×10 km×0,1 km) with the injection location away from anticline structure.
The solubility fingering was quantified with CO_2 mole fraction and Rayleigh-Darcy number; it occurred due to the instability of density differences between CO_2 saturated and connate water[1]. During injection, plume migrated into anticline dome and facilitated more interaction of CO_2 with water. Post injection period, CO_2 restricted within anticline dome structure, which can i) prevent the CO_2 migration to geological faults and cracks[2], ii) delay the solubility trapping due to stagnant CO_2 under the anticline dome[1]. This will control rapid mineralization and porosity change, which is crucial in the structural integrity of the domain[3].

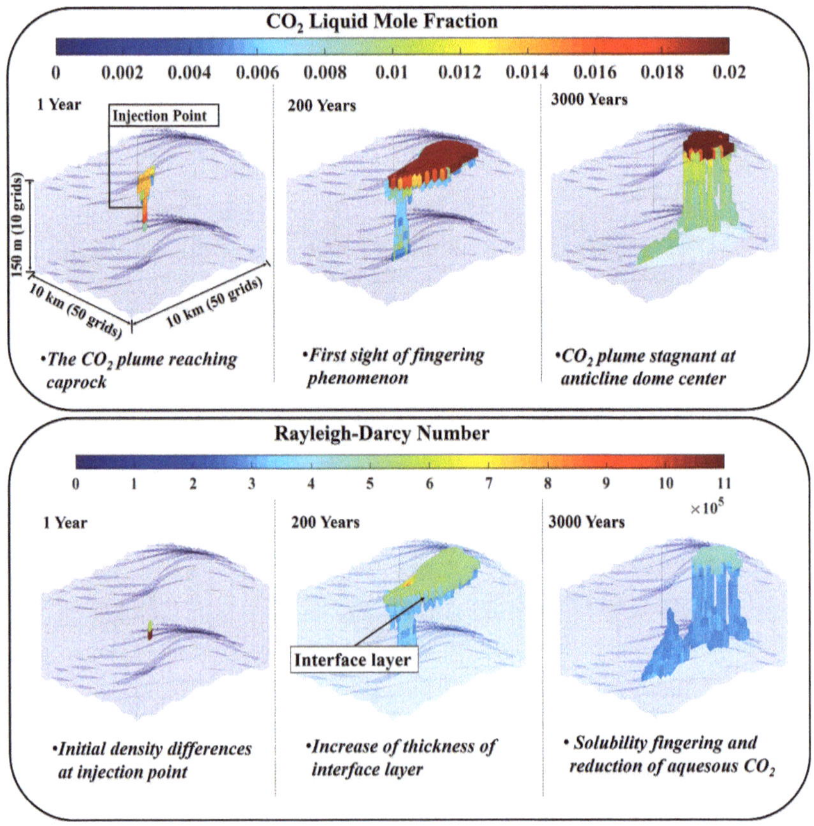

En el secuestramiento geológico de CO_2, el flujo multifásico de agua y CO_2 y la retención por solubilidad, es decir, la disolución de CO_2 en agua, desencadenan una serie de reacciones entre el CO_2 acuoso y los minerales[1]. La ubicación de la inyección y la topografía de las rocas de recubrimiento de la formación geológica controlan la migración de CO_2 y la retención por solubilidad[2]. Para dilucidar este efecto, se lleva a cabo una simulación de transporte reactivo multifásico utilizando PFLOTRAN en un dominio (10 km×10 km×0,1 km) con el emplazamiento de la inyección alejado de la estructura anticlinal.
La digitación de la solubilidad se cuantificó con una fracción molar del CO_2 y el número de Rayleigh-Darcy, la cual ocurrió debido a la inestabilidad por las diferencias de densidad entre el CO_2 saturado y el agua connata[1]. Durante la inyección, la pluma migró a una cúpula anticlinal facilitando una mayor interacción entre el CO_2 y el agua. Después del periodo de inyección, el CO_2 está restringido a la estructura de la cúpula anticlinal, lo que i) evita la migración de CO_2 a fallas y fisuras geológicas[2], ii) retrasa la retención por solubilidad debido al CO_2 estancado bajo la cúpula anticlinal[1]. Esto controlará el rápido cambio de mineralización y porosidad, lo que es crucial en la integridad estructural del dominio[3].

1. Lu C. and Lichtner P.C. (2007). *J. Phys.: Conf. Ser.*, 78:012042.
2. Nilsen H.M. *et al.* (2012). *International J. Greenhouse Gas Control*, 11:221.
3. Jayne R.S. *et al.* (2019). *Greenhouse Gas Sci Technol*, 9(5):979.

Contact: Vikranth Kumar Surasani <surasani@hyderabad.bits-pilani.ac.in>

E. F. Médici and A. D. Otero (eds.), *Album of Porous Media*,
https://doi.org/10.1007/978-3-031-23800-0_111

PORE-SCALE CONTROLS ON CALCITE DISSOLUTION USING DIRECT NUMERICAL SIMULATIONS

Sergi Molins and David Trebotich
Lawrence Berkeley National Laboratory

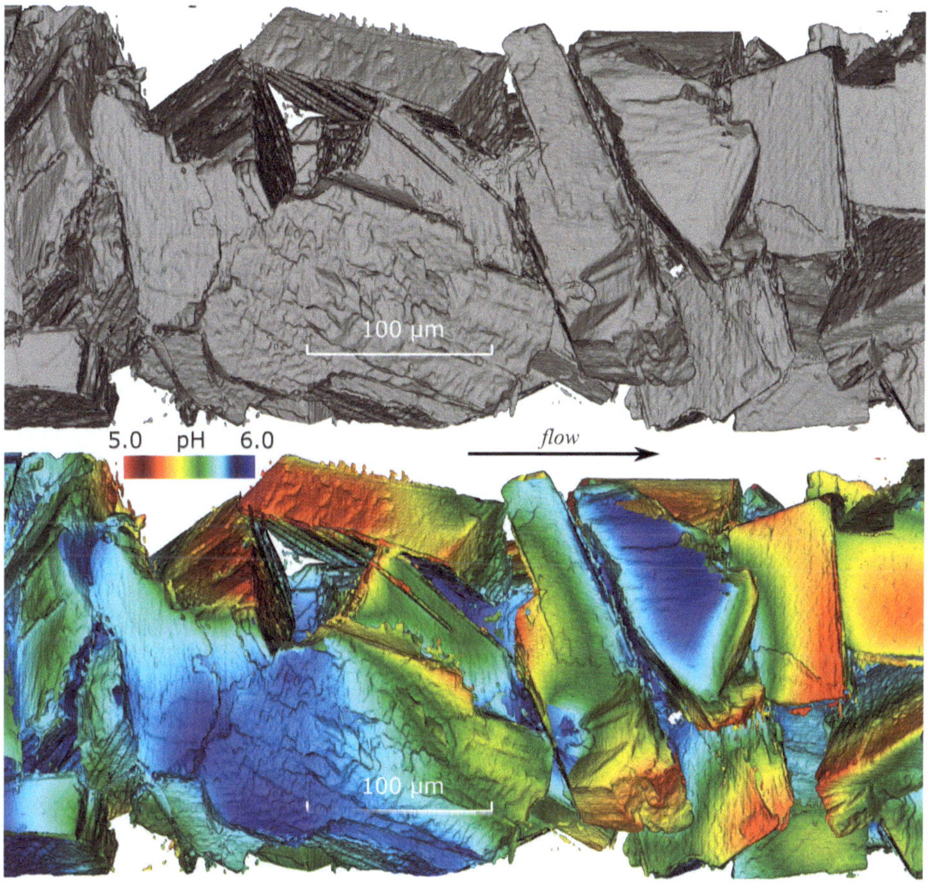

The complexity of pore geometry results in local variations of fluid velocity that affect the interplay between advective and diffusive transport. Dissolution rates are a function of the solute concentrations that are in direct contact with the mineral surfaces. In the image, the variability of pH values on the mineral surfaces indicate the existence of distinct reactive transport regimes in areas in close proximity to each other (bottom). This result was obtained by simulating calcite dissolution driven by the flow of a high-CO_2 solution using the code Chombo-Crunch. A $6144 \times 512 \times 512$ grid was used to discretize the geometry of the crushed calcite grains packed in a 0.7 cm long 0.5 mm in-diameter capillary tube with 1.16 μm resolution. The domain geometry (top) was generated from X-ray computed microtomography data. High-resolution simulations of fluid flow and reactive transport enable exploration of differences in geochemical conditions in the fluid near mineral surfaces[1, 2].

La compleja geometría de los poros resulta en variaciones locales de velocidad de fluido que afectan la relación entre el transporte advectivo y difusivo. Las tasas de disolución son una función de las concentraciones de los solutos que están en contacto directo con las superficies minerales. En la imagen, la variabilidad de los valores del pH en las superficies minerales indican la existencia de diferentes regímenes de transporte reactivo muy cercanos entre sí (abajo). Este resultado fue obtenido al simular la disolución de calcita guiada por el flujo de una solución con alto contenido de CO_2 utilizando el código Chombo-Crunch. Se utilizó una grilla de $6144 \times 512 \times 512$ para discretizar la geometría de los granos de calcita triturada y empaquetada en un tubo capilar de 0,7 cm de largo y 0,5 mm de diámetro con una resolución de 1,16 μm. La geometría del dominio (arriba) fue generada con datos de una microtomografía computarizada de rayos X. Las simulaciones de alta resolución del flujo de fluido y de transporte reactivo permiten explorar las diferencias en las condiciones geoquímicas del fluido próximo a superficies minerales[1, 2].

1. Molins S. *et al.* (2012). *Water Resour. Res.*, 48(3).
2. Molins S. *et al.* (2014). *Environ. Sci. Tech.*, 48(13):7453.

Contact: Sergi Molins <smolins@lbl.gov>

E. F. Médici and A. D. Otero (eds.), *Album of Porous Media*,
https://doi.org/10.1007/978-3-031-23800-0_112

DROPLET DETACHMENT FROM A GAS DIFFUSION LAYER OF A PEM FUEL CELL

Daniel Niblett[1], Vahid Niasar[1], Adrian Mularczyk[2], and Jens Eller[2]

[1] University of Manchester
[2] Paul Scherrer Institut

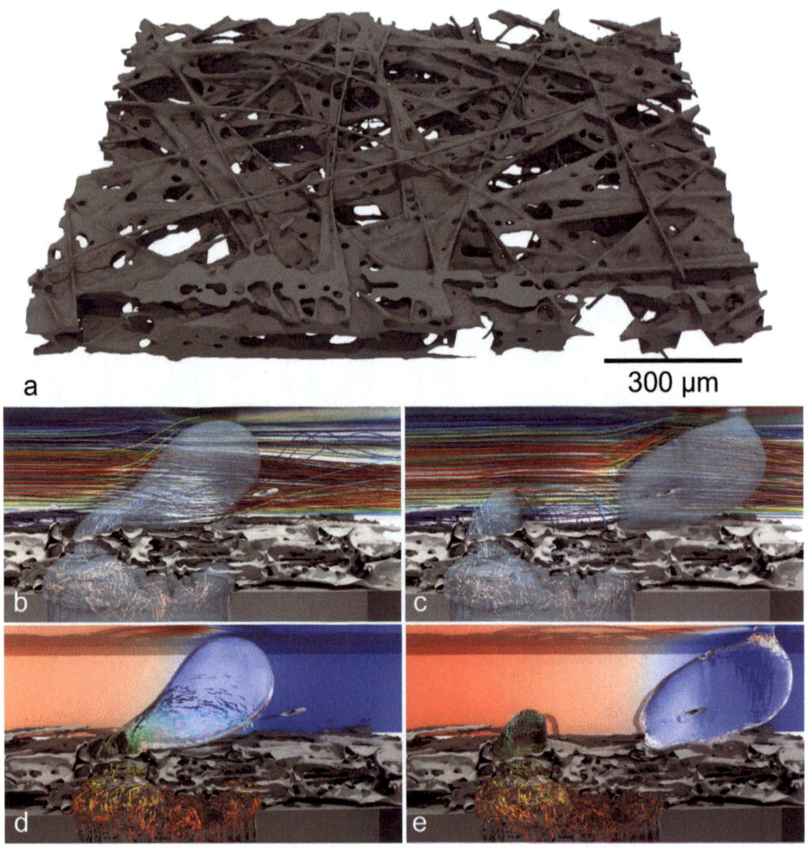

Water produced during polymer electrolyte membrane (PEM) fuel cell operation flows through the gas diffusion layer (GDL) which is simulated in an *ex situ* situation with liquid water injection from a laser perforated plate. The structure of an X-ray computed tomography microstructure SGL 25 BA can be visualised (a). Water percolates through the hydrophobic pore space based on the lowest entry capillary pressure pathway (b). The emergence of a droplet in the air supply channels decreases cluster capillary pressure (b and d), limiting further interface expansion. The presence of water in the channel causes acceleration of air around the droplet, leading to momentum exchange in the form of inertial and shear forces. This increases the air phase pressure drop (color gradient at channel walls d - e). Once these forces overcome the pinning capillary force, the droplet is detached and transported in the channel (c and e). (b - e) simulated using volume of fluid method (OpenFOAM) using the X-ray structure (a)[1, 2]. Constant flux of water (0.029 m/s) and air (15 m/s) were applied to the circular hole below the GDL and normal to the channel inlet, respectively. Rendering in Paraview.

El agua producida durante la operación de pilas de combustible de membrana de polímero electrolítico (PEM por sus siglas en inglés) fluye a través de la capa de difusión de gas (GDL, por sus siglas en inglés) simulada en una situación *ex situ* con una inyección de agua líquida desde un plato perforado por láser. La estructura de la microestructura de una SGL 25 BA puede visualizarse mediante tomografía computarizada con rayos X en (a). El agua percola a través del espacio poroso hidrofóbico siguiendo el camino de menor presión capilar de entrada (b). La aparición de una gota en los canales de suministro de aire disminuye la presión capilar que actúa sobre el agua (b y d), limitando una posterior expansión de la interfaz. La presencia de agua en el canal causa la aceleración del aire alrededor de la gota, lo que conlleva un intercambio de cantidad de movimiento en forma de fuerzas de inercia y corte. Esto aumenta la caída de la presión en la fase de aire (gradiente de colores en las paredes del canal d - e). Una vez estas fuerzas superan la fuerza capilar de anclaje, la gota se desprende y se desplaza por el canal (c y e). (b - e) Simulación mediante el método de volumen de fluido (OpenFOAM), utilizando la estructura obtenida por rayos X (a)[1, 2]. Se aplicó un flujo constante de agua (0,029 m/s) a través del orificio circular debajo de la GDL y de aire (15 m/s) en la dirección normal a la entrada del canal. Visualización con Paraview.

1. Niblett D. *et al.* (2020). *J. Power Sources*, 471:228427.
2. Mularczyk A. *et al.* (2020). *J. Electrochem. Soc.*, 167(8):084506.

Contact: Vahid Niasar <vahid.niasar@manchester.ac.uk>

E. F. Médici and A. D. Otero (eds.), *Album of Porous Media*,
https://doi.org/10.1007/978-3-031-23800-0_113

ENGLISH-SPANISH GLOSSARY OF POROUS MEDIA

Leyre Alegre-Figuero[1], *Ezequiel F. Médici*[1], *and Alejandro D. Otero*[2,3]

[1] Michigan Technological University, USA
[2] Facultad de Ingeniería, Universidad de Buenos Aires, Argentina
[3] Centro de Simulacion Computacional, CSC - CONICET, Argentina

This glossary is intended as an aid to readers of this book, whether they are researchers seeking general clarification or writing about any of the topics discussed here. The *Album of Porous Media – Structure and Dynamics* encompasses a wide range of topics and applications related to porous media. It was quite a challenge to switch between the different specializations covered, such as physics, chemistry, biology, geology... Throughout these pages we have tried to keep the terminology consistent between the chapters, and thus achieve coherence between the texts and the variety of voices. The reference framework for the translation of these chapters was scientific writing in Spanish. Various sources were used, but *Cómo traducir y redactar textos científicos en español. Reglas, ideas, consejos*, from M. Gonzalo Claros Díaz [1] was of great help. This translation, of course, would not have been possible without the support, collaboration, and knowledge of the co-editors who edited, corrected, rephrased the translation and consulted with specialists, and sometimes the authors themselves, terms and concepts. This list of terms is not intended to be an exhaustive compilation of specialized porous media terminology, but rather a list of those terms for which we have had to find a solution. These are usually terms that are either from a specialized field or have sparked discussions about their use in that precise context. We have reduced the number of entries as much as possible, eliminating those terms of general knowledge. Left for a future endeavor is the more comprehensive job in which we will embody these discussions into terminological entries of greater value for future projects where lexical information, examples, synonyms, false cognates, preferences according to contexts or usage nuances in both languages are incorporated.

El presente glosario está pensado como una ayuda a los lectores de este libro tanto si son investigadores que buscan una aclaración en general, como si están escribiendo sobre alguno de los temas aquí tratados. El *Album of Porous Media – Structure and Dynamics* comprende un amplio abanico de usos y aplicaciones de los medios porosos. Fue todo un reto saltar entre las diferentes especializaciones abarcadas, como física, química, biología, geología... A lo largo de estas páginas se ha intentado mantener la consistencia de la terminología entre los capítulos, y así lograr coherencia entre los textos y la variedad de voces. El marco de referencia para la traducción de estos capítulos fue la redacción científica en español. Para ello se usaron variadas fuentes, pero *Cómo traducir y redactar textos científicos en español. Reglas, ideas, consejos*, de M. Gonzalo Claros Díaz [1], fue de gran ayuda. Esta traducción, por supuesto, no habría sido posible sin el trabajo, colaboración y conocimiento de los coeditores que editaron, corrigieron, reformularon la traducción y consultaron con especialistas, y en ocasiones los mismos autores, términos y conceptos. Esta lista de términos no pretende ser una recopilación exhaustiva de la terminología especializada de los medios porosos, sino más bien una lista de aquellos vocablos para los que hemos tenido que encontrar una solución. Suelen ser términos que o bien son de un campo especializado, o bien han provocado discusiones sobre su uso en ese preciso contexto. Hemos reducido al máximo el número de entradas, eliminando aquellos términos de uso o conocimiento general. Para otro momento queda un trabajo de más envergadura en que podamos plasmar esas discusiones en fichas terminológicas de un mayor valor para futuros proyectos en donde se incorporen información léxica, ejemplos, sinónimos, falsos amigos, preferencias según los contextos o sutilezas de usos en una u otra lengua.

1. Claros Díaz M.G. (2016). *Cómo traducir y redactar textos científicos en español.* Cuadernos de la Fundación Dr. Antoni Esteve Nro. 39.

Contact: Leyre Alegre-Figuero <lalegref@mtu.edu>

E. F. Médici and A. D. Otero (eds.), *Album of Porous Media*,
https://doi.org/10.1007/978-3-031-23800-0_114

English – Spanish

absorption: absorción
adsorption: adsorción
adhesion: adhesión
adjustable level-cut filtered Poisson field (ALCPF): campo de Poisson filtrado con corte nivelado ajustable
advection: advección
aggregates/inter-aggregate/coarse aggregate: agregados/interagregados/agregado grueso
alkane: alcano
alluvial strata: estratos aluvionales
aluminium oxide: óxido de aluminio
amorphous: amorfo
analog: análogo
anion: anión
anodization/anodized: anodización/anodizado
anticline: anticlinal
array: arreglo
artefacts: artificios
atomic layer deposition: deposición de capas atómicas
authigenic: autigénico
axial skeleton: esqueleto axial
backscatter electron (BSE): retrodispersión de electrones
back-scattered X-ray: emisión de rayos X retrodispersados
based brine: solución salina
beaker: vaso
beam: haz
bed strike: arrumbamiento
benchtop: de mesa
Bentheimer sandstone: arenisca de Bentheimer
Berea sandstone: arenisca de Berea
binarization filtering technique: técnica del filtrado binario
biofilm: biopelícula
biogenic silica: sílice biogénico
blob: mancha
body force: fuerza volumétrica
brine: solución salina
brittle faults: fracturas frágiles
byproduct: subproducto
calcium carbonate: carbonato de calcio
calcium chloride: cloruro de calcio
canola oil: aceite de colza
capillary fingering: digitación capilar
carbon sputter coated: pulverización con carbono
carbonate: carbonato
catalyst: catalizador
cathode: cátodo
cement paste: pasta de cemento
cerebrospinal fluid (CSF): líquido cefalorraquídeo (LCR)
chain force: cadena cinemática
chalk: tiza
chemical species: especies químicas
chemical vapor deposition (CVD): deposición química de vapor
chip: circuito integrado
chlorite: clorito
chromatography: cromatografía
clogging: obstrucción
cluster: cúmulos
Cnidaria: *Cnidaria*
coating: recubrimiento
coenenchym: cenénquime
collagen: colágeno
collapse: colapso
colormap: escala de colores
comminution: conminución
computational fluid dynamics (CFD): dinámica de fluidos computacional
concrete: hormigón
condensed: condensado
confocal microscopy: microscopía confocal
conformal coatings: recubrimientos conformados
connote water: agua connata
contact angle: ángulo de contacto
contaminant tracing: rastreo de contaminantes
continuum models: modelos continuos
convective heat transfer: transferencia de calor por convección
core plug: núcleo
creeping flow: flujo de arrastre
cross sectional: transversal
crust: corteza
crystallographic: cristalográfico
cutout: recorte
dark-field: campo oscuro
data pipeline: secuencia de procesamiento de datos
decane: decano
defending: desplazado
degassing: desgasificación
deionized: desionizado
densification: densificado
deviation: desviación
diatom: diatomeas
diatomaceous rock: roca de diatomea
diatomite: diatomita
diffusion/diffusivity: difusión/difusividad
dipping: inmerso
direct numerical simulation (DNS): simulación numérica directa
discotic liquid crystals (DLC): cristales líquidos discóticos
discretised: discretizado
dispersion: dispersión
DI-water: agua desionizada
dolomite: dolomita
doping: dopaje
drainage: drenado
drying: secado
dye: tinta
earthquake: sismo
electroactuation: electroaccionamiento
electrochemical: electroquímico
electrode: electrodo
electrolyte: electrolito
electron: electrón
electropolymerization: electropolimerización

electrospinning/electrospun: electrohilado
ellipsoid: elipsoide
emulsion: emulsion
energizing: energizado
energy-dispersive X-ray (EDX): rayos X por energía dispersiva (EDX por sus siglas en inglés)
enhanced oil recovery: recuperación mejorada del petróleo
environmental scanning electron microscopy (ESEM): microscopio electrónico de barrido ambiental (MEBA)
epoxy resin: resina epoxídica
eruption: erupción
etching: grabado
Euclidean distance: distancia euclidiana
evaporation: evaporación
extra-cellular matrix (ECM): matriz extracelular
fast Fourier transform: transformada rápida de Fourier
fault: falla
feldspar: feldespato
felt: fieltro
ferrofluid: ferrofluido
fiber: fibra
film: lámina
fingering: digitación
flood/flooded: inundación/inundado
flow battery: baterías de flujo
flow cell: celda de flujo
flow field: canal de flujo
flow rate: caudal
fluorescein: fluoresceína
fluorescent: fluorescente
fluorinert: fluorinert
fluorophore: fluorocromo
flux/flow: flujo
foam: espuma
focal plane: plano focal
focused ion beam: haz de iones enfocados
Foraminifera: *Foraminifera*
force-strain curve: curva de esfuerzo-deformación
fractal: fractal
fracture: fractura
fracture-cut: líneas de corte
fragmentation: fragmentación
frictional finger labyrinths: laberintos de digitaciones por fricción
frustules: frústulas
fuel cells: pilas de combustible
funicular: funicular
furfuryl alcohol: alcohol furfurílico
gas diffusion layer (GDL): capa de difusión de gas
gas reservoir: yacimiento de gas
geo-sequestration: geosecuestro
glass: vidrio
glass bead: esfera de vidrio
glass plate: placa de vidrio
glycerine: glicerina
goethite: goethita
gorgonian/gorgonin: gorgónido/gorgonina
gradient: gradiente
grain/granular: grano/granular
grid: grilla
heat conduction: transferencia de calor por conducción
heulandite: heulanditas
hydrocarbon: hidrocarburo
hydrodynamic: hidrodinámica
hydrofluoric acid: ácido fluorhídrico
hydrogen: hidrógeno
hydrolysis: hidrólisis
hydrophilic: hidrofílico
hydrophobic: hidrofóbico
hydrous: hidratado
ignimbrite: ignimbrita
illite: illita
imbibition: imbibición
immiscible: inmiscible
infrared (IR): infrarrojo
infused: impregnado
inhomogeneity: inhomogeneidad
injection/injected: inyección/inyectado
interaction: interacción
interface: interfaz
interfacial tension: tensión superficial
interfacial transition zone (ITZ): zona de transición interfacial
interstitial fluid: fluido intersticial
invading: inyectado
ionomer: ionómero
kaolinite: caolinita
kinematic: cinemática
labyrinth: laberinto
laser/laser pulses/laser-ultrasonic: láser/láser pulsante/láser ultrasónico
Lattice Boltzmann method: método Lattice Boltzmann
layer : capa
light expanded clay aggregate (LECA): arlita
linear-elastic: elástico lineal
lumina: lumen
lymphatic vessels: vasos linfáticos
machine learning: aprendizaje automático
magma: magma
magnesium: magnesio
magnetic resonnance (MR): resonancia magnética (RM)
maltodextrin: maltodextrina
manifold-microchannels: colectores de microcanales
measurement: medida/medición
melt: material fundido
membrane: membrana
mesh: malla
mesophase: mesofase
micelles: micelas
micrograph: micrografía
microscope: microscopio
microstylolite: microestilolito
migration: migración
miscible: miscibles
mixing/mixture: mezclado/mezcla
mole fraction: fracción molar
molecular dynamics: dinámica molecular
monolayer: de una sola capa
monzonite: monzonita

multiphase: multifásico
nanopore: nanoporo
n-dodecane: n-dodecano
Newtonian fluid: fluido newtoniano
n-heptane: n-heptano
nitrogen: nitrógeno
no flow boundary conditions: condiciones de contorno de flujo nulo
non-Fickian: no fickiano
nucleate: nucleación
oblate: achatado
octahedron: octaedro
OOIP: petróleo original in situ
open boundary condition: condiciones de contorno de tipo abierto
open-source software: software de código abierto
optic: óptica
organic: orgánico
outcrop: afloramiento rocoso
outgassing: desgasificación
oxidation: oxidación
oxygen: oxígeno
pack/packing: empaquetamiento
packstone: caliza
particle image velocimetry (PIV): velocimetría de imagen de partículas
path: camino
pellets: pellas
pendular state: estado pendular
penetration: penetración
pentafluoropropane: pentafluoropropano
perchloric acid: ácido perclórico
percolation: percolación
performance: rendimiento
permeability: permeabilidad
phases: fases
photonic crystal: cristal fotónico
photoresist: fotoresistente
pillars: pilares
pixel: píxel
plastic/plasticity: plástico/plasticidad
plug: cilindro
plume: pluma
polarisation: polarización
polycarbonate: policarbonato
polydimethylsiloxane (PDMS): polidimetilsiloxano (PDMS)
polymer: polímero
polymer electrolyte membrane (PEM): membranas de electrolito polimérico
polymerization: polimerización
polymethylmethacrylate (PMMA): polimetilmetacrilato (PMMA)
polypeptide: polipéptido
polypyrrole: polipirrol
polystyrene: poliestireno
pore network: red de poros
pore space: espacio poroso
pore throat : garganta del poro
porous material: material poroso
porous media/ medium: medio poroso
positron emission tomography (PET): tomografía por emisión de positrones
precipitation: precipitación
printing: impresión
process: proceso
prolate: oblongo
propane: propano
pull phase: fase de tracción
push phase: fase de empuje
push-pull: empujar-traccionar
pyroxene: piroxeno
pyrromethene: pirrometeno
quartz: cuarzo
quasi-2D: cuasibidimensional
quiver: flecha
radioactive : radioactivo
radionuclide fluorodeoxyglucose 18F-FDG: radioisótopo de fluorodeoxyglucosa 18F-FDG
radio-opaque: radioopaco
random: aleatorio
ratio: razón
reactant: reactivo
reconstruction: reconstrucción
refractive index matching/index-matched: ajuste del índice de refracción/coincidencia del índice de refracción
remobilization: removilización
repellent: repelente
resolution: resolución
rheology: reología
rhodamine 6G: rodamina 6G
rock: roca
Rotliegendes: Rotliegendes
saline solution: solución salina
sand: arena
sandstone: arenisca
sap: savia
scale: escala
scanning electron microscopy (SEM): microscopio electrónico de barrido (MEB)
sclerites: escleritos
scuba diving: buceo
sedimentary: sedimentario
shale: lutita
shear: corte
shear-thinning: pseudoplasticidad
sheet: lámina
silica: sílice
siliceous: silíceo
silicon: silicio
silicon oil: aceite de silicona
silicon wafe: oblea de silicio
silt: limo
simplified marching cube technique: técnica de Cubos de Marcha Simplificada
simulation: simulación
size: tamaño
slice: corte transversal
slip-boundary condition: condición de contorno de deslizamiento libre
slope angle: ángulo de inclinación
smectite: esmectita
sodium salt: sal sódica
soil: suelo

solid crystals: cristales sólidos
solubility: solubilidad
solute: solutos
solvent: solvente
source rock: roca madre
spinodal phase separation: fase de descomposición espinodal
spotsize: spotsize
sputtered: pulverizado
stick-slip: oscilación de relajación
stochastic: estocástico
stoichiometry: estequiometría
storage: almacenamiento
streamlines: líneas de corriente
streamwise: dirección de la corriente
stress: tensión
stress-strain curve: curva de tensión-deformación
strike: vena
strip: banda
structure: estructura
sublimation: sublimación
subsurface: subsuelo
subtraction: sustracción
surface: superficie
surface tension: tensión superficial
surfactant: surfactante
synchrotron: sincrotrón
syringe pump: bomba de infusión de jeringa
taïga: taiga
template: plantilla
tensile failure: falla por tensión
tensile strength: resistencia a la tracción
tetraethoxysilane: tetraetoxisilano
thermal signature: característica térmica
thin: delgado
tissue scaffolds: soporte de tejido
titanium dioxide: dióxido de titanio
tracer: trazador
transducer: transductor
transmission electron microscope (TEM): microscopio electrónico de transmisión
trapping: retención
triaxial stress: tensiones triaxiales
two phase flow: flujo de dos fases
ultra violet (UV): ultra violeta (UV)
ultramicrotome: ultramicrotomo
underground: subterráneo
unsharp masking filters: filtros de máscara de enfoque
upwards fining: granocreciente
urea: urea
velocity field: campo de velocidades
vessel: recipiente
vial: vial
vibration frequency: frecuencia de vibración
viscosity: viscosidad
viscous fingering: digitación viscosa
void: hueco
voxel: vóxel
walk: camino
water phase: fase acuosa
watershed segmentation: transformación divisoria
wave velocity: velocidad de onda
wavelength: longitud de onda
wettability/wet/non-wetting: mojabilidad/mojado/no monjante
whisker: filamento
wicking: absorción
Xanthan gum: goma xantana
xenon: xenón
X-ray (micro/μ/nano) computed tomography (CT): (micro/μ/nano) tomografía computarizada (TC) de rayos X
yield: fluencia
zeolite: zeolita

Español – Inglés

absorción: absorption, wicking
adsorción: adsorption
aceite de colza: canola oil
aceite de silicona: silicon oil
achatado: oblate
ácido fluorhídrico: hydrofluoric acid
ácido perclórico: perchloric acid
adhesión: adhesion
advección: advection
afloramiento rocoso: outcrop
agregados/interagregados/agregado grueso: aggregates/inter-aggregate/coarse aggregate
agua connata: connote water
agua desionizada: DI-water
ajuste del índice de refracción/coincidencia del índice de refracción: refractive index matching/index-matched
alcano: alkane
alcohol furfurílico: furfuryl alcohol
aleatorio: random
almacenamiento: storage
amorfo: amorphous
análogo: analog
ángulo de contacto: contact angle
ángulo de inclinación: slope angle
anión: anion
anodización/anodizado: anodization/anodized
anticlinal: anticline
aprendizaje automático: machine learning
arena: sand
arenisca: sandstone
arenisca de Bentheimer: Bentheimer sandstone
arenisca de Berea: Berea sandstone
arlita: light expanded clay aggregate (LECA)
arreglo: array
arrumbamiento: bed strike
artificios: artefacts
autigénico: authigenic
banda: strip
baterías de flujo: flow battery
biopelícula: biofilm
bomba de infusión de jeringa: syringe pump
buceo: scuba diving
cadena cinemática: chain force
caliza: packstone
camino: path
camino: walk
campo de Poisson filtrado con corte nivelado ajustable: adjustable level-cut filtered Poisson field (ALCPF)
campo de velocidades: velocity field
campo oscuro: dark-field
canal de flujo: flow field
caolinita: kaolinite
capa: layer
capa de difusión de gas: gas diffusion layer (GDL)
característica térmica: thermal signature
carbonato: carbonate
carbonato de calcio: calcium carbonate
catalizador: catalyst
cátodo: cathode
caudal: flow rate
celda de flujo: flow cell
cenénquime: coenenchym
cilindro: plug
cinemática: kinematic
circuito integrado: chip
clorito: chlorite
cloruro de calcio: calcium chloride
Cnidaria: Cnidaria
colágeno: collagen
colapso: collapse
colectores de microcanales: manifold-microchannels
condensado: condensed
condición de contorno de deslizamiento libre: slip-boundary condition
condiciones de contorno de flujo nulo: no flow boundary conditions
condiciones de contorno de tipo abierto: open boundary condition
conminución: comminution
corte: shear
corte transversal: slice
corteza: crust
cristal fotónico: photonic crystal
cristales líquidos discóticos: discotic liquid crystals (DLC)
cristales sólidos: solid crystals
cristalográfico: crystallographic
cromatografía: chromatography
cuarzo: quartz
cuasibidimensional: quasi-2D
cúmulos: cluster
curva de esfuerzo-deformación: force-strain curve
curva de tensión-deformación: stress-strain curve
de mesa: benchtop
de una sola capa: monolayer
decano: decane
delgado: thin
densificado: densification
deposición de capas atómicas: atomic layer deposition
deposición química de vapor: chemical vapor deposition (CVD)
desgasificación: degassing, outgassing
desionizado: deionized
desplazado: defending
desviación: deviation
diatomeas: diatom
diatomita: diatomite
difusión/difusividad: diffusion/diffusivity
digitación: fingering
digitación capilar: capillary fingering
digitación viscosa: viscous fingering
dinámica de fluidos computacional : computational fluid dynamics (CFD)
dinámica molecular: molecular dynamics
dióxido de titanio: titanium dioxide
dirección de la corriente: streamwise
discretizado: discretised
dispersión: dispersion

distancia euclidiana: Euclidean distance
dolomita: dolomite
dopaje: doping
drenado: drainage
elástico lineal: linear-elastic
electroaccionamiento: electroactuation
electrodo: electrode
electrohilado: electrospinning/electrospun
electrolito: electrolyte
electrón: electron
electropolimerización: electropolymerization
electroquímico: electrochemical
elipsoide: ellipsoid
emisión de rayos X retrodispersados: back-scattered X-ray
empaquetamiento: pack/packing
empujar-traccionar: push-pull
emulsion: emulsions
energizado: energizing
erupción: eruption
escala: scale
escala de colores: colormap
escleritos: sclerites
esfera de vidrio: glass bead
esmectita: smectite
espacio poroso: pore space
especies químicas: chemical species
espuma: foam
esqueleto axial: axial skeleton
estado pendular: pendular state
estequiometría: stoichiometry
estocástico: stochastic
estratos aluvionales: alluvial strata
estructura: structure
evaporación: evaporation
falla: fault
falla por tensión: tensile failure
fase acuosa: water phase
fase de descomposición espinodal: spinodal phase separation
fase de empuje: push phase
fase de tracción: pull phase
fases: phases
feldespato: feldspar
ferrofluido: ferrofluid
fibra: fiber
fieltro: felt
filamento: whisker
filtros de máscara de enfoque: unsharp masking filters
flecha: quiver
fluencia: yield
fluido intersticial: interstitial fluid
fluido newtoniano: Newtonian fluid
flujo: flux/flow
flujo de arrastre: creeping flow
flujo de dos fases: two phase flow
fluoresceína: fluorescein
fluorescente: fluorescent
fluorinert: fluorinert
fluorocromo: fluorophore
Foraminifera: Foraminifera
fotoresistente: photoresist
fracción molar: mole fraction
fractal: fractal
fractura: fracture
fracturas frágiles: brittle faults
fragmentación: fragmentation
frecuencia de vibración: vibration frequency
frústulas: frustules
fuerza volumétrica: body force
funicular: funicular
garganta del poro: pore throat
geosecuestro: geo-sequestration
glicerina: glycerine
goethita: goethite
goma xantana: Xanthan gum
gorgónido/gorgonina: gorgonian/gorgonin
grabado: etching
gradiente: gradient
grano/granular: grain/granular
granocreciente: upwards fining
grilla: grid
haz: beam
haz de iones enfocados: focused ion beam
heulanditas: heulandite
hidratado: hydrous
hidrocarburo: hydrocarbon
hidrodinámica: hydrodynamic
hidrofílico: hydrophilic
hidrofóbico: hydrophobic
hidrógeno: hydrogen
hidrólisis : hydrolysis
hormigón: concrete
hueco: void
ignimbrita: ignimbrite
illita: illite
imbibición: imbibition
impregnado: infused
impresión: printing
infrarrojo: infrared (IR)
inhomogeneidades: inhomogeneity
inmerso: dipping
inmiscible: immiscible
interacción: interaction
interfaz: interface
inundación/inundado: flood/flooded
inyección/inyectado: injection/injected
inyectado: invading
ionómero: ionomer
laberinto: labyrinth
laberintos de digitaciones por fricción: frictional finger labyrinths
lámina: film, sheet
láser/láser pulsante/láser ultrasónico: laser/laser pulses/laser-ultrasonic
limo: silt
líneas de corriente: streamlines
líneas de corte: fracture-cut
líquido cefalorraquídeo (LCR): cerebrospinal fluid (CSF)
longitud de onda: wavelength
lumen: lumina
lutita: shale
magma: magma
magnesio: magnesium

malla: mesh
maltodextrina: maltodextrin
mancha: blob
material fundido: melt
material poroso: porous material
matriz extracelular: extra-cellular matrix (ECM)
medida/medición: measurement
medio poroso: porous media/ medium
membrana: membrane
membranas de electrolito polimérico: polymer electrolyte membrane (PEM)
mesofase: mesophase
método Lattice Boltzmann: Lattice Boltzmann method
mezclado/mezcla: mixing/mixture
micelas: micelles
microestilolito: microstylolite
micrografía: micrograph
microscopía confocal: confocal microscopy
microscopio: microscope
microscopio electrónico de barrido (MEB): scanning electron microscopy (SEM)
microscopio electrónico de barrido ambiental (MEBA): environmental scanning electron microscopy (ESEM)
microscopio electrónico de transmisión: transmission electron microscope (TEM)
migración: migration
miscible: miscible
modelos continuos: continuum models
mojabilidad/mojado/no monjante: wettability/wet/non-wetting
monzonita: monzonite
multifásico: multiphase
n-dodecano: n-dodecane
n-heptano: n-heptane
nanoporos: nanopores
nitrógeno: nitrogen
no fickiano: non-Fickian
nucleación: nucleate
núcleo: core plug
oblea de silicio: silicon wafe
oblongo: prolate
obstrucción: clogging
octaedro: octahedron
óptica: optic
orgánico: organic
oscilación de relajación: stick-slip
oxidación: oxidation
óxido de aluminio: aluminium oxide
oxígeno: oxygen
pasta de cemento: cement paste
pellas: pellets
penetración: penetration
pentafluoropropano: pentafluoropropane
percolación: percolation
permeabilidad: permeability
petróleo original in situ: OOIP
pilares: pillars
pilas de combustible: fuel cells
piroxeno: pyroxene
pirrometeno: pyrromethene
píxel: pixel
placa de vidrio: glass plate
plano focal: focal plane
plantilla: template
plástico/plasticidad: plastic/plasticity
pluma: plume
polarización: polarisation
policarbonato: polycarbonate
polidimetilsiloxano (PDMS): polydimethylsiloxane (PDMS)
poliestireno: polystyrene
polimerización: polymerization
polímero: polymer
polimetilmetacrilato (PMMA): polymethylmethacrylate (PMMA)
polipéptido: polypeptide
polipirrol: polypyrrole
precipitación: precipitation
proceso: process
propano: propane
pseudoplasticidad: shear-thinning
pulverización con carbono: carbon sputter coated
pulverizado: sputtered
radioactivo: radioactive
radioisótopo de fluorodeoxyglucosa 18F-FDG: radionuclide fluorodeoxyglucose 18F-FDG
radioopaco: radio-opaque
rastreo de contaminantes: contaminant tracing
rayos X por energía dispersiva (EDX por sus siglas en inglés) : energy-dispersive X-ray (EDX)
razón: ratio
reactivo: reactant
recipiente: vessel
reconstrucción: reconstruction
recorte: cutout
recubrimiento: coating
recubrimientos conformados: conformal coatings
recuperación mejorada del petróleo: enhanced oil recovery
red de poros: pore network
removilización: remobilization
rendimiento: performance
reología: rheology
repelente: repellent
resina epoxídica: epoxy resin
resistencia a la tracción: tensile strength
resolución: resolution
resonancia magnética (RM): magnetic resonnance (MR)
retención: trapping
retrodispersión de electrones: backscatter electron (BSE)
roca: rock
roca de diatomea: diatomaceous rock
roca madre: source rock
rodamina 6G: rhodamine 6G
Rotliegendes: Rotliegendes
sal sódica: sodium salt
savia: sap
secado: drying
secuencia de procesamiento de datos: data pipeline
sedimentario: sedimentary
sílice: silica

sílice biogénico: biogenic silica
silíceo: siliceous
silicio: silicon
simulación: simulation
simulación numérica directa: direct numerical simulation (DNS)
sincrotrón: synchrotron
sismo: earthquake
software de código abierto: open-source software
solubilidad: solubility
solución salina: based brine, brine, saline solution
solutos: solute
solvente: solvent
soporte de tejido: tissue scaffolds
spotsize: spotsize
sublimación: sublimation
subproducto: byproduct
subsuelo: subsurface
subterráneo: underground
suelo: soil
superficie: surface
surfactante : surfactant
sustracción: subtraction
taiga: taïga
tamaño: size
técnica de Cubos de Marcha Simplificada: simplified marching cube technique
técnica del filtrado binario: binarization filtering technique
tensión: stress
tensión superficial: interfacial tension, surface tension
tensiones triaxiales: triaxial stress
tetraetoxisilano: tetraethoxysilane
tinta: dye
tiza: chalk
(micro/μ/nano) tomografía computarizada (TC) de rayos X: X-ray (micro/μ/nano) computed tomography (CT)
tomografía por emisión de positrones: positron emission tomography (PET)
transductor: transducer
transferencia de calor por conducción: heat conduction
transferencia de calor por convección: convective heat transfer
transformación divisoria: watershed segmentation
transformada rápida de Fourier: fast Fourier transform
transversal: cross sectional
trazador: tracer
ultra violeta (UV): ultra violet (UV)
ultramicrotomo: ultramicrotome
urea: urea
vaso: beaker
vasos linfáticos: lymphatic vessels
velocidad de onda: wave velocity
velocimetría de imagen de partículas: particle image velocimetry (PIV)
vena: strike
vial: vial
vidrio: glass
viscosidad: viscosity
vóxel: voxel
xenón: xenon
yacimiento de gas: gas reservoir
zeolita: zeolite
zona de transición interfacial: interfacial transition zone (ITZ)

Correction to: Album of Porous Media

Ezequiel F. Médici and Alejandro D. Otero

Correction to:
E. F. Médici and A. D. Otero (eds.), *Album of Porous Media*,
https://doi.org/10.1007/978-3-031-23800-0

In the original version of the book, the following belated corrections have been incorporated:

1. In chapter 47, the city and country in the affiliation of Professor Richard Dawe has been changed from "Imperial College, University of the West Indies, Kingston, India" to "Imperial College, University of the West Indies, St. Augustine, Trinidad, WI" in the chapter XML.
2. The incorrect author's name has been published inadvertently in the online version of the below chapters. Now, it has be updated with the below changes.
   - In Chapter 49: From "Laura Gallardo, Juan Sanchez, Hugo Bellezza, Olivier Vincent, and Patrick Huber" to "Armin Afrough and Karen Feilberg".
   - In Chapter 50: From "Laura Gallardo, Juan Sanchez, Hugo Bellezza, Olivier Vincent, and Patrick Huber" to "Uri Nachshon".
   - In Chapter 51: From "Laura Gallardo, Juan Sanchez, Hugo Bellezza, Olivier Vincent, and Patrick Huber" to "Divyansh Pandey, Kaustubh Desai, Debashis Panda, Abdolreza Kharaghani, and Vikranth Kumar Surasani".
   - In Chapter 52: From "Laura Gallardo, Juan Sanchez, Hugo Bellezza, Olivier Vincent, and Patrick Huber" to "Navneet Kumar and Jaywant H. Arakeri".
   - In Chapter 53: From "Laura Gallardo, Juan Sanchez, Hugo Bellezza, Olivier Vincent, and Patrick Huber" to "Zahra Noori O'Connor, Jamal Yagoobi, and Burt Tilley".
   - In Chapter 54: From "Laura Gallardo, Juan Sanchez, Hugo Bellezza, Olivier Vincent, and Patrick Huber" to "Bjørnar Sandnes, James Campbell, and Deren Ozturk".
   - In Chapter 55: From "Laura Gallardo, Juan Sanchez, Hugo Bellezza, Olivier Vincent, and Patrick Huber" to "Bjørnar Sandnes, Knut Jørgen Måløy, and Eirik Grude Flekkøy".
   - In Chapter 56: From "Laura Gallardo, Juan Sanchez, Hugo Bellezza, Olivier Vincent, and Patrick Huber" to "Kristian Stølevik Olsen, Eirik Grude Flekkøy, Knut Jørgen Måløy, and James Campbell".

The book and the chapters have been updated with the changes.

The updated versions of these chapters can be found at
https://doi.org/10.1007/978-3-031-23800-0_47
https://doi.org/10.1007/978-3-031-23800-0_49
https://doi.org/10.1007/978-3-031-23800-0_50
https://doi.org/10.1007/978-3-031-23800-0_51
https://doi.org/10.1007/978-3-031-23800-0_52
https://doi.org/10.1007/978-3-031-23800-0_53
https://doi.org/10.1007/978-3-031-23800-0_54
https://doi.org/10.1007/978-3-031-23800-0_55
https://doi.org/10.1007/978-3-031-23800-0_56

E. F. Médici and A. D. Otero (eds.), *Album of Porous Media*,
https://doi.org/10.1007/978-3-031-23800-0_115

MIX
Papier aus verantwortungsvollen Quellen
Paper from responsible sources
FSC® C105338

If you have any concerns about our products,
you can contact us on
ProductSafety@springernature.com

In case Publisher is established outside the EU,
the EU authorized representative is:
Springer Nature Customer Service Center GmbH
Europaplatz 3, 69115 Heidelberg, Germany

Printed by Libri Plureos GmbH
in Hamburg, Germany